Greek and Roman Military Manuals

This volume explores the enigmatic primary source known as the ancient military manual. In particular, the volume explores the extent to which these diverse texts constitute a genre (sometimes unsatisfactorily classified as 'technical literature'), and the degree to which they reflect the practice of warfare.

With contributions from a diverse group of scholars, the chapters examine military manuals from early Archaic Greece to the Byzantine period, covering a wide range of topics including readership, siege warfare, mercenaries, defeat, textual history, and religion. Coverage includes most of the major contemporary siege manual writers, including Xenophon, Frontinus, Vegetius, and Maurice. Close examination of these texts serve to reveal the complex ways in which ancient Greeks, Romans, and Byzantines sought to understand better, and impose order upon, the seemingly irrational phenomenon known as war.

Providing insight into the multifaceted collection of texts that constituted military manuals, this volume is a key resource for students and scholars of warfare and military literature in the classical and Byzantine periods.

James T. Chlup is Associate Professor of Ancient History at the University of Manitoba, Canada.

Conor Whately is Associate Professor of Classics at the University of Winnipeg, Canada.

Routledge Monographs in Classical Studies

Titles include

Un-Roman Sex
Gender, Sexuality, and Lovemaking in the Roman Provinces and Frontiers, 1st Edition
Edited by Tatiana Ivleva and Rob Collins

Robert E. Sherwood and the Classical Tradition
The Muses in America
Robert J. Rabel

Text and Intertext in Greek Epic and Drama
Essays in Honor of Margalit Finkelberg
Edited by Jonathan Price and Rachel Zelnick-Abramovitz

Animals in Ancient Greek Religion
Edited by Julia Kindt

Classicising Crisis
The Modern Age of Revolutions and the Greco-Roman Repertoire
Edited by Barbara Goff and Michael Simpson

Epigraphic Culture in the Eastern Mediterranean in Antiquity
Edited by Krzysztof Nawotka

Proclus and the *Chaldean Oracles*
A Study on *Proclean Exegesis*, with a Translation and Commentary of Proclus' Treatise *On Chaldean Philosophy*
Nicola Spanu

Greek and Roman Military Manuals
Genre and History
Edited by James T. Chlup and Conor Whately

For more information on this series, visit: www.routledge.com/classicalstudies/series/RMCS

Greek and Roman Military Manuals

Genre and History

Edited by James T. Chlup and Conor Whately

Routledge
Taylor & Francis Group

LONDON AND NEW YORK

First published 2021
by Routledge
2 Park Square, Milton Park, Abingdon, Oxon OX14 4RN

and by Routledge
52 Vanderbilt Avenue, New York, NY 10017

Routledge is an imprint of the Taylor & Francis Group, an informa business

British Library Cataloguing-in-Publication Data
A catalogue record for this book is available from the British Library

Library of Congress Cataloging-in-Publication Data
A catalog record has been requested for this book

ISBN: 978-1-138-33514-1 (hbk)
ISBN: 978-0-429-44397-8 (ebk)

Typeset in Times New Roman
by Apex CoVantage, LLC

Contents

Contributors

Aaron L. Beek is a Postdoctoral Research Fellow at North-West University. His research focus is on banditry, piracy, and mercenaries in the ancient world, but he has also published on naval history, Mithridates, Josephus, and Cassius Dio.

Craig H. Caldwell is Associate Professor of History at Appalachian State University. His research has focused on the Balkans in late antiquity, late Roman usurpations and civil wars, military history, and numismatics.

James T. Chlup is Associate Professor of Ancient History at the University of Manitoba. With a focus on the history of the Roman Middle and Late Republic, he has published articles on Caesar, Livy, Plutarch, Onasander, and the Scottish novelist and biographer John Buchan.

Lucy Felmingham-Cockburn is currently studying for a Ph.D. at the University of Warwick. Her doctoral thesis explores sociopolitical metaphor in Xenophon's *Peri Hippikes*.

Murray Dahm is a New Zealand-born ancient historian living in Australia. His research encompasses various aspects of didactic military handbooks and ancient military history. He has written on Alexander the Great, the Peloponnesian War, the Gothic Wars, and the Battle of Leuctra for Osprey Publishing and is the assistant editor of *Ancient Warfare Magazine*.

Immacolata Eramo is a researcher and lecturer in Classical Philology at the University of Bari. She is the author of several publications on Greek, Roman, and Byzantine military literature and its tradition, including two critical editions: Syrianus, *Rhetorica militaris* (Bari, 2010) and the anonymous *De militari scientia* (Besançon 2018).

Clemens Koehn is Senior Lecturer in Classics and Ancient History at the University of New England in Armidale, Australia. He has widely published on ancient Greece and Rome with a focus on military history, including *Justinian und die Armee des frühen Byzanz* (De Gruyter, 2018). For many years, he was involved in a large research project reconstructing ancient catapults.

Meredith L. D. Riedel is a historian specialising in the religious and military history of the medieval Byzantine empire. At Duke University's Divinity School in Durham, North Carolina, she teaches graduate courses on the history of Eastern Orthodoxy, the Crusades, and Byzantium and Islam. She is the author of *Leo VI and the Transformation of Byzantine Christian Identity: Writings of an Unexpected Emperor* (Cambridge, 2018).

Jeffrey Rop is Associate Professor of History at the University of Minnesota, Duluth. His publications include *Greek Military Service in the Ancient Near East, 401–330 BCE* (Cambridge, 2019), and articles on the military and political history and historiography of ancient Greece and Persia.

Hans Michael Schellenberg studied ancient history and political sciences at Heinrich-Heine-University Düsseldorf. His master thesis received the Förderpreis of the Werner-Hahlweg-Award for military history and military sciences. He is currently writing his Ph.D. with the title *Critical studies and materials on the so-called ancient military writers* (400 BCE–600 CE).

Nicholas Sekunda is Head of the Department of Mediterranean Archaeology at the University of Gdansk. He has participated in excavations in England, Poland, Iran, Greece, Syria, and Jordan, and now co-directs excavations at Negotino Gradište in the Republic of North Macedonia. He is the author of a number of books concerning Greek Warfare, including (co-edited with Philip Rance) *Greek Taktika: Ancient Military Writing and Its Heritage* (Gdansk, 2017).

Jonathan Warner is a Ph.D. candidate in the Classics Department at Cornell University. In additional to Greek and Latin military literature, his research focuses on late Roman society and imperial administration. His dissertation, *Soldiers of Caesar and Christ: Martial Imagery and the Ethos of Church and State Service in Late Antiquity*, studies epistolography to reappraise the militarisation of the late empire.

Conor Whately is Associate Professor of Classics at the University of Winnipeg. He has published widely on Roman and late antique military history. His books include *Exercitus Moesiae* (BAR, 2016), *Battles and Generals: Combat, Culture, and Didacticism in Procopius' Wars* (Brill, 2016), and *An Introduction to the Roman Military* (Wiley, 2020).

Nadya Williams is Professor of History at the University of West Georgia. She is a military historian of the ancient world, and the co-editor of *Civilians and Warfare in World History* (Routledge, 2017).

Graham Wrightson is Associate Professor of History at South Dakota State University. He has published extensively on Fourth century and Hellenistic warfare. He is the author of *Combined Arms warfare in Ancient Greece: From Homer to Alexander the Great and his Successors* (Routledge, 2019).

Acknowledgements

This volume originates from a conference on ancient military manuals jointly held by the Universities of Manitoba and Winnipeg in October 2016. The conference arose from discussions the editors have had about courses on ancient (primarily Roman) warfare and the appropriate place for military manuals as primary sources therein. Some of the chapters in this volume are revised and expanded versions of papers presented at the conference; a few appear at the invitation of the editors. Our goal in preparing this volume is to provide a resource for scholars and students interested in ancient warfare and ancient military writing and to demonstrate the versatility of the genre. The authors constitute established, emerging, and new scholars from around the world whose work intersects with the field of military literature. The volume is very much designed with a view to opening up new lines of enquiry on ancient military manuals.

First and foremost, the editors thank the conference participants for braving the (in some instances quite long) journey to Winnipeg. Gratitude is due also to those contributors who were willing to share ideas and provide guidance on matters directly related and peripheral to this volume: Aaron L. Beek, Craig H. Caldwell, Jeffrey Rop, Nadya Williams, and Graham Wrightson. The editors most warmly express their appreciation to their home institutions for generous financial and other support for the hosting of the conference and the preparation of this collection, especially our Department Heads, Lea Stirling and Matt Gibbs, the Faculty of Arts at both institutions, Tourism Winnipeg for some promotional materials, and those colleagues from both institutions who provided support in other ways, whether by attending the conference or in conversation. Additionally, Ruth Dickinson, the students of the Roman Warfare and Imperialism course at the University of Manitoba in fall term 2016, and Oliver Zhou provided help at the conference or with the preparation of the volume. A significant debt of gratitude is owed to Immacolata Eramo for agreeing to read the draft chapters and write the epilogue. Finally, the editors thank the team at Routledge for their assistance in the publication of this volume.

James T. Chlup
Conor Whately

Note on translations

Where feasible and accepted by scholars, authors use English titles for texts. For the benefit of students and other readers without knowledge of ancient Greek and Latin, the authors quote English translations in the main text, and the original in a footnote (if necessary). Unless otherwise indicated by individual authors, translations of the primary source texts in this book are as follows:

Aelian, *Tactics*	The Tactics of Aelian (Pen and Sword, Christopher Matthews, translator, Barnsley, 2012)
Aeneas Tacticus, *How to Survive under Siege*	Aeneas Tacticus, Asclepiodotus, Onasander (Loeb Classical Library edition, Illinois Greek Club, translator, Cambridge, MA, 1923)
Asclepiodotus, *Tactics*	Aeneas Tacticus, Asclepiodotus, Onasander (Loeb Classical Library edition, Illinois Greek Club, translator, Cambridge, MA, 1923)
Frontinus, *Stratagems*	Frontinus, *Stratagems* and *Aqueducts of Rome* (Loeb Classical Library edition, Charles E. Bennett, translator, Cambridge, MA, 1925)
Leo, *Tactics*	*The Taktika of Leo VI* (Dumbarton Oaks edition (revised), George T. Dennis, translator, Washington, 2014)
Maurice, *Strategikon*	*Maurice's Strategikon: Handbook of Byzantine Military Strategy* (University of Pennsylvania Press edition, George T. Dennis, translator, Philadelphia, 2001)
Onasander, *The General*	Aeneas Tacticus, Asclepiodotus, Onasander (Loeb Classical Library edition, Illinois Greek Club, translator, Cambridge, MA, 1923)
Vegetius, *Epitome of Military Science*	Vegetius, *Epitome of Military Science* (N. P. Milner, translator, Liverpool, 1993)
Xenophon, *The Cavalry Commander*	Xenophon, *Scripta Minora* (Minor works) (Loeb Classical Library edition, E. C. Marchant and G. W. Bowersock, translators, Cambridge, MA, 1960)

Xenophon, *The Educa-tion of Cyrus*	Xenophon, *Cyropaedia* (2 vols) (Loeb Classical Library edition, Walter Miller, translator, Cambridge, MA, 1914)
Xenophon, *On Horsemanship*	Xenophon, *Scripta Minora* (Minor works) (Loeb Classical Library edition, E. C. Marchant and G. W. Bowersock, translators, Cambridge, MA, 1960)

For non-military authors, all translations come from the Loeb Classical Library edition (except where otherwise noted by the author).

Introduction

The ancient military treatise, genre, and history

James T. Chlup and Conor Whately

To point out that war continues to be a unifying human experience would seem a *sine qua non* in a book that seeks to explore an aspect of warfare, regardless of its temporal frame. Every person, regardless of historical period or geographic location, has a connection, direct or indirect, to this irrational and destructive phenomenon; it cuts unapologetically across time, geography, and culture. In our modern world, even those fortunate enough not to have experienced war as a soldier or a civilian victim still face it in the media on a near-daily basis, as television series, films, video games, and literature frequently feature war. It is also one of the main facets of the news: at the time of this volume going to press, the Syrian civil war, the campaign against DAESH, the war in Yemen, the war in Afghanistan, and the poorly named 'war on terror' are (still) ongoing; and some of these wars are directly connected, as one war feeds into, and off of, another. Most critically, scholars and pundits debate – most nervously, a few enthusiastically – the probability of future war (between the United States and China, the United States, Israel, and Iran, or the NATO alliance and Russia, for instance).[1]

In the ancient world, war was an intrinsic part of the identity of the community.[2] Heraclitus' comment that 'war is the father and king of everything' possibly reveals a disturbing paradox: war can nurture as a father and rule (one presumes harshly) like a king.[3] A prescient comment from Plato in his *Laws* confirms that the ancient Greeks understood war as the default condition for humanity, resistance against which is futile: 'For (as he would say) "peace", as the term is commonly employed, is nothing more than a name, the truth being that every State is, according to nature, engaged perpetually in an informal war with every other State'.[4] Simply put, Plato argues that war is the natural, and by extension peace the unnatural, state of human affairs. Aristotle, on the other hand, advances the belief that there is a symbiotic relationship between war and peace, whereby one needs the former in order to enjoy the latter: 'we make war in order that we may live in peace'.[5] Other prominent Greek authors readily admitted war as a negative human event.[6]

'It is not possible to fail twice at war', according to Arsenius – at least one should so hope.[7] While this proverb may mean that to fail once was enough – it could mean the loss of one's life – it may also mean that one could learn from mistakes. In other words, war could be improved upon (it is surely impolitic to call it 'progress'). This could be what Demosthenes has in mind when he remarks in his *Third Philippic* that 'in my view, while practically all matters have made a great advance and we are living today in a very different world from the old one, I consider nothing has been revolutionalised more than the practices of war'.[8]

Unfortunately, Demosthenes does not specify what the advances in warfare were or – much more critically, one might argue – *how* individuals and communities came to reflect upon the changes. However, a generous reading between the lines of Demosthenes' comment is that there was a thought process that entailed careful reflection upon past practice with a view to avoiding repetition of past mistakes and ensuring the recurrence of past successes, the overall result of which would be a war better waged to a significant degree.[9] This would seem to create an intellectual discursive space for the emergence of a literary genre – or (sub) genre – to assist in this process. Aelian, who himself in reply to the passage from Plato quoted previously, adds: 'If this is the case, then what discipline is to be more esteemed, or more necessary to a man's life, than the art of war?'[10]

By the time Demosthenes made his remark in the mid-fourth century BCE, this process had begun; in fact, it had been under way for some time, moving step-by-step alongside a defining event of Greek culture. Oliver Spaulding not quite a century ago opens his study of ancient military writing by noting that 'for the Greek military system, we must of course begin with Homer, that source of everything Greek'.[11] The entry for the art of war in ancient Greece in the most recent edition of the *Oxford Classical Dictionary* confirms that scholars still recognise the Homeric *Iliad* as the earliest account of warfare in the ancient Greek world. A very recent publication on Greek warfare concurs: 'Homer . . . composed a work that had an incalculable impact on every aspect of Greek life, including, *perhaps most of all*, war'.[12] This is, of course, simply an acknowledgment of what ancient military authors thought. Aelian, for instance, unequivocally declares that 'Homer wrote first about the theories of strategies in warfare'.[13] Polyaenus concurs when in his preface he provides no fewer than ten quotations from the Homeric poems. Alexander the Great, for instance, considered the *Iliad* to be a military guide-book *par excellence*; this is surely a qualified endorsement.[14] The role of the *Iliad* as a foundational text of 'Western Culture' (artificial and problematic construct though that may be) also means that the military knowledge that its author(s) seek to impart forms at least part of its cultural legacy.[15] If – and surely this need not be a big 'if' – it is the first military manual, or a proto-manual, then it establishes an impressive pedigree for this genre.[16] In other words, it was the textbook of generations of military leaders, some of whom in turn became objects of study in the practice of warfare.

I. The military manual and the study of ancient warfare

Despite the pervading interest in ancient warfare by scholars and the general public, a most unfortunate reality is the unconscious marginalisation of ancient military writing in scholarship and popular literature. In the university lecture hall, courses on ancient warfare are increasingly common, but these do not normally entail students engaging deeply with military manuals.[17] This is rather unfortunate, since these texts provide valuable insight into ancient Greeks' and Romans' thinking (or thinking about thinking) about warfare. In other words, one should not seek to study the practice of warfare on its own (for example, the minutiae of weapons, training, specific campaigns) only; one should also seek to capture the flavour of how ancient experts *and* amateurs *thought* about warfare.[18]

The confidence of the ancients did not translate into the interest of the modern, whereby scholarship, lamentably, does not appear to agree, at least not fully, declining to afford military manuals full attention that some feel they merit.[19] Yvon Garland's chapter on warfare in the fourth century BCE in the *Cambridge Ancient History* notes the professionalisation of the army, which 'provoked the appearance . . . probably under the influence of the sophists, of a body of technical literature', that is, the military treatise, which itself marks the rise of military 'experts' – here he might mean authors of military texts and their readers.[20] Some of the other chapters across the series also draw upon and reference military treatises.[21] In English, *The Cambridge History of Greek and Roman Warfare* is an invaluable resource to the scholar and student alike, but it eschews a specific chapter or substantial discussion on ancient military texts as a whole, though specific references to individual military authors and texts appear *passim* in citations.[22] Paul Erdkamp's *A Companion to the Roman Army*, an important resource especially for students, utilises historians much more than military authors, with only a few citations of the latter.[23] Finally, Blackwell's *Encyclopedia of Ancient Battles* provides references to primary sources on individual battles, but these are exclusively to historical texts.[24] This is a particular shame, since some military texts – those of Frontinus and Polyaenus, for instance – consist of tactics employed in historical battles.[25] One might argue here that the exception is J. E. Lendon's masterful *Soldiers and Ghosts*, which ascribes to military manuals a place in its discussion.[26]

This, in part, may be the unfortunate result of the (r)evolution of the study of warfare over the past century. As Victor Hanson observes, the publication of John Keegan's *The Face of Battle* encouraged scholars to think beyond tactics and begin to consider 'what battle was actually like for the men who did the fighting and the dying'.[27] This focus on the front-line soldier's perspective would initially seem incompatible with the top-down general's perspective that seems to be the focus of the military treatise; historical narrative offers a better line-of-sight in this instance, conveying (sometimes) the point of view of soldiers in the front-line.

These examples should be sufficient to highlight that ancient military manuals, unfortunately, rest farther from the centre of ancient warfare studies than perhaps they ought. These should not be obscure texts in the purview of only a small, eccentric cadre of scholars who seek to explore very narrow questions or engage in philological analysis.[28] The Loeb Classical Library, which is perhaps the most widely consulted series, provides English translations of five military authors: Aeneas Tacticus, Asclepiodotus, and Onasander appear together in a volume; Xenophon's *Cavalry Commander* (*Hipparchus*) and *On Horsemanship* (*Peri Hippikes*) appear in the volume of his shorter works; and Frontinus, whose

Stratagems (*Strategemata*) appears with his *On the Aqueducts of the City of Rome* (*De aquaeductu urbis Romae*).[29] Liverpool University Press' series of texts and commentaries includes what is arguably the most famous military treatise: Vegetius' *Epitome of Military Science* (*Epitoma rei militaris*). Brian Campbell's source book in translation also makes these texts accessible to readers without knowledge of ancient Greek or Latin. Translations of these texts, either as stand-alone editions or as part of commentaries, are also available in other languages.[30] And good commentaries – a common facet of classical scholarship – are available for some texts.[31] Students and even scholars without philological expertise should therefore not feel intimidated to become better acquainted with these authors.

II. Genre and history

> I say that this is bullshit and I say the hell with it. 'Genre', if it means anything at all, is a restrictive commercial requirement.
>
> Norman Spinrad

The chapters in this volume concern themselves with two not unrelated, though not necessarily interdependent, topics that pertain to the military treatise: genre and history.[32] The former – a structuralist fetish to some – concerns the extent to which authors agree to uphold a set of delineated textual characteristics to which a text ought to adhere; the latter relates to the seemingly close, if not (potentially uncomfortably) intimate, relationship between the military treatise and the events upon which it draws its lessons, both the actual event itself and the (re)creation of the event in a historical narrative. Here one could (should?) take 'history' as meaning 'reality', that is, a realistic representation of warfare as it occurred in the past and the rational belief in how it may unfold in the present and future.

Ancient authors and readers embraced their structuralist leanings in their understanding the conventions of literary genres. The epic poet, for instance, provides, in dactylic hexametre, an invocation of the muse, speeches, ecphrasis, involvement of divine beings in a multifaceted narrative on a grand theme – quite oftentimes war.[33] Each poet understood that he worked within a tradition of the genre. The same holds for historical narrative: a programmatic preface, ethnographic digressions, speeches, and last but not least engaging descriptions of battles as the garnish to, if not the main course of, foreign affairs (*res externae*).[34]

Heretofore, there has not been substantial agreement on the conventions of the military manual as a genre – or even if it is a genre in its own right.[35] For instance, the tendency has been to situate military texts under the broad and quite bland terms of 'technical writing' (sometimes including also the term 'scientific') or 'technical literature'.[36] On the one hand, the term 'technical' does serve to convey that these texts seek to impart practical knowledge or craft (τέχνη, μάθημα, θεωρία, ἐπιστήμη; *scientia* or *ars* in Latin), as if to plan how to fight a battle is an activity similar to the design and maintenance of an aqueduct, for instance.[37] And the Greeks had a word that could convey 'art of war': πολεμικός.[38] The possible classification of military texts as 'literature' is intriguing, since it allows for the possibility that one may choose to read a military text for pleasure; that is, they are not the exclusive prevue of the professional soldier or general, but the casual enthusiast also.[39] If – hopefully this is not too big an 'if' – war is an art that one studies and then seeks to replicate, then surely the writing about war is a literary 'art'.

What is in a name? One of the challenges in beginning to contemplate these texts comes from the difficulties in agreeing upon appropriate nomenclature: that is, 'manual', 'treatise', 'monograph', or 'military literature'; these would seem to allow for varying frames of exegesis.[40] The first term – manual – is, arguably, (the most) optimistic, but sadly also potentially the most problematic: it suggests – implies? insists? – that one reads in order to learn and later to emulate; therefore, there is a great deal at stake in using this term with regards to the estimation of the qualitative value of the texts.[41] The author writes and the reader studies in some kind of implicit transactional relationship by which the former promises the latter that his advice guarantees success. The second and third seem to reflect more their status and length: short(ish), self-standing texts, perhaps more an 'exposition of enquiry' (λόγου ἀπόδεξις) than something of specific utility.

If – and hopefully this is not necessarily a significant 'if' – the military manual does merit consideration as a genre in its own right, how would one (begin to) define its parametres? Herein a not insubstantial challenge presents itself, since the extant texts appear to fall into two very broad, yet seemingly distinct categories: the theoretical and the specific. The former describes battle and war in general terms, providing suggestions for situations that come up in most, but admittedly not all, conflicts; in some ways it imagines a 'perfect' battle or war, where the optimal conditions exist for the reader-general to put into action exactly the advice proffered by the author. Aeneas Tacticus, Onasander, Vegetius, and Maurice are examples of this type. The latter consists of examples from historical battles: Frontinus and Polyaenus are examples of this type – though in the case of the former, it was probably a supplementary volume to a more generalised treatise like the first category. Both categories would seem to have merits and shortcomings. While both types of texts clearly (ought to) belong to the same group or genre, they in fact may seem complementary: this may be why Frontinus wrote a theoretical text and a exempla-based treatise, with the latter designed to demonstrate the principles of the former. Of course, an author from one category does not face a prohibition from having his text draw upon aspects of the other category: for example, Aeneas Tacitus refers to specific historical episodes in support of stratagems.[42] Be that as it may, the question of genre may have to include the contemplation of division of its texts into subgenres. Or – to put forward a more positive 'spin' – acknowledge that the wide range of texts is evidence of the genre's multivalence: the differences between individual manuals reveals that ancient Greek and Romans believed there were many ways to teach, and therefore to learn about, war.

What they appear to share in common is the desire to convey a transhistorical and transcultural perspective of ancient warfare. The theoretical manual presumes that one can establish commonalities to warfare – situations, actions, and reactions – that occur in every conflict; the *exempla*-style treatise describes a stratagem employed successfully and often counteracted with an equal degree of success, several times across centuries by generals of different communities.[43] In both instances, one may surely reasonably argue, later cultures (that is, Roman, Byzantine) could understand an additional context through which they could – and ought – to establish a connection to the past.

One might suppose that the relationship between the military manual and historical narrative may be a fraught one, since they both lay claim as an, if not *the*, authoritative voice for the practice of war.[44] After all, the first great historians – Herodotus, Thucydides, and Xenophon – made war a prominent, if not the organising, subject of their works; the latter two also had first-hand experience of war in positions of command.[45] Herodotus and Thucydides, for instance, use the same word – ἐπολέμησαν ~ 'they waged war' – in their opening sentences, making war the *sine qua non* in their narratives, if not of historical narrative henceforth.[46] The aorist (past tense) marks the war as complete, an organic (w)hole, available for thorough necropsy. Moreover, Thucydides' and Xenophon's penchant for describing instances of stratagems and ingenious generalship would seem to necessitate its exploration in a more focused text.[47] Xenophon's exploration of aspects of warfare across so many different texts earned him the title 'the first military writer'.[48]

The Roman historians, understandably, worked within this narrow rubric; this was not an area in which they sought to chart a new course. Caesar set things in motion with great aplomb, with narratives on foreign *and* domestic war. Sallust chose a war as the focus for one of his monographs: the *Jugurthine War* (*Bellum Iugurthinum*); though the war does 'spill over' into politics (or does politics spill over into the war?).[49] Livy uses war as a means to structure his narrative: some pentads or decades centre around a particular conflict.[50] To the Romans, war was an equal partner in the narrative (*res*) in the record of events past (*memoriae rerum gestarum*).[51] At the other end of antiquity, Procopius produced an interconnected series of histories on wars.[52] One may argue that military manuals that contain historical *exempla* serve to direct (back) their readers to historical narratives that describe and contextualise those military events. The strong focus of historical narrative on war makes it an invaluable source for scholarly studies.

From the perspective of authors of military manuals, there was possible cognisance of potential criticism for regurgitating what appears in historical texts. Frontinus' preface in his *Stratagems* clearly betrays this concern: 'I neither ignore nor deny the fact that historians have included in the compass of their works this feature also, nor that authors have already recorded in some fashion all famous examples'.[53] Frontinus confesses that his reader might ponder whether it is more fruitful to read, say, Livy or Polybius instead of his treatise. In the next sentence, however, he claims that the medium is the message: his text is a more efficient means of acquiring information on the most effective means to win a war: 'I ought, I think, out of consideration for busy men, to have regard to brevity. For it is a tedious business to hunt out separate examples scattered over the vast body of history'.[54] A reader with more time for study, however, may choose to read military text in the first instance, and coming to a basic understanding of the principle of a particular stratagem, then desire to appreciate it in its full(er) historical context by reading a historical narrative. Therefore, one might recognise a symbiotic relationship between the military manual and historical narrative, with the latter providing additional insight into the principles delineated in the former. They can exist comfortably side-by-side under the broad term 'military literature'.[55]

The fact that some authors – Xenophon and Polybius are two particularly noteworthy examples – wrote historical *and* military texts (and, in one instance, possibly inserted an example of the latter in one of the former) surely encourages this line of enquiry.[56] And then there is Xenophon's genre-transgressing (or genre-busting) *Education of Cyrus* (*Cyropaedia*), 'the amusement of his later years, the vehicle for his military fancies'.[57] It was a text that proved *very* popular with ancient generals: Scipio Aemilianus kept a copy with him; Caesar and Cicero also may have read it.[58] This surely establishes the *bona fides* of this text *qua* military manual. However, one must then surely engage in a chicken-and-the-egg scenario: did Xenophon and Polybius write history because they wanted to explore warfare in its broad(er) contexts, or did they write military manuals out of a concern to rationalise, or perhaps simplify, the war experience as they tried to understand it when writing history?[59]

III

Like Caesar's Gaul, this volume is in three parts. The first three chapters delineate contexts for beginning to think (or to begin to think about thinking) about ancient Greek and Roman military manuals. **Conor Whately** provides a broad introduction to the military manual, noting its possible origins, intended scope, and current scholarly approaches. In a bid to broaden the range of approaches applied to military manuals, this chapter focuses on the issues of genre and audience, and returns to the relationship between manuals and historiography. **Hans Michael Schellenberg** offers an emphatic caveat on how one reads – or perhaps ought not to read – military manuals. Whereas ancient readers and scholars may want to consider the description of a military tactic as historically accurate, and therefore the manual as a whole as an historical source, this may not be the case. A degree of uncertainty surrounds our understanding of ancient warfare: thus the ability of authors to fully and accurately represent warfare is limited. It may seem counterintuitive (an 'own goal', as one says in football) to include in this volume a work that problematises one of the main scholarly approaches to engaging with military manuals, and, in fact, is the *modi legendi* of some of the subsequent chapters. The reader will (hopefully) acquire an appreciation of the difference of opinion that exists in the study of the military manual, and future scholarship will understand the need to reflect upon, and potentially begin to contemplate a means to reconcile, this difference of opinion.

Nadya Williams addresses an important topic for any genre, and one necessary for contemplation of 'genre': the frames of reference of the author and reader of a military manual. This is an especially important topic in this instance, because the nature of military 'knowledge' is such that one may presume an author to be an 'expert' and the reader to be someone in search of 'expertise'. However, this is not always the case, since some authors and readers are amateur enthusiasts.

Then follow eleven studies that examine different aspects of the ancient military manual across a period of approximately a millennium and a half, from Archaic Greece to the post-Roman world, covering many (but admittedly not all) of the major authors of the 'genre': Xenophon, Aeneas Tacticus, Frontinus, Polyaenus, Vegetius, Maurice, and Leo. The chapters boldly delineate the military manual as a multivalent genre in that authors appear to be offering more than advice on how to win a battle or war only: that is, ancient military authors sought to make broader, qualitative points about warfare – and their communities. As with most collections of essays that examine different aspects of a variety of texts, the goal is not to provide definitive answers, but rather to open up lines of enquiry and to convey an appreciation of the multifaceted nature of the military text. And if, as the quotation at the start of this introduction suggests, war is a form of tyranny, then perhaps the military manual is a democratic discourse in that it seeks to provide the knowledge on ways to the most practical path to bringing about its successful conclusion.

Nicholas Sekunda considers a unique, form of military writing: Homeric tactical treatises. On the one hand, this chapter underscores the continuing relevance throughout ancient Greek and Roman history of the Homeric epics to the formulation of military thought and writing. However, the nature of what these texts articulated about warfare in Homeric verse, and in what sense the authors sought to apply those 'tactics' to real-world scenarios, is not easy to discern.

Graham Wrightson appears to employ the scholarly approach against which **Schellenberg** cautions by exploring a historical military event side-by-side with potentially pertinent military texts. This is not an accident: the presence of these chapters serves to provoke reflection on a persistent approach to using military texts. At the same time, **Wrightson** explores how military manuals across time can provide insight into a particular kind of warfare: the siege. Demetrius' siege of Rhodes was *sine dubio* one of the most famous sieges of antiquity. Ancient texts that provide advice on sieges do so from the position of the community that seeks to endure the siege. This chapter inverts that perspective by mining information of siege defence to examine how it could inform an *offensive* siege – or when examined alongside an example of an unsuccessful assault, inform how *not* to conduct a siege.

Mercenary armies were, to the chagrin of many, an enduring feature of ancient warfare, presenting both clear advantages and disadvantages, but they do not receive much attention in military manuals. **Aaron L. Beek** examines how mercenary warfare features in Polyaenus' *Stratagems*. In so doing, Polyaenus makes a broader political and cultural argument about the general (who in this case could be the Roman emperor) and the army. In this instance, an author composes a military manual that appears to work outside its narrow rubric – providing examples of military scenarios to emulate and avoid – to maintain the status quo, to which the use of mercenary forces may represent a threat.

Few doubt Xenophon's contribution to the development of the genre.[60] The chapters of **Lucy Felmingham-Cockburn** and **Jeffrey Rop** examine one of Xenophon's tactical treatises – *On Horsemanship* – and a work that, while not a military manual *stricto sensu*, evinces aspects of that genre: the *Education of Cyrus*. In *On Horsemanship*, Xenophon's vocabulary establishes a myriad of connections between the military treatise and other literary works of almost every major literary genre. The implication is that a military treatise does not exist in a literary vacuum, but exists in a relationship to other texts, contributing to wider political and cultural discourses. As noted earlier, the *Education of Cyrus* is a peculiarly enigmatic text, appearing as a unique hybrid of many genres; one might perhaps proffer to describe it as 'intergeneric'. One genre to which it may belong is that of a military manual, with a focus on the qualities of being a good general, a topic that surely is central to military study. It is for this reason that several famous generals regarded it very highly.

The next four chapters discuss Roman military treatises. **Murray Dahm** reviews evidence for a possible lost military manual from the late Republic, which M. Tullius Cicero may have received. The possible relationship(s) between this lost text and earlier (some extant, some missing) treatises further enhances awareness of how a treatise is not a stand-alone work but exists within a broader literary tradition – at least that may be how those working outside of the genre thought of it. It also raises potential answers for why some manuals come into being: in response to a military defeat, which causes a need for these texts to allow possible military commanders to refresh their knowledge. **James T. Chlup** seeks to explain Frontinus' presentation of Hannibal's victory and, interestingly for a Roman military text, the Romans' defeat at Cannae. On the one hand, Frontinus shows a willingness to do what few authors of military texts do: explore Roman defeat, which entails affording the Romans' greatest enemy a prominent place in the text. On the other hand, the way in which Frontinus contextualises the *exempla* that feature Hannibal and the Romans at Cannae dilutes Hannibal's victory and mitigates Roman defeat.

To be sure, Vegetius casts a *very* long shadow across the genre of the military treatise, where his immense popularity in the medieval, pre-modern, and modern worlds makes his *Epitome of Military Science* (*Epitoma Rei Militaris*) the 'most widely read military book in Europe until Carl von Clausewitz's *On War*'.[61] Two chapters on this important author discuss different aspects of his text. **Jonathan Warner** explores the relationship between the collection of maxims at the end of book 3 and the manual as a whole, as well as contextualising this section with the broader literary-historical list tradition (a possible literary 'sub' genre?). The latter creates a further point of contact between the military manual and ancient literature. **Craig H. Caldwell** considers the possible relationship between Vegetius' naval appendix and the naval battle between Licinius and Constantine, which placed the latter as sole emperor. While Vegetius does not explore specific *exempla* in the manner of Frontinus and Polyaenus, a recent historical event – an *exemplum* implied or inferred – may impact interpretation of the *Epitome*.

The final two chapters explore Byzantine manuals and warfare. **Clemens Koehn** explores the extent to which military manuals of the period reflect the changing practice of war, in this instance evidence for Justinian initiating a new way of waging war, that is, that there was a shift from the 'western' to a 'Byzantine' way of fighting, especially as conveyed through Maurice's military manual. **Meredith L. D. Riedel** examines the military manual of the Byzantine emperor Leo the Wise. It is a unique work of consolidation and innovation: while Leo's *Tactics* draws upon earlier military texts, he also imbues it with Christian ideals – that is, he presents a Christianised general. This balance of imitation and innovation would appear to give a new lease of life to the military manual. These chapters appear to agree that one aspect of Byzantine military thinking is the focus on how to conceptualise the enemy. Whereas the Greeks and Romans considered themselves as superior fighters to their foreign (barbarian) enemies, by the Byzantine period there appears to be a recognition that this was no longer the case. In this sense, Maurice and Leo balance between following the generic traditions of the military manual and creating texts that are relevant to the shifting realities of warfare in that period.

The volume concludes with an epilogue by **Immacolata Eramo**, whose contribution surveys some early modern editions of military treatises, the history of the genre in the ancient Mediterranean, and then the role of medieval Byzantium in bringing additional definition to the genre. One of the many important themes covered is the relationship between history (wars especially), historiography, and the military literature. **Eramo** also aims to provide a set of criteria that delineate what constitutes military literature in the ancient Mediterranean world, and revisits the issue of readership, discussed in previous chapters.

Notes

1 On the possibility of the first of these, at least one scholar uses an ancient war (and an ancient interpretation of that war) as a filter: Allison (2017). We do not claim originality for beginning a study of ancient military writing with reference to the pernicious presence of war in the modern world: note the opening of Spaulding (1937) 1, who claims that there will be no more wars!

2 As demonstrated by Pritchett (1971–79); see also Hanson (1994); Dillon and Welch (2006); and Chlup (2012). More broadly, see Gat (2005).

3 πόλεμος πάντων μὲν πατήρ ἐστι, πάντων δὲ βασιλεύς (53 DK = 29 Markovich). See also the Epilogue (268).

4 626a, ἣν γὰρ καλοῦσιν οἱ πλεῖστοι τῶν ἀνθρώπων εἰρήνην, τοῦτ᾽ εἶναι μόνον ὄνομα, τῷ δ᾽ ἔργῳ πάσαις πρὸς πάσας τὰς πόλεις ἀεὶ πόλεμον ἀκήρυκτον κατὰ φύσιν εἶναι. At *Republic* 372d–373e and *Phaedo* 66c, he observes that the desire to accumulate wealth is the prime motivator for war. See also Sage (1996) xi.

5 *Nicomachean Ethics* 1177b4, πολεμοῦμεν ἵν᾽ εἰρήνην ἄγωμεν. Aristotle goes on to observe that no one willingly engages in warfare for its own sake: 'no one desires to be at war for the sake of being at war, nor deliberately takes steps to cause a war: a man would be thought an utterly bloodthirsty character if he declared war on a friendly state for the sake of causing battles and massacres' ~ αἱ μὲν πολεμικαὶ καὶ παντελῶς (οὐδεὶς γὰρ αἱρεῖται τὸ πολεμεῖν τοῦ πολεμεῖν ἕνεκα, οὐδὲ παρασκευάζει πόλεμον: δόξαι γὰρ ἂν παντελῶς μιαιφόνος τις εἶναι, εἰ τοὺς φίλους πολεμίους ποιοῖτο, ἵνα μάχαι καὶ φόνοι γίνοιντο).

6 For example, Polybius (4.31.3), Thucydides (4.59.2), and Euripides (*Trojan Women* 400): see also Chapter 13 (240 n.51).

7 13 39h, οὔκ ἐστιν ἐν πολέμωι δὶς ἁμαρτάνειν.

8 §47, ἐγὼ δ᾽ ἁπάντων ὡς ἔπος εἰπεῖν πολλὴν εἰληφότων ἐπίδοσιν, καὶ οὐδὲν ὁμοίων ὄντων τῶν νῦν τοῖς πρότερον, οὐδὲν ἡγοῦμαι πλέον ἢ τὰ τοῦ πολέμου κεκινῆσθαι κἀπιδεδωκέναι. The verb κεκινῆσθαι may be a double entendre, since it can also mean to set an armed force in motion, that is, to order the battle-line to advance.

9 See Echeverría Rey (2010) and Isaac (2017) 82–98.

10 πῶς οὖν ἄλλο τις ἡγήσεται προὐργιαίτερον μάθημα ἢ τὸ βιβλίον χρειωδέστερον τούτου. On μάθημα, see later, 4.

11 Spaulding (1933) 658 and (1937) 7.

12 Sears (2019) 1–2 (italics are mine). The first chapter of Sage (1996) focuses on Homeric warfare; and note how heavily Krentz (2007) draws upon the *Iliad* in his survey of warfare in the Greek Archaic period.

13 Ὅμηρος πρῶτος περὶ τῆς ἐν τοῖς πολέμοις τακτικῆς θεωρίας ἔγραψεν. So too Niceratus in Xenophon's *Symposium*: 'the sage Homer has written about practically everything pertaining to man. Any one of you, therefore, who wishes to acquire the art of the householder, the political leader, or the general' ~ Ὅμηρος ὁ σοφώτατος πεποίηκε σχεδὸν περὶ πάντων τῶν ἀνθρωπίνων. ὅστις ἂν οὖν ὑμῶν βούληται ἢ οἰκονομικὸς ἢ δημηγορικὸς ἢ στρατηγικὸς γενέσθαι (4.6). Vela Tejada (2004) provides a succinct analysis of the evolution (emergence) of military writing; it is a good starting point for anyone new to the genre. He discusses Homeric poetry specifically at 130–132. See also Lendon (2005) 20–38; Dueck (2011); and in this volume, Chapter 1 (18), Chapter 4, and the Epilogue (265–266). On war as an 'art', see later, 4; and the Epilogue (265).

14 Plutarch, *Alexander* 8.2: 'he reflected upon and called the *Iliad* a ways and means of the art of warfare' ~ καὶ τὴν μὲν Ἰλιάδα τῆς πολεμικῆς ἀρετῆς ἐφόδιον καὶ νομίζων καὶ ὀνομάζων. Note Plutarch's use of ἀρετή – not a standard word in this context, to be sure.

15 Lest one not consider potentially texts from other ancient cultures, a useful comparandum in this instance is the influence of the *Mahabharata* and other Sanskrit literature on ancient Indian warfare: see Singh (1965) and Chakravarti (1972). So too one could postulate that ancient Israelite texts (the Book of Joshua in the Hebrew Bible and the War Scroll (1QM), for instance) serve as military manuals of a kind. In fact, the latter may have a connection to Greek and Roman military manuals: Yadin (1962) regards it as a military treatise; Duhaime (1988) 134–137 agrees, who explores parallels between the War Scroll and the treatises of Asclepiodotus, Aelian, and Arrian (137–151). A cross-cultural study of ancient military treatises (here one should extend discussion to ancient Chinese texts) is a desideratum.

16 A honourable mention should go to Tyrtaeus, a military lyric poet: see Luginbill (2002) and Vela Tejada (2004) 132–133, who notes that Tyrtaeus and the Homeric poet share 'an agonal ideal'.

17 Several contributors to this volume have developed courses (and in one case a textbook) on ancient warfare, and in so doing continue to reflect upon the appropriate role of military manuals. The *communis opinio* is that students should familiarise themselves with military manuals, but the extent to which students should study them in-depth is a matter of instructor preference.

18 That is, it is important to consider writers of military texts who had direct experience of warfare and those who appear to have written about it for academic interest. See Chapter 3 in this volume for discussion of the latter.

19 The reader should not take what appears in this and the next few paragraphs as a back-of-the-hand slap of criticism on previous scholarship on ancient warfare. It merely seeks to raise awareness of the military manual's unfortunate position too close to the periphery of broad studies on warfare. See also Chapter 1 in this volume, specifically 21–22. As Singh (1965) and Chakravarti (1972) make clear, studies of ancient Indian

warfare place literary texts – including military manuals – very much indeed front and centre of their discussion.

20 Garlan (1994) 679–680. He lists Xenophon's *Cavalry Commander, On Horseman-ship*, and Aeneas Tacticus' *How to Survive under Siege* as examples, to which he adds pseudo-Aristotle's *Oikonomikos* due to its discussion of 'strategies' for the payment (or not) of mercenaries.

21 Keppie (1996) and Campbell (2005) rely more on historical texts, inscriptions, and papyri; Lee (1997) and especially Whitby (2001) make use of military texts.

22 The only focused discussions appear to be in Catherine M. Gilliver's and Philip Rance's chapters on Roman battle in the Republican and late antique periods, respec-tively (124–125; 343–348). Hans Michael Schellenberg makes a similar observation in his chapter: 51 n.44.

23 James Thorne's chapter on battle and tactics in the West is perhaps an exception in its use of Onasander in discussing battle from the general's perspective. The companion as a whole is, unfortunately, slightly frustrating in that the general index and the index locorum do not include all citations of military texts that appear in the chapters.

24 There is, however, *some* discussion of military texts in one of the introductory chapters: in discussion of sources on ancient siege warfare, for example, Harry Sidebottom cata-logues and comments briefly upon the subgenre of siege manuals (42, 44); however, he then notes, perhaps as an oblique *modus scribendi* for the work as a whole, 'more important than the specialist literature are the ancient historians' (44); by 'specialist literature', he surely means military treatises.

25 And possibly Vegetius too, though obliquely: see Chapter 12.

26 Lendon (2005) 279–285, for instance.

27 Hanson (1994) 24–25. See Rance (2017) 41–42.

28 See Rance (2017) 29–36.

29 These volumes are approaching their centenary; fortunately, the lapse of their copy-right means that they are now also available to the general public online (for example, LacusCurtius (http://penelope.uchicago.edu/Thayer/E/Roman/home.html) provides texts for Aeneas Tacticus, Asclepiodotus, Frontinus, and Onasander). Revised editions are surely very much desirable; volumes of Aelian, Arrian, Polyaenus, and Vegetius are highly desired as future additions. We note also that there used to be (at least there was a few years ago) a website devoted to Aeneas.

30 For instance, De Gruyter is currently issuing new German translations by Kai Broder-sen of several authors: Arrian, Asclepiodotus, Aeneas Tacticus, Polyaenus, and Xeno-phon's *On Horsemanship* have appeared thus far. In the Budé series, there are editions of Aeneas Tacticus and Xenophon's *On Horsemanship* and *Cavalry Commander*.

31 For example, Milner (1993); Whitehead (2002); Petrocelli (2008).

32 For further discussion of these topics and the military treatise, see Whately (2015) and Chapter 2; Rance (2017) 23–27. On genre more broadly, including the ongoing debate in critical theory, see Duff (1999) 1–24; Bawarshi and Reiff (2010); and Frow (2014).

33 In addition to the *Iliad*, Lucan's *Pharsalia* and Silius Italicus' *Punica* are especially pertinent examples; though mythological in its subject, one might consider also Sta-tius' *Thebaid*. Their historical rather than mythological settings of Lucan's and Silius' poems, where they can be read side-by-side with a relevant historical narrative, may lend themselves also to consideration as poetic military manuals, though admittedly the former presents challenges, though not without precedent, in providing potential areas of imitation from civil war: Frontinus includes stratagems from Roman civil war, not necessarily unproblematically: see König (2018).

34 The study of battle descriptions in historical narrative has understandably received some scholarly attention: for example, see Levene (2010) 261–300 on Livy, with addi-tional bibliography; Whately (2016) on Procopius; and Lendon (2017a) and (2017b).

35 Conor Whately in his discussion of Xenophon's *Education of Cyrus* 1.6.41–43 notes what some might take as a rubric *par excellence* for the military treatise (18–20).

36 The latter term is the category under which Frontinus' works appear in the *Cambridge History of Classical Literature*. For a broad theoretical approach to technical writing, see Formisano (2017).

37 Some examples include: τέχνη: Leo, *Tactics* 1.2. μάθημα: Aelian 1. θεωρία: Onasander Preface 1; Aelian, Preface ἐπιστήμη: Polyaenus, Preface 2., Syrianus Magister 14, Leo, *Tactics* 1.1 (also a term favoured by Thucydides with specific reference to skill in war – 1.121.4, 6.72.4, 7.62.2). *Scientia*: Frontinus, Preface Book 1; *Ars*: Vegetius Preface Book 1. Onasander appears to stand apart in extant military texts by using in one instance λόγος – this is clearly a (not wholly unwelcome) carry over from his philosophy. ἱππική (for instance, at Xenophon, *On Horsemanship* 1.1) implies τέχνη (per LSJ); this is surely what Xenophon means in his *Cavalry Commander* and *On Horsemanship*. To move forward to the Renaissance, Machiavelli refers to war as an 'art' (*arte*) in his *On the Art of War* (*Dell'arte della Guerra*) and elsewhere. More generally, see Derla (1996) with additional bibliography; Heuser (2016).

38 As in Plato, *Sophist* 222C.

39 In fact, as Chapter 3 in this volume considers, some authors of military texts may have had the lay reader specifically in mind.

40 The terms listed previously are not various English translations of a single word in ancient Greek or Latin. Ancient authors did not limit themselves to a single term: in Greek, for instance, two terms – στρατηγικά and τακτικά – appear; in Latin, *praecepta* seems to occur (for example, Vegetius 2.3 and 4 Preface). See also Rance (2017) 23–24. On 'military literature', see Chapter 2 and n.2 in this volume. Schellenberg also notes the corresponding German terms.

41 See also Formisano (2011) 43.

42 In this instance, see Chapter 2.

43 Chapter 10 discusses the example of Hannibal and Romans in this scenario.

44 To which one should add biography: Gazzano and Traina (2014) and Jacobs (2018). Xenophon's *Agesilaus* and Tacitus' *Agricola* are additional possible examples.

45 Thucydides mentions his naval command of 424 (4.104.4ff), while Xenophon features prominently in his *Anabasis*. On historiography, see Vela Tejada (2004) 135–140. Sage (1996) and (2008) provide almost exclusively historical authors; Campbell (2004) includes salient extracts from historical authors such as Thucydides, Polybius, Caesar, Josephus, and Cassius Dio. See his rationale in his preface (xxi). Note Schellenberg's criticism of this in his chapter.

46 See Petrocelli (2012) 16–19. See also the Epilogue in this volume (269–270).

47 Wheeler (2010) 435.

48 Spaulding (1933) 658–659.

49 See Syme (1964) 138–177; Kraus and Woodman (1997) 21–30; and Kraus (1999).

50 For example, books 21–30 focusing on the Hannibalic War. The 'preface' to book 31 makes clear how his narrative almost seamlessly proceeds from one war to the next: the Macedonian War followed the Punic Peace' ~ *pacem Punicam bellum Macedonicum excepit*. Here the Latin is quite instructive: the wars appear side-by-side, in the order in which they occurred.

51 For example, as Livy writes in his preface: 'The subjects to which I would ask each of my readers to devote his earnest attention are these: what the character and the morals of the community were, through which men and by which skills at home and abroad the empire was won and extended' ~ *ad illa mihi pro se quisque acriter intendat animum, quae vita, qui mores fuerint, per quos viros quibusque artibus domi militiaeque et partum et auctum imperium sit* (§9). To Livy, military endeavour, both in self-defense and in the construction of an empire (that is, warfare) is an *ars*.

52 On Procopius, see Cameron (1985); Kaldellis (2004); Whately (2016); Lillington-Martin and Turquois (2017); and Greatrex and Janniard (2018).

53 *illud neque ignoro neque infitior, et rerum gestarum scriptores indagine operis sui hanc quoque partem esse complexos et ab auctoribus exemplorum.*

54 *sed, ut opinor, occupatis velocitate consuli debet. longum est enim singula et sparsa per immensum corpus historiarum persequi.* A few sentences later he acknowledges that the reader may fault his work for *omitting* an expected stratagem. On this passage, see also later, 204.

55 Schellenberg suggests this term (40–41). He argues that the accurate representation of the full scope of war is a challenge that both manuals and historical narrative share.

56 On Polybius, see Chapter 1 (27–28) and the Epilogue (270).

57 Spaulding (1933) 659.

58 Cicero, *Letters to Quintus* 1.1.23; *Tusculan Disputations* 2.62; Suetonius, *Life of Caesar* 87. See Astin (1967) 16, 21, 118. Cicero also claimed to keep a personal copy: *Letters to Friends* 9.25/114: see Chapter 9. And Alexander too may have read it: McGroarty (2006) with additional bibliography. On the *Education of Cyrus*, see Chapter 1 (18–20), Chapter 2 (45), Chapter 8, and the Epilogue (268).

59 While Xenophon's *Cavalry Commander* and *On Horsemanship* clearly fall into the category of a military text, one might also postulate that his *Agesilaus* and *Education of Cyrus* (see the previous footnote) evince characteristics of the genre: the catalogue of the king's positive character traits in the former work, for instance, aligns broadly with Onasander's catalogue of the characteristics of the ideal general. See Vela Tejada (2004) 141 on Xenophon.

60 Wheeler (2010) 435.

61 Wheeler (2010) 436.

Works cited

Allison, G. 2017. *Destined for War: Can America and China Escape Thucydides's Trap?* New York: Houghton Mifflin Harcourt.

Astin, A. E. 1967. *Scipio Aemilianus*. Oxford: Oxford University Press.

Bawarshi, A. S. and Reiff, M. J. 2010. *Genre: An Introduction to History, Theory, Research, and Pedagogy*. West Lafayette, IN: Parlor Press.

Cameron, A. 1985. *Procopius and the Sixth Century*. London: Duckworth.

Campbell, B. 1987. 'Teach Yourself How to Be a General'. *JRS* 77. 13–29.

——— 2004. *Greek and Roman Military Writers: Selected Readings*. London: Routledge.

——— 2005. 'The Army', in A. K. Bowman, P. Garnsey, and A. Cameron, eds., *The Cambridge Ancient History Volume XII: The Crisis of Empire, A.D. 193–337*. Cambridge: Cambridge University Press. 110–130.

Chakravarti, P. C. 1972. *The Art of War in Ancient India*. New Delhi: Oriental Publishers.

Chlup, J. T. 2012. 'Identity and the Representation of War in Ancient Rome', in E. V. Baraban, S. Jaeger, and A. Muller, eds., *Fighting Words and Images: Representing War across the Disciplines*. Toronto: University of Toronto Press. 209–232.

Derla, L. 1996. 'Machiavelli: La guerra come opera d'arte'. *Aevum* 70. 597–617.

Dillon, S. and Welch, K. E., eds. 2006. *Representations of War in Ancient Rome*. Cambridge: Cambridge University Press.

Dueck, D. 2011. 'Poetry and Roman Technical Writing: Agriculture, Architecture, Tactics'. *Klio* 93. 369–384.

Duff, D. 1999. *Modern Genre Theory*. London: Pearson Longman.

Duhaime, J. 1988. 'The *War Scroll* from Qumran and the Greco-Roman Tactical Treatises'. *Revue de Qûmran* 13. 133–151.

Echeverría Rey, F. 2010. 'Weapons, Technological Determinism, and Ancient Warfare', in G. Fagan and M. Trundle, eds., *New Perspectives on Ancient Warfare*. Leiden: Brill. 21–56.

Foote, N. and Williams, N., eds. 2017. *Civilians and Warfare in World History*. London: Routledge.

Formisano, M. 2011. 'The Strategikós of Onasander: Taking Military Texts Seriously'. *Technai* 2. 39–52.

——— 2017. 'Introduction: The Poetics of Knowledge', in P. van der Eijk and *idem*, eds., *Knowledge, Text and Practice in Ancient Technical Writing*. Cambridge: Cambridge University Press. 12–26.

Frow, J. 2014. *Genre*. Second edition. London.

Garlan, Y. 1972. *La Guerre dans l'Antiquité*. Paris: F. Nathan.

——— 1994. 'Warfare', in D. M. Lewis, J. Boardman, S. Hornblower, and M. Ostwald, eds., *The Cambridge Ancient History Volume VI: The Fourth Century B.C.* Second edition. Cambridge: Cambridge University Press. 678–692.

Gat, A. 2005. *War in Human Civilization*. Oxford: Oxford University Press.

Gazzano, F. and Traina, G. 2014. 'Plutarque, historien militaire?'. *Ktèma* 39. 347–370.

Greatrex, G. and Janniard, S., eds. 2018. *Le Monde de Procope – The World of Procopius*. Paris: Éditions de Boccard.

Hanson, V. D. 2009 [1994]. *The Western Way of War: Infantry Battle in Classical Greece*. With a preface by John Keegan. Berkeley: University of California Press.

———, ed. 2012. *The Makers of Ancient Strategy*. Princeton: Princeton University Press.

Heuser, B. 2016. 'Theory and Practice, Art and Science in Warfare: An Etymological Note', in D. Marston and T. Leahy, eds., *War, Strategy and History: Essays in Honour of Robert O'Neill*. Canberra: Australian National University Press. 179–196.

Isaac, B. 2017. *Empire and Ideology in the Greco-Roman World*. Cambridge: Cambridge University Press.

Jacobs, S. G. 2018. *Plutarch's Pragmatic Biographies*. Leiden: Brill.

Kaldellis, A. 2004. *Procopius of Caesarea*. Philadelphia: University of Pennsylvania Press.

Keegan, J. 1976. *The Face of Battle*. London: Jonathan Cape.

——— 1987. *The Mask of Command*. London: Jonathan Cape.

Keppie, L. 1996. 'The Army and the Navy', in A. K. Bowman, E. Champlin, and A. Lintott, eds., *The Cambridge History Volume X: The Augustan Empire, 43 B.C – A.D. 69*. Cambridge: Cambridge University Press. 371–396.

König, A. 2018. 'Reading Civil War in Frontinus' *Strategemata*: A Case-Study for Flavian Literary Studies', in L. D. Ginsburg and D. A. Krasne, eds., *After 96 CE: Writing Civil War in Flavian Rome*. Berlin: de Gruyter. 145–177.

Kraus, C. S. 1999. 'Jugurthine Disorder', in *eadem*, ed., *The Limits of Historiography: Genre and Narrative in Ancient Historical Texts*. Leiden: Brill. 217–247.

——— and Woodman, A. J. 1997. *Latin Historians*. Oxford: Oxford University Press.

Krentz, P. 2007. 'Warfare and Hoplites', in H. A. Shapiro, ed., *The Cambridge Companion to Archaic Greece*. Cambridge: Cambridge University Press. 61–84.

Lee, A. D. 1997. 'The Army', in A. Cameron and P. Garnsey, eds., *The Cambridge Ancient History Volume XIII: The Late Empire, A.D. 337–425*. Cambridge: Cambridge University Press. 211–237.

Lendon, J. E. 2005. *Soldiers and Ghosts: A History of Battle in Classical Antiquity*. New Haven: Yale University Press.

——— 2017a. 'Battle Descriptions in the Ancient Historians, Part I: Structure, Array, and Fighting'. *G & R* 64. 39–64.

——— 2017b. 'Battle Descriptions in the Ancient Historians, Part II: Speeches, Results, and Sea Battles'. *Greece & Rome* 64. 145–167.

Levene, D. S. 2010. *Livy on the Hannibalic War*. Oxford: Oxford University Press.

Lillington-Martin, C. and Turquois, E., eds. 2017. *Procopius of Caesarea: Literary and Historical Interpretations*. London: Routledge.

Luginbill, R. D. 2002. 'Tyrtaeus 12 West: Come Join the Spartan Army'. *CQ* 52. 405–414.

McGroarty, K. 2006. 'Did Alexander the Great Read Xenophon?'. *Hermathena* 181. 105–124.

Milner, N. P. 1993. *Vegetius: Epitome of Military Science*. Liverpool: Liverpool University Press.

Petrocelli, C. 2008. *Onasandro: Il generale*. Bari: Edizioni Dedalo.

Pretzler, M. and Barley, N., eds. 2017. *Brill's Companion to Aineias Tacticus*. Leiden: Brill.

Pritchett, W. K. 1971–79. *The Greek State at War*. 4 vols. Berkeley: University of California Press.

Rance, P. 2017. 'Introduction', in *idem* and N. V. Sekunda, eds., *Greek Taktika: Ancient Military Writing and Its Heritage*. Gdansk: Foundation for the Development of Gdansk University for the Department of Mediterranean Archeology. 9–64.

Sabin, P., van Wees, H., and Whitby, M., eds. 2007. *The Cambridge History of Greek and Roman Warfare*. Cambridge: Cambridge University Press.

Sage, M. M. 1996. *Warfare in Ancient Greece: A Sourcebook*. London: Routledge.

———— 2008. *The Republican Roman Army: A Sourcebook*. London: Routledge.

Sears, M. A. 2019. *Understanding Greek Warfare*. London: Routledge.

Singh, S. D. 1965. *Ancient Indian Warfare with Special Reference to the Vedic Period*. Leiden: Brill.

Spaulding, O. L. 1933. 'The Ancient Military Writers'. *CJ* 28. 657–669.

———— 1937. *Pen and Sword in Greece and Rome*. Princeton: Princeton University Press.

Syme, R. 1964. *Sallust*. Berkeley: University of California Press.

Vela Tejada, J. 2004. 'Warfare, History and Literature in the Archaic and Classical Periods: The Development of Greek Military Treatises'. *Historia* 53. 129–146.

Whately, C. 2015. 'The Genre and Purpose of Military Manuals in Late Antiquity', in G. Greatrex, H. Elton, and L. McMahon, eds., *Shifting Genres in Late Antiquity*. Burlington, VT: Ashgate. 249–262.

———— 2016. *Battles and Generals: Combat, Culture, and Didacticism in Procopius' Wars*. Leiden: Brill.

Wheeler, E. L. 2010. 'Military Treatises', in M. Gagarin, ed., *The Oxford Encyclopedia of Ancient Greece and Rome*. Vol. 4. Oxford: Oxford University Press. 434–438.

Whitby, M. 2001. 'The Army, c. 420–602', in A. Cameron, B. Ward-Perkins, and *idem*, eds., *The Cambridge Ancient History Volume XIV: Late Antiquity: Empire and Successors, A.D. 425–600*. Cambridge: Cambridge University Press. 288–314.

Whitehead, D. 2002 [1994]. *Aineias the Tactician: How to Survive Under Siege*. Bristol: Bristol Classical Press.

Yadin, Y. 1962. The Scroll of the War of the Sons of Light against the Sons of Darkness. Engl. trans. Oxford: Oxford University Press.

1 Military manuals from Aeneas Tacticus to Maurice

Origins, scholarship, genre, audience, and history

Conor Whately

One of the most important sources for the military history of the classical world is the enigmatic military manual. Both Greek and Latin handbooks survive, with the earliest dating to the fourth century BCE, and the latest to the end of antiquity (c. 590 CE) and beyond (Byzantine Middle Ages, c. 600–1453 CE). Although usually lumped together, their diversity belies simple classification.[1] While authors and works like Aeneas Tacticus' *How to Survive a Siege* (*Poliorketika*), Arrian's *Expedition Against the Alans*, and Maurice's *Strategikon* demonstrate real practical utility, others, like Asclepiodotus' *Tactics* and Frontinus' *Stratagems* (*Strategemata*) seem, on the surface, to be more antiquarian in outlook.[2] Some authors, like Polybius and Arrian, were seasoned generals with real military experience, and so their works would seem to have been the most useful to would-be commanders. And yet, Polybius' *Tactics* does not survive, while one later reader of Apollodorus Mechanicus' very technical *On Siege Engines* (*Poliorketika*) found it useful enough to correct it where necessary, and Vegetius' outwardly antiquarian *Epitome on Military Matters* (*Epitoma Rei Militaris*) was one of the most popular ancient texts in the medieval west. Indeed, while the interest expressed by later Roman-era authors like Asclepiodotus and Aelian in the phalanx had once seemed antiquarian at best and anachronistic at worst, a good case has been made for the continued employment of that formation, the phalanx, when needed in Roman combat. Such a contention underscores the complexity of these manuals and shows how little understood they remain despite their prominence and regular use as sources for ancient warfare and ancient history in general.

This chapter provides a necessarily select overview of the surviving manuals and the scholarship, addresses the hitherto overlooked questions of genre and who read military manuals, and looks at their use as historical sources and their often close relationships to works of ancient history. This chapter will provide a frame of reference for the chapters that follow, while also providing an introduction to the genre as a whole.

I. Origins

It is not much of a surprise that military manuals emerged in such a bellicose and highly literate (at least on some levels) world. War was ubiquitous in the ancient Mediterranean, and its impact and regularity were such that it had a marked impact not just on more obvious matters, like economic and political concerns, but less obvious ones too, like a state's cultural output.[3] As the introduction notes, the poetry of Homer and the histories of Herodotus and Thucydides used war as their subjects and in turn influenced how later authors wrote about war. In their world, war was usually waged either by means of battles, sieges, or some combination thereof. It is not surprising, then, that most of the surviving military manuals, especially those in Greek, focus on some aspect of open battle or siege warfare.

One author and one war has had a bigger impact than any of the others, and that is Homer and his account of the Trojan war in the *Iliad*. Indeed, Homer has cast such a large shadow over Graeco-Roman military thinking that much military theory can be traced back to him.[4] To many a writer from the Mediterranean world, Homer's two epic poems, and in particular the *Iliad*, provide the basis for all future discussions of the nature of warfare. Early in his treatise on generalship, which dates to the first century CE, Onasander includes a reference to Homer, which alludes to heroic leadership when he discusses the selection of a general.[5] This trend is even more pronounced in the Second Sophistic with Polyaenus' *Stratagems* and Aelian's *Tactics*. Polyaenus, who published a collection of stratagems, possibly in different stages, opens book one with – no surprise – Homer, and provides several examples of the exploits of Odysseus, especially from the *Odyssey*.[6] The second century theoretician Aelian claims, in his *Tactics*, that Homer was the first to write about tactical theory in war.[7] Maurice, the late-sixth-century-CE writer of the *Strategikon*, includes a quotation from Homer in his collection of maxims.[8] This should not surprise. The *Iliad* is dominated by warfare but does not refer to one aspect of warfare only: in fact, it provides examples of all sorts of military matters from single combat, battle between massed formations, and the use of stratagems and ambushes, to the speeches of generals before battle, their subsequent attempts to array the battle-lines, and the vast array of different weaponry from bows to spears.[9] On the other hand, the *Odyssey*, though not specifically concerned with battle, is replete with stratagems, as Polyaenus' references attest, and examples of, from a Greek perspective, effective leadership. Moreover, the two works provide the reader with two distinctive types of general: the Achilles, the fight-from-the-front-line archetype, and the Odysseus, who deploys stratagems or ruses. Indeed, regardless of whether a historian had a predilection for presenting one kind of general or another, or any type of warfare, with the exception of naval combat, he need only trawl the pages of the *Iliad* or, less likely, the *Odyssey*, to find a Homeric precedent. With the exception of generalship, more often than not, single combat and massed 'phalanx-like' combat were what they remembered and chose.

Whether Aelian was right, and Homer was the first person to write about military theory, it is more difficult to assert that he was the first writer of military manuals. So what was the first manual *stricto sensu*? A good case could be made that one or two of Xenophon's works might constitute the first such text, particularly his *On the Cavalry Commander* (*Hipparchus*), or even all or part of his *On Horsemanship* (*Peri Hippikes*) and *The Education of Cyrus* (*Cyropaedia*).[10] The latter is representative of the military doctrine that he espouses, and delves into key issues, like the importance of tactics and morale, and it presents his ideal characteristics for a commander; battle and generalship are key components of later manuals.[11]

In one particular episode of the text, the main characters cover a host of pertinent material, and it is centred round a discussion involving Cyrus and his father Cambyses about education in the arts of war. During that discussion, an unnamed man had trained Cyrus on how to be a general and he taught him a number of things. After referring to logistics and health Cyrus says the following:

> he professed to have been teaching me generalship. And thereupon I answered, 'tactics'. And you laughed and went through it all, explaining point by point, as you asked of what conceivable use tactics could be to an army, without provisions and health, and of what use it could be without the knowledge of the arts invented for warfare and without obedience.[12]
>
> (1.6.14)

Tactics, morale, obedience, and generalship are stressed here.[13] A little later, Cambyses advocates the use of stratagems; it is not enough to defeat one's foe on the field in pitched battle alone (1.6.27–40). Then Cambyses turns to battle itself (a long passage, to be sure, but worth quoting in full):

> But if it is ever necessary – as it may well be – to join battle in the open field, in plain sight, with both armies in full array, why, in such a case, my son, the advantages that have been long since secured are of much avail; by that I mean, if your soldiers are physically in good training, if their hearts are well steeled and the arts of war well studied. Besides, you must remember well that all those from whom you expect obedience to you will, on their part, expect you to take thought for them. So never be careless, but think out at night what your men are to do for you when day comes, and in the daytime think about how the arrangements for the night may best be made. But how you ought to draw up an army in battle array, or how you ought to lead it by day or by night, by narrow ways or broad, over mountains or plains, or how you should pitch camp, or how station your sentinels by night or by day, or how you should advance against the enemy or retreat before them, or how you should lead past a hostile city, or how attack a fortification or withdraw from it, or how you should cross ravines or rivers, or how you should protect yourself against cavalry or spearmen or bowmen, and if the enemy should suddenly come in sight while you are leading on in column, how you should form and take your stand against them, and if they should come in sight from any other quarter than in front as you are marching in phalanx, how you should form and face them, or how any one might best find out the enemy's plans or how the enemy might be least likely to learn his [Cambyses does not complete his thought]. . . . I think, then, that you should turn this knowledge to account according to circumstances, as each item of it may appear serviceable to you.
>
> (1.6.41–43; Miller's translation, slightly modified)[14]

Although this list includes some general comments, it reads like a table of contents for any of the military treatises that appear in the fourth century BCE through to the high Byzantine period: thus one may ascribe to Xenophon the role of the formal founder of the military manual.[15] In sections such as these, there is little doubt that the *Education of Cyrus* comes across as a proto-theoretical treatise. Cambyses' words cover just about any military engagement and contingency that Cyrus' army would likely have faced. Clearly, for Xenophon, a great deal was involved in warfare.[16] For all of its fictional character, Xenophon's *Education of Cyrus* has a lot to say about Greek military theory.[17] Most importantly, many of the points discussed by Xenophon in the text find their way into the 'proper' treatises which were to follow in the fourth century and later.

The earliest full-fledged military manual that survives is Aeneas Tacticus' treatise on siege warfare, possibly written at the time Xenophon was writing his works (that is, the mid-fourth century BCE), possibly a bit later.[18] Aelian considered Aeneas to be the first writer on war *after* Homer, and he seems to have written a number of texts, of which unfortunately only one survives: the *How to Survive under Siege*.[19] Before Aeneas goes through his points, and following the preface, he writes the following:

> The disposition of troops should be made in the light of the size of the state, the topography of its urban centre, the posting of sentries and patrols and other public functions which call for troops; these are the criteria which must determine the allocations. Thus the expeditionary forces should be organized with regard to the terrain through which they will have to march – bearing in mind danger-spots, strongholds, defiles, plains, commanding heights, ambush-points, river-crossings, and the formation of battlefronts in such circumstances. There is no call, on the other hand, to organize on these lines the troops which will remain within the walls and guard the citizens, but rather in accordance with the configuration of the city and immediacy of danger.
>
> (1.1–3; Whitehead's translation)[20]

Aeneas is setting his work within the conceptual framework that stresses the importance of tactics and the order of battle. Even though the treatise is not concerned with pitched battle, Aeneas will still discuss the issue in those terms. In fact, much of the treatise is concerned with engaging the enemy in the field, and in the positioning of sentinels along the fortifications much as the later theoreticians like Aelian and Arrian – and Polybius – are concerned with the positioning of the men in formation, and usually in a phalanx.[21] Aeneas advocates the use of stratagems when engaging an enemy who is approaching one's territory.[22] There are references to punishment and morale, and not just of the morale of the soldiers, but of the inhabitants of the besieged cities.[23] As the manual is concerned primarily with sieges, Aeneas also discusses quite a bit about how to prevent treachery that could lead to the capture of the besieged city; in the context of that discussion, Aeneas also refers to discipline and the capabilities of the commanders as well in some detail. Preparation is a key to success when besieged. Communications are important.[24] Aeneas treats stratagems, fire, and countermeasures to be used against an enemy whether they are using machinery or mines.[25] Morale is also important to Aeneas: he provides some suggestions on how to act if the army's morale takes a hit.[26] 'Panic' is an interesting component of Aeneas' discussion of morale. It can arise in camps or cities suddenly and without warning; suffering a defeat is not a prerequisite: thus he considers not just the actions and counteractions in warfare, but also their emotional frame – physical action and emotional reaction are important factors in warfare.[27] In sum, Aeneas' treatise discusses a wide range of issues so covering most of the contingencies that an ancient commander could expect to encounter; despite his stress on morale and stratagems, it is tactics and formations that predominate.

II. Scholarship

Although military manuals have a long tradition, they have, in many instances, failed to capture the attention of the academic community.[28] Some scholars have often treated the writers of manuals as parts of larger studies. For example, John Anderson provides a useful overview of the Spartan military system, Xenophon, and military theory and practice in the fourth century.[29] William Pritchett's far-reaching studies on the Greek state at war are also important and delve into Greek military theory, with volume II particularly noteworthy.[30] Eric Marsden's detailed analysis of Greek and Roman artillery draws heavily on the writers of manuals, and the second volume in particular includes the texts and translations of several Greek and Latin authors.[31] Finally, J. E. Lendon's *Soldiers and Ghosts: A History of Battle in Classical Antiquity* – a very important work on ancient warfare, to be sure – discusses military texts specifically and utilises them as evidence.[32]

Perhaps the best indication of the lack of scholarship on this group of authors is the nature of work that has already been completed. Of the 246 entries for Arrian in *l'Année philologique*, for instance, only a handful are concerned specifically, or even partially, with his military manuals, the *Tacticon* and the *Against the Alans*.[33] Searches of research on the work of other writers of military manuals reveal comparable results.[34] Most of the scholarship on Aelian concentrates on the other Aelian, whether they survive (*On the Nature of Animals*) or they do not (*Varia Historia*), not the Aelian of the *Tactica*.[35] Vegetius' *Epitome* has been comparably well served, which matches his later reputation in the medieval west.[36] The scholarship has ranged from points of detail, such as the type of trumpets he refers to in individual passages,[37] to one of the more popular topics, the date for his work,[38] though translations and his work's transmission are also popular.[39] A very small number have ranged wider, with one of Michael Charles' papers delving into questions of masculinity.[40] Vegetius, quite simply, is the exception. As I say, however, a great many of the publications are concerned with the fundamentals: editions and translations.

It is also the case that some scholars have devoted themselves to systematic study of a particular author. Charles has published widely on Vegetius with papers on Vegetius' comments on armour, Vegetius and the Ioviani and Herculiani legions, Vegetius and Liburnian galleys, Vegetius' reference to the legions using weighted darts, the Mattiobarbuli, and the date and context of the *Epitome of Military Matters*.[41] Much of Immacolata Eramo's work too has concentrated on the works of one author, in her case Syrianus Magister. Besides writing an edition, translation, and commentary of Syrianus Magister, she has published widely on many aspects of Syrianus' writing, including the relationship between his three works, and the date of his work.[42] Both Everett Wheeler and Phil Rance, very much leading scholars in the field of military texts, have published several articles focused on philological issues and ancient tactics. Wheeler's scholarship on military theory, drawn largely from the military manuals, has ranged from Classical Greece to late antiquity.[43] Rance's publications have concentrated on the late antique and Byzantine worlds, with a particular emphasis on Maurice's *Strategikon*.[44] Rance has also examined the reception of earlier treatises in the Byzantine period, and a host of philological issues based, to a significant degree, on the material in the military manuals.[45] David Whitehead, on the other hand, has concentrated primarily on translations and commentaries of military manuals, with editions of Aeneas Tacticus, Athenaeus Mechanicus, Apollodorus Mechanicus, and Philo Mechanicus, and some noteworthy papers.[46] What we lack, by and large, are studies that concentrate on literary and textual issues.

III. Genre

Classical military manuals and questions of genre do not often go hand-in-hand. Though scholars of classical literature have discussed the vagaries of genre, both ancient and modern, for some time, this discussion has had virtually no impact on the reading of these manuals.[47] To some degree, this is unsurprising: ancient literary theorists, like Aristotle, either predated the first real manuals, or later ones, like Lucian, were too interested in traditional texts to bother analysing some of the more recent ones. Indeed, you could argue that their very emergence, at least in Greek, after the classical era doomed their literary study, at least to some degree. Given the lack of scholarship on literary and textual issues, it behooves us to begin to explore the issue of genre here.

After Homer, Xenophon, and Aeneas, the development of Greek military theory continued into the Hellenistic age and beyond. Indeed, a good case has been made that much of Greek military theory was formulated in the Hellenistic age, and Ted Lendon, for instance, has treated its Hellenistic development.[48] This was due, in large part, to the significant progress in the realm of military technology, with some going so far as to characterise the age as a period of military revolution.[49]

Despite the advancements in military technology, we do not have a matching collection of military treatises, with many no longer extant. Some Hellenistic examples include a lost *Strategika* of Demetrius (second half of the fourth century BC) and a lost *Tactics* of Evangelus (third century BC). One of the best-known Hellenistic military writers is Polybius, and besides writing his history of the rise of Rome, Polybius also wrote a tactical treatise, which, regrettably, has also not survived.[50] Despite this, it is possible to tease out some of Polybius' views of warfare from the pages of his history.[51] Polybius did not hesitate to make his views heard, and there are more than a few military-themed digressions. His emphasis on tactics and battle-order in his *Histories* belies the focus of his lost *Taktics* and supports the claims that the later works of Aelian and Arrian drew heavily on the lost text.

Although Polybius' *Taktics* is lost, we can get some sense of the kind of material it might have contained by looking more closely at some of his digressions, particularly the famous digression on the Roman political and military system in book six, and his comparison of the Macedonian phalanx and Roman legion in book eighteen. Starting with the former, Polybius discusses the means by which the Romans encourage their soldiers to face danger. He notes that some rewards are given out for merit, and the first person responsible for boosting morale and emboldening the troops is the man in charge.[52] The distribution of rewards in fact follows the laudatory words of the general. After giving a precis of the rewards, Polybius goes on to say that 'by such incentives they excite to emulation and rivalry in the field not only the men who are present and listen to their words, but those who remain at home also'.[53] It is not only rewards, however, that are effective means of improving the soldiery; even before rewards, Polybius discusses punishment.

Tactics and battle-order play a big part in Polybius' conception of warfare, and this is brought out clearly in his digression on the differences between the Roman legion and the Macedonian phalanx. It is perhaps here that we get our best sense of the views Polybius might have expounded in his *Tactics*. In this digression, Polybius' premise is that the reason for the Romans' string of successes against the Macedonians was their superior battle-order.[54] Polybius notes that the phalanx is unstoppable in ideal conditions, especially frontal assaults; however, the role of the general is pervasive, for suitable terrain is needed for the phalanx to be effectively employed.[55] One of the principal conclusions which Polybius draws from his comparison is tied to his belief that it is imperative for a general to be well prepared: the phalanx is only really useful under certain fixed conditions, and since Polybius knows that anything is possible while on campaign, it is the legion's adaptability that makes it so effective.[56] When Polybius turns to describe the battle itself, it is the tactical advantage, which the Romans enjoy, that turns out to be the decisive factor.[57] While morale is important, and Polybius does refer to its role in combat, it is tactics that rule the field of battle, and it is up to the well-prepared general to determine which ones suit the occasion.

It is perhaps with the development of military theory as a science that we emergence of two seemingly divergent strands of military manuals: the theoretical, like Onasander's *The General*, and the practical like Arrian's *Against the Alans*. Onasander, a philosopher like two other authors of manuals, Asclepiodotus and Aelian, wrote a manual on generalship which puts great emphasis on tactics.[58] The sixth chapter, in particular, is concerned with the importance of maintaining one's formation.[59] Only a few books later, Onasander returns to battle-order and essentially summarises many of the points made by Asclepiodotus, Aelian, and Arrian.[60] Onasander discusses making camps and passing through narrow defiles, and dealing with spies. The need to be cautious in pursuit is stressed, as is the importance of the general taking matters into his own hands when the morale of his troops drops.[61] We also find that Onasander shares some of the views expressed by Polybius; he says that a general must not do anything rash, nor must he enter battle himself.[62] Towards the end of his guide, Onasander discusses sieges.[63] Despite the breadth of coverage, there is no explicit section devoted to the stratagem; yet, in the preface, Onasander makes it abundantly clear that strategems are of vital importance to generalship.[64] And so, much like Xenophon's general, Onasander's is able to act in a variety of situations, is adept at using tactical formations, is aware of his troops' morale, and is not averse to using stratagems when needed. What is more, Onasander's general is also broadly similar to the well-prepared general characterised by Polybius.

Maurice's *Strategikon*, a practical work, which manages to describe late Roman warfare while still adhering to many of the practices of the genre, is a text of considerable importance with an obvious slant towards tactics. Its focus is didactic, and it is aimed at the would-be general, though it is meant to supplement training in the field rather than to replace it.[65] Onasander's treatise was certainly a major influence, though it is more likely to have provided the impetus for Maurice, rather than the explicit framework and material for the work.[66] Cavalry manoeuvres make up a considerable portion of the work, in part a reflection of the sort of warfare that the army was engaged in, in part because the section on infantry seems to have been added later as a supplement.[67] Like its sixth-century predecessors, the *Strategikon* betrays a real interest in stratagems and opportunism on the field of battle; besides the sections scattered throughout devoted to ambushes, the subject warrants an entire chapter.[68] There is also a chapter devoted to surprise attacks.[69] One of the most innovative features of the *Strategikon* is the inclusion of a chapter devoted to the types of enemy that the army is likely to face, something which has no literary precedent in the genre of Greek military writing.[70] As noted, there is considerable emphasis on tactics and formation. Significantly, however, Maurice is concerned not only with the formations themselves, but with how they are created, how they operate, and the human efforts behind their employment. Communication, discipline, morale, and training are all rightly regarded as essential to a unit's ability to carry out any of the tactical manoeuvres described, or any of the other actions for that matter. This conflation of the two strands of tactics and morale marks out the *Strategikon* as unusual among military manuals, though not histories. Indeed, as Rance notes, 'It also reveals an acute understanding of the realities of combat and an insight into the psychological preoccupation of both generals and troops'.[71] In many ways, and as we might expect given the text's practical purpose, as evidenced, in part, by its deliberately simple language, it marks a fitting final text since it incorporates the two strands of military thought referred to in this survey of Greek military theory, order, and morale, as well as the equally important issues of generalship and the use of stratagems.

There is another strain of thought – a subgenre, if you will – that is only vaguely discernable in many of those works, one which was to have a marked impact on later Byzantine warfare, and this is an emphasis on avoiding battle if at all possible, and using whatever means were at one's disposal to gain an advantage over one's foe.

This dichotomy goes back to Homer and is best described by Wheeler's two terms the 'Achilles ethos' and the 'Odysseus ethos'. In Wheeler's own words: 'The Achilles ethos promotes chivalry, face-to-face confrontation, open battle, and the use of force, while the Odysseus ethos asserts the superiority of trickery, deceit, indirect means, and the avoidance of pitched battle, although not the denial of the use of force or battle if advantageous'.[72] There are two authors who have left us fairly complete and extensive treatises on the stratagem, Frontinus and Polyaenus.[73] Frontinus was a first-century-CE author perhaps best known for his work on aqueducts, and who also wrote a now lost work on tactics and a work called the *Stratagems*, largely on stratagems. That last work covers a range of topics, much of which are concerned with how a general should approach war, campaigning, and battle. Works like Onasander's *Strategikos* and Maurice's *Strategikon* cover similar themes, though they approach it in different ways, for the former are much more explicit in that they provide clear directions for generals in the moment. Frontinus, on the other hand, chose to illustrate his instructions with historical episodes, with little in the way of explication. In the section on choosing a site for battle, for example, Frontinus opens with a reference to the Second Punic War and a case in which Publius Scipio, upon learning that the Carthaginians had already had breakfast, delayed engaging Hasdrubal until the afternoon to wear them out.[74] Each section, then, is a collection of *exempla* that both serve as a stratagems and seek to explain the section under review.

The organisation of Frontinus' *Stratagems,* such as we have it, contrasts starkly with Polyaenus' collection of stratagems, which is quite extensive and comprises historical *exempla* from mythological, classical, and Hellenistic Greece. The order is somewhat haphazard, and loosely chronological; though it was composed in the second century CE, the Romans have only a supporting role, and a quite minor one at that, which is not unsurprising, given its Second Sophistic context. The stratagems which he discusses cover a wide variety of different possibilities.[75] There is a stratagem that advocates supplying one's foe with misleading information about what is happening in battle, a stratagem that cautions against encirclement, a stratagem that shows how psychology can be used to help overcome a numerical disadvantage, a stratagem that shows how one's formation and daring can cause a foe to flee, and a stratagem involving a dust cloud.[76] What that ever-so-brief list should show is that there is an incredible number and variety of stratagems covered, and that they incorporate many of the issues discussed previously, including the importance of tactics and morale. Plus, the size of Polyaenus' work, and its popularity, suggests that there was no misapprehension among commanders about using stratagems in battle.[77]

IV. Audience

Who read the military manuals in antiquity? Well for one, other military theorists: military authors were themselves probably avid consumers of military literature, especially other manuals.[78] Some texts demonstrate exceptionally close alignment in discussion that suggests that military manuals are more closely tethered together than other genres. In other words, one might postulate that, say, Vegetius and Frontinus have a closer relationship than Virgil and Homer. Asclepiodotus, Aelian, and Arrian all wrote works specifically devoted to tactics, and precisely in that order. Asclepiodotus (first century BCE) wrote an account of the phalanx; Aelian (early second century CE) wrote a manual on tactics devoted to the phalanx; Arrian (middle of the second century CE) also wrote a manual on tactics centred on the phalanx. For these theoreticians, the formation of the battle-line is the decisive factor in battle. Though an emphasis on the phalanx might seem anachronistic or at least antiquarian, some have argued that it continued to be in use by the Romans into the middle of the second century CE.[79] The organisation of these works is similar, as is the variety and number of issues which they discuss. There are twelve books in Asclepiodotus' work, and the topics range from the branches of the phalanx, and the names of the various subdivisions, to the intervals between soldiers and the character of the arms used.[80] Aelian's *Tactics*' voluminous chapters cover a wide range of different tactical matters. The topics range from the different varieties of mounted troops (on horse or on elephant), the division of the infantry line into three parts, how to arm the hoplites, the size of the arms and spears, and the formations of elephants and chariots, to a plethora of definitions such as what a πρόσταξις, an ἔνταξις, and a ὑπόταξις are, what a φάλαγξ ἀμφίστομος is, and how the phalanx becomes a rectangle or a square.[81] Arrian's *Tactics*, or at least what we have of it, is arranged into fewer books. The topics range from the division of the armies and their respective names, the equipment of the various troops, the intervals between the arranged troops, and the tactical manoeuvres used by the formations, and the orders given.[82]

All three manuals used material from earlier and now lost Hellenistic manuals, including, perhaps, Polybius' lost *Taktics*. The issues which they discuss are similar; this is in part the result of the somewhat derivative nature of the genre.[83] P.A. Stadter notes that while both Arrian and Aelian were influenced by the same common source, Asclepiodotus was influenced by some other sources. It is also possible that all three works go back to the lost treatise of Posidonius.[84] On the other hand, the parallels in arrangement between Aelian and Arrian in particular are striking, as noted by H. Köchly in his edition of Aelian's work, and Stadter in his article on Arrian's influences.[85] A quick comparison on *TLG* reveals some of their shared content,[86] and this short survey reveals that the readership of these three authors alone was in part made up of other authors of military manuals.

But this was not restricted to the classical age, or even to known authors. Whitehead's edition of Apollodorus Mechanicus' *On Siege Engines* is full of interpolations, which Whitehead himself has painstakingly revealed.[87] As Whitehead notes, Apollodorus' interpolators, besides providing clarification on difficult points, likely saw themselves as engaged in a dialogue with their predecessor. The so-called 'Heron of Byzantium', Syrianus Magister, and Leo the VI all participated in the same sort of exercise as Apollodorus' interpolators, incorporating material from earlier authors including Aelian and Arrian. The *Taktics* of Leo VI, for instance, is based on Maurice's *Strategikon*, though it also relied heavily on Onasander, Aelian, and Polyaenus.[88] In the very prologue to the text, Leo stresses that he collected and presented anything of value pertaining to war he found in ancient and more recent texts, and in other places mentions Aelian and Onasander by name.[89] It was not simply a case of slavish copying of an earlier classical text. Rather, as John Haldon argues, later Byzantine writers like Leo were bound by convention 'to use as much as [they] could of the earlier treatises, precisely on the grounds that innovation for its own sake was regarded not simply as unnecessary but positively a bad thing'.[90] This might also apply to the military writers of the imperial era and beyond, many of whom shared this penchant for extracts and paraphrases from the works of earlier writers. Historians believed that writing a narrative of events covered by an earlier author was a good thing, with one example being Cassius Dio's coverage of events already described by others, like Appian. It seems that military authors felt the same way and that repetition was the sincerest form of flattery.

V. History and historiography

One topic that has received very little attention is the degree to which the military manuals drew on other texts, and vice versa – it is a potentially fruitful topic worthy of consideration. In this final section, I want to draw attention to a few examples where histories drew on military manuals. I also want to flip things around: at the end, I will look at what value some military manuals might have as historical sources.

The writing-down of military theory, as with historiography, had a long tradition, which extended from Homeric Greece to the middle Byzantine period in medieval Constantinople. Though it seems to have lacked the popularity of works of ostensibly similar value like Greek historiography, at least among the literary elite, it is still an important part of the military thinking of the ancient world. The writers of that other, sometimes didactic, genre of military writing, historical narrative, were often impacted by the theorists, whether they were conscious of this or not.[91] As Lendon notes,

> Greek historians, heavily influenced by a tactical conception of combat, usually used that model as the structuring armature of their accounts of battles: the formation, deployment, and movement of forces tend to form the backbone of the narrative, with other material – stratagems, the brave deeds of individuals, remarkable occurrences like panics, paradoxes, and touching stories – included intermittently along the way.[92]

To give one example, Polybius' account of the Battle of Cynoscephalae emphasizes the formation and battle order of the two sides, with the Romans' much the superior.[93] Unsurprisingly, his account is followed by his comparison of the legion and phalanx.

It is also the case, however, that historians made good use of the manuals – this surely recommends their importance as texts. Polybius, the author of a manual himself, mentions and quotes approvingly a passage of Aeneas Tacticus, no mean feat for an author quick to disparage his fellow historians. Agathias refers to one in his description of the Battle of Casilinum at 2.9.2 of the *Histories*, though he is not explicit.[94] Given the lack of disclosure of sources by ancient historians, an overview of theoretical works is therefore not unwarranted.[95] Therefore, what ultimately may be at play is not that military manuals draw upon history, but rather that history through historical texts draws upon the military manual tradition.

There are late antique examples of this interaction between histories and military manuals. Both Vegetius and Maurice include military maxims in their treatises, and traces of these maxims show up in Procopius' *Wars*.[96] For example, Vegetius and Maurice discuss what to do when a general faces an unfamiliar foe, something highlighted by Procopius (via Belisarius) in the *Wars*.[97] In a different context, Belisarius gives a speech in which he stresses the value of training, a view also found in those same two treatises (Vegetius and Maurice).[98] Vegetius and Maurice were likely drawing on the same shared tradition, even if Maurice, who wrote at least a century and a half after Vegetius, did not draw specifically upon Vegetius. At the same time, other sixth-century authors shared Maurice's interest in how to succeed on the field of battle and in the importance of the general. Procopius regularly stresses the importance of generalship in the *Wars*, so following the established tradition of his classical predecessors like Thucydides and Polybius.[99] At the same time, the anonymous author of the *Dialogue on Political Science* includes an entire chapter devoted to the merits of infantry and cavalry, a sure sign of a wider, shared, intellectual milieu.[100]

Let us shift from one genre's engagement with another to the value of military manuals as historical sources. In the case of the collections of Frontinus and Polyaenus, it seems easy to see how they might serve as repositories of historical information, in their cases famous stratagems performed by famous commanders. As we saw earlier, both of their works are filled with references to historical events and evinced real antiquarian slants. In other words, they are not so much military manuals *per se*, but rather highly selective historical texts with a military focus. Reading these authors, who both draw upon both Greek, Roman, and other cultures for their examples, one gets a totalising overview of the history of warfare from the Homeric period to the author's own day.

On the other hand, the antiquarianism that we find in many of the manuals can reduce their value as repositories of historical information. Vegetius is full of references to historical features of the Roman military and its history, but his use of these is often nebulous. For example, Vegetius makes passing references to Quinctius Cincinnatus (1.3.5), Hannibal (1.28.), Gaius Marius (3.10.23), and Jugurtha (3.24.6). When discussing the type of arms and armour of the ancients (*antiqui*), Vegetius notes the impact of the Alans, Goths, and Huns (1.20.2) and makes a vague reference to the long use of cataphracts, which N. P. Milner takes to mean scale-armour.[101] In that same section, Vegetius connects those ancients with the Romans whose legions were filled with *principes, hastati*, and *triarii* (1.20.13). Indeed, the focus of most of book two of the four-book work is on a mysterious *antiqua legio*, 'ancient legion'. Yet, Sylvain Janniard has pointed out some similarities between Vegetius' comments and the legions of the Severan era, though even he admits that the date is a matter of contention.[102] The Severan legion was not organised in the same way as the three-lined, mid-republican-era legion with *principes*, *hastati*, and *triarii*. It is because of issues like these that the value of Vegetius' *Epitome* as a source for military practices in the late fourth/early fifth century has been doubted.[103] That said, though Vegetius' use of historical artefacts, episodes, and personae in general is often wrong, his antiquarianism serves other ends, for his references to this material serves to emphasise his authority.

The relationship between military manuals, works of history, and history itself, therefore, is a complex one. On the one hand, there are certainly cases where the connection between the manuals and works of historiography seems symbiotic, with the histories, for instance, describing how a general should (and has) acted in a given situation, and a manual explaining how – and when and why – a general should act in such a situation. In the realm of ancient military thought, the two genres also regularly betray a similar outlook. On the other hand, while the manuals often contain historical information, their value as sources for earlier events is often mixed, at best.

VI. Conclusion

The last military handbook of antiquity is Maurice's *Strategikon*, a text which makes a fitting place to stop for any number of reasons.[104] For one, this was the last in a long line of military treatises, at least for a few hundred years. This branch of military writing, much like that other notable branch, historiography, went into abeyance for nearly 300 years, with the next such text not appearing until the works attributed to Emperor Leo VI in the late ninth century.[105] Maurice's *Strategikon*, which dates to c. 590 CE, is aimed at the would-be general, though it is meant to supplement training in the field rather than to replace it.[106] Onasander's treatise was a major influence, though it is more likely to have provided the impetus for Maurice, rather than the explicit framework and material for the work.[107] By means of the tacit acknowledgement of earlier theoretical approaches to combat,[108] Maurice manages to combine the theoretical and the practical approach to warfare without succumbing to any of the tension between the two strands that has been seen as characteristic of the genre.[109] Along those lines, not only is there good evidence that it was consulted as a source of military expertise by the historian Theophylact, but the emperor Heraclius might also have used the text while on campaign himself.[110] On the other hand, Maurice's *Strategikon* has considerable historical value. From its emphasis on cavalry manoeuvres[111] to its interest in stratagem and opportunism on the field of battle,[112] the text betrays an interest in real-life, sixth-century warfare. With the *Strategikon*, we have a text we can date with certainty to the end of the sixth century (c. 590), which makes sparing use of historical anecdotes, is not at all ambiguous with respect to its content, reveals a real concern with practical issues, and which reflects well the combat of its age.[113] The looming publication of Rance's two-volume text, translation, and commentary of the *Strategikon* should make Maurice's oeuvre an excellent candidate for sustained literary analyses of the sort only touched on at the end here, and in the chapter in general. Whitehead's translations and commentaries of Athenaeus, Apollodorus, and Philo should offer the same opportunities for those earlier authors.

In this chapter, I have covered a lot of ground, from the origins of military manuals and some highlights of the scholarship, to questions of genre and audience and their relationship to historiography, all themes explored at some point or other by the other contributors to this book. My discussion of genre only scratched the surface, however, and my brief and superficial glance at the degree to which military manuals drew on historiography and vice versa should have revealed the wealth of potential that such work offers.

Notes

1 Note the comment by Aaron L. Beek in his chapter (120).
2 On antiquarianism, see Chapter 1 and the Epilogue; see also later, 28–29.
3 Vela Tejada (2004) 130–135.
4 Lendon (2005) 15–161, esp. 20–38. See also Chapter 4 and the epilogue (especially 265–267).
5 Onasander, *The General* 1.7.
6 Polyaenus, pref. Preface 4–12. Frontinus is very much the odd author out here: he does not provide any Homeric *exempla*, drawing upon almost entirely historical stratagems.
7 Aelian, *Tactics* 1.1: ὅτι Ὅμηρος πρῶτος περὶ τῆς ἐν τοῖς πολέμοις τακτικῆς θεωρίας ἔγραψεν ~ 'that Homer first wrote the about the theory of tactics in warfare'.
8 Maurice, *Strat.* 8.B.82. See *Iliad* 11.802–803, 16.44–45.
9 22.248–374 (the final duel between Achilles and Hector), 11.407ff (Odysseus standing firm in the battle-line), 10.247ff (Odysseus leading a night raid; to the Greeks night attacks, particularly at this time, would fall under the category of stratagem), 4.293ff (Nestor deploys a range of Greek soldiers in battle order, and gives a speech), 16.169ff (Achilles exhorts his men).
10 Xenophon has much to say on warfare: Anderson (1970). On the *On Horsemanship*, see Chapter 7. On the *Education of Cyrus*, see Due (1989); on this text as a possible military manual, see Chapter 8. On the value of the *Anabasis* for military matters, see Whitby (2004) and Lee (2007) *passim*. Cf. Lendon (1999) 290–295.
11 Vela Tejada (2004) 141, n. 38. Besides, Xenophon seems to have transposed onto his fictional Persians the characteristics and values commonly attached to contemporary (i.e. early fourth century) Sparta. For a more detailed exposition of Xenophon's views on generalship and tactics, see Hutchinson (2000).
12 τέλος δή μ᾽ ἐπήρου ὅ τι ποτὲ διδάσκων στρατηγίαν φαίη με διδάσκειν. κἀγὼ δὴ ἐνταῦθα ἀποκρίνομαι ὅτι τὰ τακτικά. καὶ σὺ γελάσας διῆλθές μοι παρατιθεὶς ἕκαστον τί εἴη ὄφελος στρατιᾷ τακτικῶν ἄνευ τῶν ἐπιτηδείων, τί δ᾽ ἄνευ τοῦ ὑγιαίνειν, τί δ᾽ ἄνευ τοῦ ἐπίστασθαι τὰς ηὑρημένας εἰς πόλεμον τέχνας, . . . τί δ᾽ ἄνευ τοῦ πείθεσθαι.
13 Xenophon, according to his account, was instrumental in the Greek mercenaries' escape from the Persian empire. One should not be surprised, then, that he pays so much attention to generalship, particularly morale and obedience, since in some ways it was these traits that contributed most to the Greeks' success. Cf. Lee (2007) 43–108.
14 ἦν δέ ποτε ἄρα ἀνάγκη γένηται καὶ ἐν τῷ ἰσοπέδῳ καὶ ἐκ τοῦ ἐμφανοῦς καὶ ὡπλισμένους ἀμφοτέρους μάχην συνάπτειν, ἐν τῷ τοιούτῳ δή, ὦ παῖ, αἱ ἐκ πολλοῦ παρεσκευασμέναι πλεονεξίαι μέγα δύνανται. ταύτας δὲ ἐγὼ λέγω εἶναι, ἢν τῶν στρατιωτῶν εὖ μὲν τὰ σώματα ἠσκημένα ᾖ, εὖ δὲ αἱ ψυχαὶ τεθηγμέναι, εὖ δὲ αἱ

πολεμικαὶ τέχναι μεμελετημέναι ὦσιν. εὖ δὲ χρὴ καὶ τοῦτο εἰδέναι ὅτι ὁπόσους ἂν ἀξιοῖς σοι πείθεσθαι, καὶ ἐκεῖνοι πάντες ἀξιώσουσι σὲ πρὸ ἑαυτῶν βουλεύεσθαι. μηδέποτ᾽ οὖν ἀφροντίστως ἔχε, ἀλλὰ τῆς μὲν νυκτὸς προσκόπει τί σοι ποιήσουσιν οἱ ἀρχόμενοι, ἐπειδὰν ἡμέρα γένηται, τῆς δ᾽ ἡμέρας ὅπως τὰ εἰς νύκτα κάλλιστα ἕξει. ὅπως δὲ χρὴ τάττειν εἰς μάχην στρατιὰν ἢ ὅπως ἄγειν ἡμέρας ἢ νυκτὸς ἢ στενὰς ἢ πλατείας ὁδοὺς ἢ ὀρεινὰς ἢ πεδινάς, ἢ ὅπως στρατοπεδεύεσθαι, ἢ ὅπως φυλακὰς νυκτερινὰς καὶ ἡμερινὰς καθιστάναι, ἢ ὅπως προσάγειν πρὸς πολεμίους ἢ ἀπάγειν ἀπὸ πολεμίων, ἢ ὅπως παρὰ πόλιν πολεμίαν ἄγειν ἢ ὅπως πρὸς τεῖχος ἄγειν ἢ ἀπάγειν, ἢ ὅπως νάπη ἢ ποταμοὺς διαβαίνειν, ἢ ὅπως ἱππικὸν φυλάττεσθαι ἢ ὅπως ἀκοντιστὰς ἢ τοξότας, καὶ εἴ γε δή σοι κατὰ κέρας ἄγοντι οἱ πολέμιοι ἐπιφανεῖεν, πῶς χρὴ ἀντικαθιστάναι, καὶ εἴ σοι ἐπὶ φάλαγγος ἄγοντι ἄλλοθέν ποθεν οἱ πολέμιοι φαίνοιντο ἢ κατὰ πρόσωπον, ὅπως χρὴ ἀντιπαράγειν, ἢ ὅπως τὰ τῶν πολεμίων ἄν τις μάλιστα αἰσθάνοιτο, ἢ ὅπως τὰ σὰ οἱ πολέμιοι ἥκιστα εἰδεῖεν, ταῦτα δὲ πάντα τί ἂν ἐγὼ λέγοιμί σοι; ὅσα τε γὰρ ἔγωγε ᾔδειν, πολλάκις ἀκήκοας, ἄλλος τε ὅστις ἐδόκει τι τούτων ἐπίστασθαι, οὐδενὸς αὐτῶν ἠμέληκας οὐδ᾽ ἀδαὴς γεγένησαι. δεῖ οὖν πρὸς τὰ συμβαίνοντα, οἶμαι, τούτοις χρῆσθαι ὁποῖον ἂν συμφέρειν σοι τούτων δοκῇ.

15 The following list is by no means exhaustive: Nicephorus II's *De velitatione bellica* (from the tenth century) includes two sections on watch posts (1 and 2), a section on 'the assembling and movement of the army', and a section on 'the siege of a fortified town'; Nicephorus II's *Praecepta militaria*, which includes a section entitled 'On the Encampment' (4); Syrianus' *Peri Strategias* begins with a section on setting up camp, two consecutive sections (3 and 4) on night and day sentinels, what to do when you are attacked while on the march (section 12), how to cross a river and pass through a narrow place (section 14), marching through mountain passes (19 and 20), and siege warfare (21, cf. 26, 27); and Nicephorus Ouranus' *Tactics* includes among a host of different sections one on passing through a variety of different terrains (64), and another on siege warfare (65).

16 Perhaps the most interesting discussion of 'how to build morale' comes at 2.1.25 when Xenophon describes Cyrus' decision while on campaign to use tents large enough for each company (μέγεθος δὲ ὥστε ἱκανὰς εἶναι τῇ τάξει ἑκάστῃ): 'he thought that if they tented together it would help them to get acquainted with one another. And in getting acquainted with one another . . . a feeling of considerateness was more likely to be engendered in them all' (ἐν δὲ τῷ γιγνώσκεσθαι καὶ τὸ αἰσχύνεσθαι πᾶσι δοκεῖ μᾶλλον ἐγγίγνεσθαι, οἱ δὲ ἀγνοούμενοι ῥᾳδιουργεῖν πως μᾶλλον δοκοῦσιν). Even here, although the morale would be fostered by the close proximity of the soldiers to each other, the morale of the men is dependent on the general.

17 For a battle narrative with these precepts in action, see book seven of the *Education of Cyrus*.

18 On Aeneas, see Whitehead (1990) 1–42 and the collected essays in Pretzler and Barley (2017).

19 Aelian, *Tactics*. 1.2. Cf. Polybius 10.22.

20 τὴν οὖν τῶν σωμάτων σύνταξιν σκεψαμένους πρὸς τὸ μέγεθος τῆς πόλεως καὶ τὴν διάθεσιν τοῦ ἄστεος καὶ τῶν φυλάκων τὰς καταστάσεις καὶ περιοδίας, καὶ ὅσα ἄλλα σώμασι κατὰ τὴν πόλιν χρηστέον, πρὸς ταῦτα τοὺς μερισμοὺς ποιητέον. τοὺς μὲν γὰρ ἐκπορευομένους δεῖ συντετάχθαι πρὸς τοὺς ἐν τῇ πορείᾳ τόπους, ὡς χρὴ πορεύεσθαι παρά τε τὰ ἐπικίνδυνα χωρία καὶ ἐρυμνὰ καὶ στενόπορα καὶ πεδινὰ καὶ ὑπερδέξια καὶ ἐνεδρευτικά, καὶ τὰς τῶν ποταμῶν διαβάσεις καὶ τὰς ἐκ τῶν τοιούτων παρατάξεις. τὰ δὲ τειχήρη καὶ πολιτοφυλακήσοντα πρὸς μὲν τὰ τοιαῦτα οὐδὲν δεῖ συντετάχθαι, πρὸς δὲ τοὺς ἐν τῇ πόλει τόπους καὶ τὸν παρόντα κίνδυνον.

21 Aeneas Tacticus 16, 22.

22 Aeneas Tacticus 8. According to the author of the entry in the Suda (alphaiota, 215), Polybius claims that Aeneas wrote a book on Stratagems.

23 Aeneas Tacticus 10.19.

24 Aeneas Tacticus 15, 18, 19, 20, 24, 25, 29, 31.

25 Counterdevices: Aeneas Tacticus 32, 34, 35; stratagems: Aeneas Tacticus 39; and fires: Aeneas Tacticus 33, 36, 37. To get back to the Suda (alphaiota, 215), the author claims that Aeneas was, first and foremost, a writer of fire signals.

26 Aeneas Tacticus 26.7ff.

27 Aeneas Tacticus 27.

28 See also the discussion in the Introduction, 2–4. And as the Introduction observes, these texts are not inaccessible to students.

29 Anderson (1970).

30 Pritchett (1974).

31 Marsden (1971).

32 Lendon (2005). See also the Introduction, 3.

33 Dain (1974); Stadter (1978); Wheeler (1978); Pavcovic (1988); Devine (1993) and (1995); Saxtorph (2002); Rance (2004a); and Colombo (2011).

34 Of the approximately 100 entries on Frontinus, for example, only a half dozen or so deal with his *Stratagems* specifically: most concern his treatise on aqueducts.

35 Though note the translations of Devine (1989) and Matthew (2012).

36 Whately (2015) 249.

37 Speidel (1975).

38 Barnes (1979); Birley (1985); Charles (2004), (2007a), and (2007b).

39 Reeve (2000).

40 Charles (2010). See too Bond (2016) 149–151.

41 Charles (2003), (2004), (2005), (2007a), and (2007b). On Vegetius and naval warfare, see Chapter 12.

42 Eramo (2009), (2010), and (2012).

43 Wheeler (1979), (1982), (1983), (1988a), (1988b), (1991), (2004a), and (2004b).

44 Rance (2000), (2004a).

45 Rance (2004b), (2007), (2008), (2010), and (2017). See also Rance (2007a) 343–348 for his overview of late antique military manuals.

46 Whitehead (1990), (2008a), (2008b), (2010), (2016) and Whitehead and Blyth (2004).

47 See the Introduction, 12 n.32 for a select bibliography.

48 Lendon (1999), (2005) *passim*, and (2007) 500–508. Besides the comments of Lendon, see Chaniotis' (2005) detailed treatment of Hellenistic warfare. Cf. Shipley (2000) 334–341.

49 Cuomo (2007) 40.

50 For an overview of Polybius' oeuvre, see Walbank (1972); for a detailed treatment see Walbank (1957–1979). Lendon (1999) 282–285 discusses Polybius' views of tactics. On the intellectual framework that underscores Polybius, see Champion (2004). Cf. Sacks (1981) 125–132.

51 As Polybius puts it, 'About this matter I have entered into greater detail in my notes on tactics'. (9.20.4; Patton's translation from his Loeb text).

52 Polybius 6.39.2.

53 Polybius 6.39.8, trans. Paton.

54 Interestingly, Polybius prefaces his detailed exposition of the differences (and following the introduction just mentioned) with some comments on the importance of generalship. As important as tactics are, generalship trumps all else. Livy, who used Polybius in his description of Cynoscephalae, also alludes to this tactical difference, and the importance of level ground. Cf. Livy 33.4.1–4.

55 Polybius 18.29. We also find, and as we saw earlier rather unsurprisingly, a reference to Homer and the tight ranks described at *Iliad* 13.131.

56 Polybius 18.32.1ff.

57 Polybius 18.21–26. In his description of the battle, Livy modifies Polybius by stressing the importance of morale and tactics (33.8–9).

58 For an introduction to Onasander, see Smith (1998) 151–166. Cf. Schellenberg (2007).

59 Onasander 6.

60 Onasander 15–21, cf. 27, 30, 31.
61 Onasander 11, 13, 14.2.
62 Onasander 32, 33. Onasander also has a brief discussion of the importance of rewards which ought to match their action, another element that echoes Polybius (34).
63 Onasander 40–41.
64 Onasander 7–10.
65 See Maurice, Preface 21–27.
66 Cf. Rance (1993) 99ff.
67 In other words, this work was first conceived of as a treatise on cavalry warfare, and only later was the infantry component added. Chapter 12, the later addition, is something of a mixed bag lacking the unity of chapters 1 through 11, including, as it does, points which, in the earlier parts of the work, had been covered under separate chapters, such as the types of formation (12.A.1–7), armament (12.B.4–5), and the crossing of various types of terrain (12.B.18–21). Thus, it does not provide evidence for the inferiority of late-sixth-century infantry, on which see Rance (2004a), (2005) 427–443 and (2007a) 348–359.
68 Maurice 2.5, 3.16, 4.1–5 (the chapter devoted to ambushes – Περὶ Ἐνέδρας).
69 Maurice 9.1–5. cf. Maurice 7.A.12.
70 On these 'ethnika', see Wiita (1977).
71 Rance (2007a) 347.
72 Wheeler (1988a) xiv. Cf. Brizzi (2004) 15–41. The 'Achilles ethos' is the same as the 'western way of war' which is championed by Keegan (1994) and Hanson (2000). Despite their strong assertions, however, western forces have long used stratagems and have often sought to avoid pitched battle. Cf. the discussion of Lynn (2003) 12–25 and Gat (2005) 391–400, 505–508, 609–618.
73 We might put Xenophon in the Odysseus camp and Polybius in the Achilles camp along with the theoreticians discussed previously.
74 Frontinus 2.1.1.
75 On stratagems in general, see Wheeler (1988a).
76 Polyaenus 1.35, 2.1.22 (the whole of section 2.1 is concerned with Agesilaus and includes both historical discussion and a few stratagems; Polyaenus appears to play armchair historian in this instance: 2.3.12, 3.9.63, and 4.19.
77 Polyaenus started to become popular in the fifth century CE, and this continued well into the Byzantine period. See Krentz and Wheeler (1994) xvi–xx for a good overview.
78 On readers, see also Chapter 3, which suggests a possibly different – that is, non-specialist – audience.
79 Of the three authors, Arrian was a commander in the Roman military and so had practical military experience, which suggests, though by no means proves, that his treatise might have had practical application. See Wheeler (1979), (2004a), and (2004b).
80 Asclepiodotus Book 1: concerning the different parts of the phalanx; Book 2: concerning the number and names of the divisions of the phalanx of hoplites; Book 3: concerning the disposition of the men both in the entire phalanx and in its divisions; Book 4: concerning the intervals between them; Book 5: concerning the size and shape of the arms; Book 6: concerning the phalanx of light infantry (*psiloi*) and peltasts and the names of the dispositions and divisions; Book 7: concerning the phalanx of cavalry and the names of the whole and of its divisions; Book 8: concerning chariots; Book 9: concerning elephants; Book 10: concerning the common names for military manoeuvres; Book 11: concerning the arrangements by divisions on expeditions; Book 12: concerning the commands for their military manoeuvres.
81 Aelian, *Tactics* 6, 7, 49, 75, 93, 94, 95, 106, and 109. Many of these points are later noted by Syrianus (31.3–9).
82 Arrian, *Tactics* 2, 3–4, 12, 20–27, and 31–32.
83 This is not meant as a criticism; rather, this is a feature of much ancient literature and is no less true of other genres like historiography and epic poetry.

84 Stadter (1978) 117–118.
85 Köchly presented Aelian's Greek text in tandem with Arrian's Greek text to empha-
    size the similarity. Cf. Stadter (1978) 120–121.
86 Compare, for instance, Aelian, *Tactics* C.24.1, **Τί ἐστι δεξιὸν κέρας καὶ κεφαλὴ καὶ
    τί εὐώνυμον κέρας καὶ οὐρά**, 'what is the right wing and the head and the left wing
    and the rear', with Arrian *Tactics* 8.4.1, **ἀριστερᾷ εὐώνυμον κέρας καὶ οὐρά**, 'the
    honoured left wing and the rear'; Aelian C.35.2, **σημειοφόρος, οὐραγός, σαλπιγκτής,
    ὑπηρέτης, στρατοκῆρυξ**, 'a standard-bearer, a file-closer, a trumpeter, an aide, an
    army herald', with Arrian 10.4.2, **σημειοφόρος, οὐραγός, σαλπιγκτής, ὑπηρέτης,
    στρατοκῆρυξ**, 'a standard-bearer, a file-closer, a trumpeter, an aide, an army her-
    ald'; and Aelian, *Tactics* C.58, **Ὅτι καθ' ἑκάστην ἑκατονταρχίαν ἔκτακτοι ἄνδρες
    τάσσονται πέντε· σημειοφόρος, οὐραγός, σαλπιγκτής, ὑπηρέτης, στρατοκῆρυξ**,
    'that to each heacatonarchy are deployed five men; a standard-bearer, a file-closer,
    a trumpeter, an aide, and an army herald' with Arrian *Tactics* 14.2, **καθ' ἑκάστην δὲ
    ἑκατονταρχίαν ἔκτακτοι ἄνδρες τέσσαρες ἔστων, σημειοφόρος καὶ σαλπιγκτὴς
    καὶ ὑπηρέτης καὶ στρατοκῆρυξ**, 'that to each hecatonarchy there were four men,
    standard-bearer, trumpeter, an aide, and an army herald', give but three examples.
87 Whitehead (2010).
88 Dain and de Foucault (1967) 327–330, 333–335; Haldon (2014) 39, 46–53.
89 Leo Preface 6; 6.25 (Aelian), 14.98 (Onasander).
90 Haldon (2014) 52–53.
91 Lendon (1999) 276.
92 Lendon (1999) 316.
93 Polybius 18.21–26; Lendon (1999) 284–285. On the Greek tactical vocabulary, see
    Pritchett (1985) 44–93.
94 In this particular instance, Agathias writes, 'as the tacticians say' (ὡς ἂν οἱ τακτικοὶ
    ὀνομάσαιεν).
95 Cf. Whitby (1988) 94–105 on the points of contact between the combat described by
    Theophylact and the treatise of Maurice.
96 Vegetius 3.26; Maurice 8. For parallels, see the discussion in Whately (2016) 145–151.
97 Vegetius 3.9.4–7; Maurice 7.A.1.25–27; Procopius, *Wars* 3.19.11–13. Whately (2016)
    146, 150.
98 Procopius, *Wars* 3.12.12–14; Vegetius 2.23.6–10; Maurice 53; Whately (2016) 146.
99 Thucydides 1.22.4; Polybius 1.1, 2.35, 12.25.
100 Anonymous, *On Political Science* 4. See too Bell (2009) 53–54.
101 Milner (1993) 19, n. 5.
102 Janniard (2011) 231, n. 717.
103 Meißner (1999) 251–252, 284–292, 335–339. Janniard (2008) makes a strong case to
    the contrary.
104 For an overview, see Rance (2000) 230–233. See too Rance (2004a), (2004b) and
    Luttwak (2009) 266–303. A comprehensive study will be found in Rance's (forth-
    coming) two-volume text, translation, and analysis of the *Strategikon* (Rance
    forthcoming).
105 A new text by Dennis (2014) and commentary by Haldon (2014) have been published
    recently of Leo's *Taktics*.
106 See Maurice pr. 21–27.
107 Rance (1993) 99ff.
108 Other influences on Maurice include the manuals of Aelian and Arrian, on which see
    Rance (forthcoming).
109 Formisano 2017.
110 Such is the supposition of Kaegi (2003) 130.
111 In other words, this work was first conceived of as a treatise on cavalry warfare, and
    only later was the infantry component added. It does not provide evidence for the

inferiority of late-sixth-century infantry, on which see Rance (2004a), (2005) 427–443, (2007a) 348–359).

112 Maurice 2.5, 3.16, 4.1–5 (the chapter devoted to ambushes – Περὶ Ἐνέδρας).

113 On the last point, note the comments of Janniard (2011) 19.

Bibliography

Anderson, J. K. 1970. *Military Theory and Practice in the Age of Xenophon*. Berkeley: University of California Press.

Barnes, T. D. 1979. 'The Date of Vegetius'. *Phoenix* 35. 254–257.

Bell, P. 2009. *Three Political Voices from the Age of Justinian*. Liverpool: Liverpool University Press.

Birley, E. 1985. 'The Dating of Vegetius and the Historia Augusta'. *Bonner Historia-Augusta Colloquium 1982–83*. 57–67.

Bond, S. 2016. *Trade and Taboo: Disreputable Professions in the Roman Mediterranean*. Ann Arbor, MI: University of Michigan Press.

Brizzi, G. 2004. *Le guerrier de l'Antiquité classique, de l'hoplite au legionnaire*. Paris: Éditions du Rocher.

Champion, C. 2004. *Cultural Politics in Polybius' Histories*. Berkeley: University of California Press.

Chaniotis, A. 2005. *War in the Hellenistic World: A Social and Cultural History*. Oxford: Blackwell.

Charles, M. B. 2003. 'Vegetius on Armour: The "Pedites Nudati" of the "Epitoma Rei Militaris"'. *Ancient Society* 33. 127–167.

——— 2004. '"Mattiobarbuli" in Vegetius' "Epitoma Rei Militaris": The "Iouiani" and the "Herculiani"'. *Ancient Society* 18. 109–121.

——— 2005. 'Vegetius on "Liburnae": Naval Terminology in the Late Roman Period'. *SCI* 24. 181–193.

——— 2007a. *Vegetius in Context: Establishing the Date of the Epitoma Rei Militaris*. Stuttgart: Franz Steiner.

——— 2007b. 'A Regimental Nickname from Late Antiquity: Vegetius and the "Mattiobarbuli" Again'. *AHB* 21. 89–94.

——— 2010. 'Unseemly Professions and Recruitment in Late Antiquity: *Piscatores* and Vegetius *Epitoma* 1.7.1–2'. *AJP* 131. 101–120.

Colombo, M. 2011. 'La "lancea", i "lanciarii", il "pilum" e l'"acies" di Arriano: un contributo alla storia dell'esercito romano'. *Historia* 60. 158–190.

Cuomo, S. 2007. *Technology and Culture in Greek and Roman Antiquity*. Cambridge: Cambridge University Press.

Dain, A. and Foucault, J. A. 1967. 'Les stratégistes Byzantins'. *T & M Byz* 2. 317–392.

——— 1974. 'Arrian's Array'. *History To-Day* 24. 570–574.

Dennis, G. 2014. *The Taktika of Leo VI: Revised Edition*. Washington, DC: Dumbarton Oaks.

Devine, A. M. 1989. 'Aelian's Manual of Hellenistic Military Tactics: A New Translation from the Greek with an Introduction'. *The Ancient World* 19. 31–64.

——— 1993. 'Arrian's Tactica'. *ANRW* 34. 312–337.

——— 1995. 'Polybius' Lost Tactica: The Ultimate Source for the Tactical Manuals of Asclepiodotus, Aelian, and Arrian?'. *AHB* 9. 40–44.

Due, B. 1989. *The Cyropaedia: Xenophon's Aims and Methods*. Aarhus: Aarhus University Press.

Eramo, I. 2009. 'Ῥωμαῖοι e Ἄραβες a battaglia?: nota al "De re strategica" di Siriano Μάγιστρος'. *Invigilata Lucernis* 31. 95–104.

———— 2010. 'Retorica militare fra tradizione protrettica e pensiero strategico'. *Talla Dixit* 5. 25–44.

———— 2012. 'Composition and Structure of Syrianus Magister's Military "Compendium"'. *Classica et Christina* 7. 97–116.

Formisano, M. 2017. 'Fragile Expertise and the Authority of the Past: The "Roman Art of War"', in J. König and G. Woolf, eds., *Authorities and Expertise in Ancient Scientific Culture*. Cambridge: Cambridge University Press. 129–152.

Gat, A. 2005. *War in Human Civilization*. Oxford: Oxford University Press.

Haldon, J. 2014. *A Critical Commentary on the Taktika of Leo VI*. Washington, DC: Dumbarton Oaks.

Hanson, V. D. 2000. *The Western Way of War: Infantry Battle in Classical Greece*. Second edition. Berkeley: University of California Press.

Hutchinson, G. 2000. *Xenophon and the Art of Command*. London.

Janniard, S. 2008. 'Végèce et les transformations de l'art de la guerre aux IVe et Ve siècles après J.-C.'. *Antiquité Tardive* 16. 19–36.

———— 2011. *Les Transformations de l'Armée Romano-Byzantine (IIIe – VIe Siècles apr. J.-C.): Le Paradigme de la Bataille Rangée*. Unpublished PhD Thesis. L'Atelier du Centre de Recherches Historiques. Paris.

Kaegi, W. 2003. *Heraclius, Emperor of Byzantium*. Cambridge: Cambridge University Press.

Keegan, J. 1994. *A History of Warfare*. London: Random House.

Krentz, P. and Wheeler, E. L., trans. 1994. *Polyaenus: Stratagems of War*. Chicago: Ares Publishers.

Lee, J. W. I. 2007. *A Greek Army on the March: Soldiers and Survival in Xenophon's Anabasis*. Cambridge: Cambridge University Press.

Lendon, J. E. 1999. 'The Rhetoric of Combat: Greek Military Theory and Roman Culture in Julius Caesar's Battle Descriptions'. *CA* 18. 273–329.

———— 2005. *Soldiers and Ghosts*. New Haven: Yale University Press.

———— 2007. 'War and Society [the Hellenistic World and the Roman Republic]', in P. Sabin, H. Van Wees, and M. Whitby, eds., *The Cambridge History of Greek and Roman Warfare Volume I: Greece, the Hellenistic World and the Rise of Rome*. Cambridge: Cambridge University Press. 498–516.

Luttwak, E. 2009. *The Grand Strategy of the Byzantine Empire*. Cambridge, MA: Harvard University Press.

Lynn, J. A. 2003. *Battle: A History of Combat and Culture*. Boulder: Westview Press.

Marsden, E. W. 1971. *Greek and Roman Artillery*. Oxford: Oxford University Press.

Matthew, C. 2012. *The Tactics of Aelian*. Barnsley, UK: Pen and Sword.

Meißner, B. 1999. *Die technologische Fachliteratur der Antike: Struktur, Überlieferung und Wirkung technischen Wissens in der Antike: (ca. 400 v. Chr.-ca. 500 n. Chr.)*. Berlin: De Gruyter.

Milner, N. P. 1993. *Vegetius. Epitoma Rei Militaris*. Liverpool: Liverpool University Press.

Pavcovic, M. 1988. 'A Note on Arrian's Ἔκταξις κατα Ἀλανῶν'. *AHB* 2. 21–23.

Pretzler, M. and Barley, N., eds. 2017. *Brill's Companion to Aineias Tacticus*. Leiden: Brill.

Pritchett, W. K. 1974. *The Greek State at War, Part II*. Berkeley: University of California Press.

—— 1985. *The Greek State at War*. Vol. 4. Berkeley: University of California Press.

Rance, P. 1993. *Tactics and Tactica in the Sixth Century: Tradition and Originality*. Unpublished PhD Dissertation. University of St. Andrews. St Andrews UK.

—— 2000. '*Simulacra Pugnae*: The Literary and Historical Tradition of Mock Battles in the Roman and Early Byzantine Army'. *GRBS* 41. 223–275.

—— 2004a. 'The *fulcum*, the Late Roman and Byzantine *testudo*: The Germanisation of Roman Infantry Tactics'. *GRBS* 44. 265–326.

—— 2004b. '*Drungus*, δρου'γγο, and δρουγγιστις: A Gallicism and Continuity in Roman Cavalry Tactics'. *Phoenix* 58. 96–130.

—— 2005. 'Narses and the Battle of Taginae (Busta Gallorum): Procopius and Sixth-Century Warfare'. *Historia* 54. 424–472.

—— 2007a. 'Battle [the Later Roman Empire]', in P. Sabin, H. Van Wees, and M. Whitby, eds., *The Cambridge History of Greek and Roman Warfare Volume II: Rome from the Late Republic to the Late Empire*. Cambridge: Cambridge University Press. 342–378.

—— 2007b. 'The Date of the Military Compendium of Syrianus Magister (Formerly the Sixth-Century Anonymous Byzantinus)'. *BZ* 100. 701–737.

—— 2008. 'Noumera or Mounera: A Parallel Philological Problem in De Cerimoniis and Maurice's Strategikon'. *Jahrbuch der Österreichischen Byzantinistik* 58. 121–129.

—— 2010. 'The De Militari Scientia or Müller Fragment as a Philological Resource: Latin in the East Roman Army and Two New Loanwords in Greek: Palmarium and *Recala*'. *Glotta* 86. 63–92.

—— 2017. 'The Reception of Aineias' *Poliorketika* in Byzantine Military Literature', in M. Pretzler, M. and N. Barley, eds., *Brill's Companion to Aineias Tacticus*. Leiden: Brill. 290–374.

—— forthcoming. *The Roman Art of War in Late Antiquity: The Strategikon of the Emperor Maurice*. London: Routledge.

Reeve, M. 2000. 'The Transmission of Vegetius' "Epitoma Rei Militaris"'. *Aevum* 74. 243–354.

Sacks, K. S. 1981. *Polybius on the Writing of History*. Berkeley: University of California Press.

Saxtorph, N. and C. G. Tortzen. 2002. '"Acies Contra Alanos": Arrian on Military Tactics', in K. Ascani, V. Gabrielsen, K. Kvist, and A. Rasmusen, eds., *Ancient History Matters*. Rome: L'Erma di Bretschneider. 221–226.

Schellenberg, H. M. 2007. 'Einige Bemerkungen zum "Strategikos" des Onasandros', in L. De Blois, E. Lo Cascio, O. Hekster, and G. De Kleijn, eds., *Impact of the Roman Army (200 BC–AD 476): Economic, Social, Political, Religious, and Cultural Aspects*. Leiden: Brill. 181–191.

Shipley, G. 2000. *The Greek World after Alexander, 323–30BC*. London: Routledge.

Smith, C. 1998. 'Onasander on How to Be a General', in M. Austin, J. Harries, and C. Smith, eds., *Modus Operandi: Essays in Honour of Geoffrey Rickman*. London: Bulletin of the Institute of Classical Studies. 151–166.

Speidel, M. P. 1975. 'Vegetius (3.5) on Trumpets'. *Acta Classica* 17. 153–155.

Stadter, P. 1978. 'The Ars Tactica of Arrian: Tradition and Originality'. *CP* 73. 117–128.

Vela Tejada, J. 2004. 'Warfare, History and Literature in the Archaic and Classical Periods: The Development of Greek Military Treatises'. *Historia* 53. 129–146.

Walbank, F. 1957–1979. *A Historical Commentary on Polybius*. Oxford: Oxford University Press.

—— 1972. *Polybius*. Berkeley: California University Press.

Whately, C. 2015. 'The Genre and Purpose of Military Manuals in Late Antiquity', in G. Greatrex and H. Elton, eds., *Shifting Genres in Late Antiquity*. Abingdon, UK: Ashgate. 249–261.

——— 2016. *Battles and Generals*. Leiden: Brill.

Wheeler, E. 1978. 'The Occasion of Arrian's Tactica'. *GRBS* 19. 351–365.

——— 1979. 'The Legion as Phalanx'. *Chrion* 9. 303–318.

——— 1982. '*Hoplomachia* and Greek Dances in Arms'. *GRBS* 22. 223–233.

——— 1983. 'The *Hoplomachia* and Vegetius' Spartan Drillmasters'. *Chiron* 13. 1–20.

——— 1988a. *Stratagem and the Vocabulary of Military Trickery*. Leiden: Brill.

——— 1988b. 'Sapiens and Stratagems: The Neglected Meaning of a Cognomen'. *Historia* 37. 166–195.

——— 1991. 'The General as Hoplite', in V. D. Hanson, ed., *Hoplites: The Classical Greek Battle Experience*. London: Routledge. 121–170.

——— 2004a. 'The Legion as Phalanx in the Late Empire (I)', in Y. Le Bohec and C. Wolff, eds., *L'armée romaine de Dioclétien á Valentinien 1er*. Paris: de Boccard. 309–358.

——— 2004b. 'The Legion as Phalanx in the Late Empire (II)'. *Revue des Etudes Militaires Anciennes* 1. 147–175.

Whitby, M. 1998. *The Emperor Maurice and His Historian: Theophylact Simocatta on Persian and Balkan Warfare*. Oxford: Oxford University Press.

——— 2004. 'Xenophon's Ten Thousand as a Fighting Force', in R. Lane Fox, ed., *The Long March: Xenophon and the Ten Thousand*. New Haven: Yale University Press. 215–242.

Whitehead, D. 1990. *Aineias the Tactician, How to Survive a City under Siege*. Oxford: Oxford University Press.

——— 2008a. 'Fact and Fantasy in Greek Military Writers'. *Acta Antiqua Academiae Scientarium Hungaricae* 48. 139–155.

——— 2008b. 'Apollodorus' "Poliorketika": Author, Date, Dedicatee', in H. M. Schellenberg, V.-E. Hirschmann, and A. Krieckhaus, eds., *A Roman Miscellany: Essays in Honour of Anthony R. Birley on His Seventieth Birthday*. Gdansk: Foundation for the Development of Gdansk University. 204–212.

——— 2010. *Apoloodrus Mechanicus, Siege-Matters*. Stuttgart: Franz Steiner.

——— 2016. *Philo Mechanicus, On Sieges*. Stuttgart: Franz Steiner.

Whitehead, D. and Blyth, P. 2004. *Athenaeus Mechanicus: On Machines*. Stuttgart: Franz Steiner.

Wiita, J. E. 1977. *The Ethnika in Byzantine Military Treatises*. Unpublished PhD Dissertation. University of Minnesota. Minneapolis.

2 The limited source value of works of military literature

Hans Michael Schellenberg

Ancient texts dealing with the art of generalship, tactics, the construction of artillery, and machines of war, as well as the layout of fortifications and of military camps, and even military law, are currently lumped together in scholarly discourse in a single group of texts.[1] This group of texts was already fully established in the nineteenth century. The labels given to this group and their authors vary considerably in the past and present scholarly literature. A commonly accepted and convincing definition of the group of texts and the authors who belong to it is lacking. For this reason, it is not entirely clear which authors belong to this group and by contrast, why other authors do *not* belong to it. One possible label for this group of text is 'military manuals', as chosen in the current volume. In what follows, I use the neutral term 'military literature' and the established term 'military writer'.[2] The texts of military literature, however, contain a number of epistemological problems, which have not been sufficiently considered in their use in scholarly literature. It is therefore useful to acquire awareness of the problems that can arise from using the works of military authors as evidence within historical interpretations.

At this point, I define 'historical interpretation' and what it should achieve. Its purpose is an explanation and presentation that is objective, in accordance with hermeneutic rules, and as far as possible accurate and comprehensive, of a particular historical problem or set of circumstances. For this, it must draw upon all the necessary evidence that can contribute to this explanation, and exclude whatever cannot do so. The evaluation of sources and the justification of the way they are evaluated to be used for an historical explanation are of decisive importance.[3]

I

It is desirable to keep in mind war and the conduct of war, which are after all the main subjects of military writers, in order to point out their complexity and some of the connected problems, and to make them more intelligible. The observation and conduct of military conflicts and the resultant experiences and perceptions derived from them form the foundations of a knowledge and know-how of warfare. The direct bearers of this knowledge and know-how of warfare are the members of a fighting force, however organised. Given a sufficient degree of development in literacy in a society, this knowledge of warfare can be expressed in the form of written texts.

Moreover, war and warfare are multidimensional processes involving the whole of society. It is an organised conflict between human communities, carried out by means of force. Its degree of complexity is dependent on the degree of organisation and technical development of those human communities that use these means. Human communities, because of the finite existence of their members, are incapable of being static, timeless institutions. The form and contents of mentalities and social rules of societies are changeable. A linear evolution of war and an unbroken continuity in the conduct of war are for this reason improbable. Wars, because of their extension in space and time and the number of individual participants, cannot be directly observed as such by a single individual. Discussions about the conduct of war within a society and the depiction of war and warfare in its written reports, especially historical texts, are collective acts. The extent to which they match the truth can be extremely variable or, indeed, negligible. Therefore, knowledge of war and thinking and acting in relation to the conduct of war cannot be examined from a timeless point of view. They are tied to particular societies in their respective historical context. Theories of a timeless and generally valid or universally applicable military knowledge should therefore be seen as extremely improbable.[4]

From these remarks, it should be clear that war and the conduct of war are subjects of knowledge that are complex and difficult to grasp and therefore to depict in written form, whether in a work of military literature or in a historical text. An ancient military writer who chose as the subject of his work either some part of the conduct of war or war in its entirety faced a challenge that should not be underestimated. He himself may not have been aware of the extent of this challenge or the problems that resulted from it. Modern historians do not have the luxury of this unawareness.[5]

Let one suppose that the intention of a military writer is to supply the knowledge and to inform his reader how to carry out a specific task and to what he should pay attention in the process.[6] An ancient military writer can call upon several sources of information for his subject: his own personal practical experience, of whatever kind, in military affairs; the military experience of others, of which he has been told; and written contributions on his topic. On the basis of this set of sources, he has the possibility to draw inferences from the state of affairs in the present or the past and to make generalisations about what the reader of his work should do in the future and what is necessary in order repeatedly to be successful in what he does. As far as can be seen from the surviving works of military writers, the personal qualifications that were required did not include practical military experience.[7] Hence anyone who for whatever reason assumed that he could contribute to any given discussion could write a work of military literature. It should be added at this point that each of these authors wrote his work in a private capacity. Works of military literature are therefore not to be construed as official and binding sets of instructions or seen as some sort of 'ancient military field manuals'.[8]

Basically, such works of military literature are to be seen as forecasts of the future. Forecasts are problematic, especially when they concern the behaviour of larger groups of people, as in the form of armed forces or whole societies. The problems that arise from forecasts of this kind are increased all the more the further that the forecast they make extends into the future. For a forecast of this kind to be correct requires a reliable empirical factual basis. In general, the future is open, and the way that the present develops uncertain: thus, solutions for all possible cases cannot be thought of in advance. New problems that first arise as the result of putting into practice existing solutions for problems cannot be foreseen from the start.[9]

Military writers transmit their own subjective knowledge and know-how; the supposed enrichment of this subjective knowledge with empirical random historical examples, with the intention of confirming it and making it more convincing, can only lead in a rhetorical sense to the desired result. Most importantly, that an action was carried out successfully on one occasion in the past does not mean that it can always be carried out successfully again, nor does it adequately explain why it was able to be carried out successfully in the past.[10] Thus, neither is the personal knowledge and ability of a military writer objective, nor can his recommendations for action be generalised in a modern empirical sense and seen as 'common military doctrine'.

To establish their source value for historical explanations, the writers of military literature can be roughly divided into two groups: on the one hand, authors with personal experience and personal capacity in the subject they discuss, on the other hand those lacking this experience and this capacity. To express this in a modern and simplified fashion: military experts and military laymen.[11] What needs to be taken note of when assessing a work by an expert? The only man who can be classed as an expert is one who possesses practical knowledge and ability in the subject he discusses. This expertise is confined to the judgement on, and evaluation of, military matters in the expert's lifetime and in the society to which he belongs. A man who counts as a military expert in his own lifetime cannot count as an expert on military affairs before and after his own lifetime.[12]

Reporting on purely historical affairs would fall into the category of historiography, the subject of which is not necessarily to transmit practical steps to carry out military engagements, but to explain and describe them.[13] To be an expert means neither that one understands everything in the sphere of one's experience, nor that one is free from making mistakes in one's assessments. Likewise, to be able to do something does not necessarily mean that one is able to take a considered look at one's own behaviour and is able to teach others to act according to rules in a manner comparable to one's own ability. For this, a corresponding competence as a teacher would be required on the part of the expert and the reader would, equally, need sufficient capacity to understand.[14] Consequently, every author who writes a work in the category of military literature and is not an expert in the prior sense is a layman. Not surprisingly, for the historical interpretation of military affairs, works by laymen possess less value than works by experts.[15]

What should be taken into account when assessing the work of a layman? The work of an author who seeks to transmit practical knowledge and capacity on how to wage war, without possessing such knowledge himself, and takes these instructions on how to act from written sources such as history books or, also, goes back to the works of military experts from the past and mixes such pieces of information together, produces arguably a work that can no longer be referred to a single specific past epoch.[16] A work of this kind would only say something about the military thinking of its author and not necessarily about the waging of war and the practice of his own lifetime. This statement would also apply if the author compiled his work exclusively from the writings of earlier experts.[17] The author's own work is mostly irrelevant for the interpretation of military affairs in his own lifetime. The worth of a work of this kind for historical interpretation would not be irrelevant if it could be cut up into separate sections for the purpose of *Quellenforschung* (research into sources), and individual sections of the texts by experts that had been used could be clearly ascribed to these experts.[18] This would make parts of this work relevant for the interpretation of military affairs.

To raise a general question: can someone simply become a military expert by reading a book on one's own? Reading a book on warfare can prepare for a practical action, but it is not an adequate substitute for its actual practice. Practical action and experience is in addition carried out jointly with other people in a military institution of whatever kind (such as an army) and cannot be regarded separately from this.[19]

Any reader of a book by military experts and laymen can use these books as a source of inspiration for his own ideas. In this case (source of inspiration), the truth and empirical content of these books is irrelevant. A repetition of successful past military actions cannot be achieved in this way. Reading such a work can thus lead to all kinds of ideas, but not necessarily to reliable knowledge or useful 'know-how knowledge' about the conduct of war. For this reason, the fact that a work has been handed down and has been highly regarded for centuries means precisely *nothing* for historians as far as its value as a source for the historical interpretation of military affairs in a specific epoch is concerned (see later, Leo VI 'Tactica' 7, 10).

II

Up to this point in my paper, I have tried to point out general problems concerned with works of military literature. Now I would like to illustrate the earlier-mentioned limitations by reference to two particular authors. For this purpose, I have decided on Aeneas Tacticus from the fourth century BCE and Onasander from the first century CE. The first was supposedly an expert, the second a layman: how is one to grade the value of the two writers as a source for historical interpretation of military affairs in their own lifetime?

Since Aeneas' origin is unknown and cannot be deduced from his historical examples either, the historical context of his military experience and his biography cannot be connected to a particular city.[20] No personal information about

Aeneas is available except from his own work. As far as can be recognised, he possessed personal military experience, the nature and extent of which are, however, unknown. Evidently, Aeneas concerned himself with warfare in his own lifetime in its totality and wrote a number of books about it, out of which only an incomplete one is preserved. The subject of this book concerns measures a city should take for its defence. One of his main points is the area of 'internal security'. He does not offer direct criticism of possible predecessors and their methods. However, Aeneas informs his readers thoroughly about possible problems that may occur when possible solutions that he proposed are repeated. Among his instructions as to how something should be done, Aeneas frequently uses historical examples, including those that derive from periods before his own lifetime.

Whatever the manner in which Aeneas acquired personal experience, the spectrum of his topics extends far beyond it. The greater part of his information comes from his general knowledge and from written sources, among them works of historiography. In one passage, he repeats a story from Herodotus on the siege of Barca (37.6–7), which is at least over a hundred years old by Aeneas' own lifetime. This is one of the *very few* instances in which one can identify his source that can be partly checked. In this passage, Herodotus tells how the Persians laid siege to the city of Barca in Cyrenaica, perhaps in about 512 BCE. Before this, Arcesilaus III of Cyrene and his father-in-law were murdered in Barca (Herodotus 4.165). Arcesilaus III's mother Pheretime was able to escape the city of Cyrene and fled to Egypt in search of help. The Persian satrap of Egypt, Aryandes, decided to help to retake Barca and avenge her son. He sent the full garrison force of his satrapy under the command of Amasis (4.167). The Persians were unable to take Barca by force and had to resort to trickery. After they conquered the city that way, the murderer and enemies of Arcesilaus III were impaled, and the surviving ones were deported:[21]

> Then the Persians besieged Barca for nine months, digging under ground passages leading to the walls, and making violent assaults. As for the mines, a smith discovered them by the means of a brazen shield, and this is how he found them: carrying the shield round the inner side of the walls he smote it against the ground of the city; all other places where he smote it returned but a dull sound, but where the mines were the bronze of the shield rang clear. Here the Barcaeans made a countermine and slew those Persians who were digging the earth. Thus the mines were discovered, and the assaults were beaten off by the townsmen.
>
> (Herodotus 4.200)[22]

Aeneas reworks this passage from Herodotus, reducing the story about the siege to the essential core for his advice on how to detect the undermining of city walls. For this purpose, it was probably not necessary to mention that Barca was taken by the Persians and what gruesome fate the citizens had to face afterwards:

> Now an old incident is told . . . of Amasis in his siege of Barca, when he was trying to dig a mine. The people of Barca, who were aware of the attempt of

Amasis, were concerned lest he might elude or anticipate them, until a coppersmith thought out a device. Carrying a bronze shield around inside the wall he held it against the ground above various points. And of course at all other points the parts to which he applied the bronze were without a sound, but where the digging was in progress beneath the shield became resonant. So the people of Barca dug a countermine at this point and killed many of the enemy's miners, and as a result even now men use this means of ascertaining where mines are being dug.

(Aeneas 37.6–7)[23]

Herodotus' and Aeneas' description seems unlikely. Sound pressure emitted by a given source follows the inverse-proportional law that predicts a quick decrease in value with increased distance from the source.[24] The geological location of the ancient city of Barca in what is today known as the 'Jebel Akhdar' uplands contributes further to this assessment. This plateau is composed mostly of medium dense limestone of the Upper Cretaceous and Paleogene that is not an ideal conductor for sound (pressure) waves. While sound travels better through a rock with higher density, a tunnel is easier to dig in unconsolidated earth that has even lower conductivity for sound waves. However, it has to be mentioned that a possible amplification effect of a bronze shield is unknown. But it should be noted that the thicker the metal is, the less it can function as an amplifier because its stiffness increases and therefore it is less able to vibrate. Taking everything into consideration, there is nothing that supports Aeneas' description of the use of a bronze shield as a tool to locate a siege tunnel.[25]

Aeneas comments that this method was still in use. How could he know that?[26] As the previous discussion suggests, this method does not work in the manner Herodotus and Aeneas describe. Aeneas offers no reliable and repeatable practical advice to his reader here. He offers his personal belief in this method and reveals the limits of his knowledge and personal experience in such matters.[27]

If this small part of Aeneas' work already contains such an unreliable judgement by Aeneas that results in unhelpful advice to his reader, how many more of these pieces of advice are present? For this reason, one should be careful and not assume in the first place that everything Aeneas describes is mostly sound and common military doctrine. The commonplace nature of Aeneas' military convictions and advice should not be taken for granted. His pieces of advice were developed for the military circumstances of his lifetime. They may not reasonably be used as evidence within an historical explanation about Greek warfare much outside this timespan.

What about Onasander's 'Strategikos Logos'? No certain biographical information is available about Onasander apart from what is in his work. According to his own statement, he had no personal military experience (strat. pr. 7). Onasander is thus to be categorised as a military layman. Research takes it that he dedicated his work to Quintus Veranius, consul in the year 49 and Roman governor of Britain from 57 to 58, where he died in the course of an expedition.[28]

The sources for Onasander's work are no longer definitely identifiable, but insofar as they can be recognised, they derive overwhelmingly from the period between the fourth and the second centuries BCE. It looks as if works of military literature were among his sources.[29] Onasander depicts an *ideal* expedition and of what the *ideal* commander of an army needs to be aware, and to do, in order to carry it through. The work contains no historical examples, and other than Quintus Veranius does not name any historical persons. This makes nearly all information and advice given by Onasander very hard or impossible to date.

The problems that arise when using Onasander's work in the context of a historical interpretation can be seen at 10.4. This passage is too readily taken to be evidence for the contemporary Roman army of Onasander's lifetime carrying out mock battles.[30] Onasander here describes an army which fights against itself in the field with 'staves, clods of earth' and 'leather straps'. The weaponry, and in particular his vocabulary, are so unusual that one can identify Xenophon's *Education of Cyrus* (*Cyropaedia*) as the basis for this passage.[31] Xenophon tries to convince his readers that archers are not to be feared and could easily be overwhelmed in close combat. Most parts of Xenophon's *Education of Cyrus* are fictitious, and there is no reason to believe that this is different for this passage:[32]

> And once Cyrus invited a captain and his whole company to dinner, because he had noticed him drawing up one half of the men of his company against the other half for a sham battle. Both sides had breastplates and on their left arms their shields; in the hands of the one side he placed stout cudgels, while he told the other side that they would have to pick up clods to throw. Now when they had taken their stand thus equipped, he gave the order to begin battle. Then those on the one side threw their clods, and some struck the breastplates and shields, others also struck the thighs and greaves of their opponents. But when they came into close quarters, those who had the cudgels struck the others – some upon the thighs, others upon the arms, others upon the shins; and as still others stooped to pick up clods, the cudgels came down upon their necks and backs. And finally, when the cudgel-bearers had put their opponents to flight, they pursued them laying on the blows amid shouts of laughter and merriment. And then again, changing about, the other side took the cudgels with the same result to their opponents, who in turn threw clods.
>
> (2.3.17–18)[33]

Onasander reduces this story from Xenophon to its essential aspects. He interlaces Xenophon's story and vocabulary in his advice for a mock battle. It is necessary to keep in mind that half the soldiers in Xenophon's story are armed with 'staves' (= Miller's 1914 'cudgels'), while the other one uses 'clods'. The 'staves' are used for *beating* the soldiers armed with 'clods':

> Next after dividing the army into two parts he should lead them against each other in a sham battle, armed with staves or the shafts of javelins; if there

should be any fields covered with clods, he should command them to throw clods; if they have any leather straps, the soldiers should use them in the battle. Pointing out to the soldiers ridges or hills or steep ascents, he should command them to charge and seize these places; and sometimes arming the soldiers with the weapons I have just mentioned, he should place some on the hilltops and send the others to dislodge them. He should praise those who stand firm without retreating, and those who succeed in dislodging their opponents.

(Onasander 10.4)[34]

Onasander gives here superficial advice for a mock battle. He gives no clue to his reader what kind of information he used to write this passage (10.4). He is rather vague about what to do with the 'staves' and 'leather straps' he mentions. The 'clods' are to be thrown. Armour and shields are missing in his description, but he adds some terrain and combat features. In scholarly literature, it is assumed that the 'staves' or 'shafts of javelins' are practice weapons for sword fighting. The 'shafts of javelins' are viewed as some kind of evidence for this assumption combined with a cross-reference to Arrian, *Tactics* 34.8. Indeed they are practice weapons at 34.8, but they are to be thrown without iron tips from horseback by the soldiers of the Roman cavalry at each other. They are *not* intended as practice swords. This kind of cross-reference is therefore misleading in explaining the passage from Onasander. The reader encounters the same words, but each author uses the terms differently. One cannot view Onasander's 'staves' and 'clods' independently from Xenophon's story. The 'leather straps' are especially difficult to explain. They belong to the same type of weapons as the 'staves' and 'clods', which are already highly unusual close combat weapons, even in an ancient mock battle. But, according to Onasander, they have to be weapons of some sort. The best explanation is that they are probably the ancient equivalent of 'boxing gloves'.[35] A mock battle fought with 'staves, clods' and 'leather straps'? In Onasander's account, what is described amounts to a freestyle outdoor brawl, not a mock battle. His choice of weapons and vocabulary betrays his literary source, to which he added his own special and unusual close combat weapon. There is no hint that either Greek military doctrine or Roman military doctrine was the template for 10.4. This is only Onasander's imagination of a mock battle developed from Xenophon's *Education of Cyrus*.[36]

How should one assess the value of Onasander's work for the interpretation of military affairs in his own lifetime? The work is most irrelevant for the interpretation of actual Roman warfare in the first century CE and earlier and can offer no evidence for historical interpretations concerned with it, especially because of its ideal character and the concealment of his sources. Anyone who uses Onasander's work for this very reason, at best, endangers only his own set of arguments.[37] The same applies, with modifications, to Greek warfare in previous times, insofar as the work does not refer to accounts that can be definitely dated and ascribed to their source. What can be investigated in Onasander's work is the military thought of its author in the context of his time.

Onasander's work formed one of the major templates for the military work of the Byzantine emperor Leo VI. He adopted the passage 10.4 into his own work, including 'clods, staves' and 'leather straps (Leo VI: *charzania*)'[38] and slightly modernised language and content. Leo VI himself had no military experience (that is, he was a layman) but was the commander-in-chief of all Byzantine forces. His idea of a mock battle is a much more elaborate version of what Onasander provides:

> Divide the army in two parts, then have them come together in a mock battle, the lances and, likewise, the arrows without points or, as we said, with staffs instead of swords. Or instead of lances distribute staves or reeds. If the ground on which they are drilling has clods of earth, order them to throw these at each other in practicing for battle. At times let them make use of what are called charzania (= leather straps) or similar items in their battles. Point out to the men steep hills and order them to ascend them on the run and seize them. Of course, you will have other soldiers in position on top of those hills.
>
> (Leo 7.10)[39]

The Byzantine general Nikephorus Ouranus (an expert) later modified this passage for his own military work (*Tactics* 8).[40] This a *very* rare instance in which the transmission and development of an idea within the military literature can be observed for 1350 years and traced back to its origin. However, this transmitted idea is neither evidence for common Greek and Roman nor Byzantine military doctrine.[41] It is only evidence for three authors (Onasander, Leo VI, and Nikephorus Ouranus) relying on the works of their predecessors.

III

I have sought to demonstrate that works of military literature contain a number of problems that have not been sufficiently addressed so far. These problems make their use as evidence for historical explanations difficult. This applies particularly to works that guide what should be done in the future and for this purpose rely on past experience of whatever type (whether oral or literary). For historians, this state of affairs means that it is not possible without further discussion of a text which treats the way in which a procedure should be carried out in the future to infer directly to an actual state of affairs in the present time (that is, the present or the lifetime of the author). This kind of inference, difficult enough as it is, becomes many times more complicated when the author of such a text relies not merely on contemporary experience but appeals for support to the recent or distant past in the shape of historical writing (Aeneas 37.6–7; see earlier, 43–44) or works of military literature. If this kind of appeal to the past is not clearly recognisable, the use of such a work in the context of historical interpretation is only possible to an extremely limited extent, and it is not suitable as evidence for the actual conduct of warfare in its author's lifetime. Works that have as their subject an ideal state of affairs have hardly any value as evidence for the actual conduct of war in a specific historical period.[42] That does not mean that works of military literature are sources without relevance for the history of the conduct of war in antiquity, but only that they are difficult to use to provide historical interpretations than has previously been assumed.

Hence before one cites the work of a military writer in favour or against a particular tactic, one ought to reflect whether it can contribute anything to our interpretation.[43] The works by experts can be rated as of greater value as a source than those by laymen, but the works of experts must also be subject to clear restrictions in their use. Those who adduce the work of a military writer in the context of their interpretation as evidence for facts that belong to a period before or after the lifetime of this military writer must explain why this can be done.[44]

Notes

1 Quotations from ancient sources (lightly modified) come from the Loeb Classical Library editions, unless otherwise noted. I would like to thank Professor Anthony R. Birley for translating this paper. Any remaining errors are my own responsibility. I also would like to thank Torsten René Sander, a geologist, for his contribution to the discussion of Aeneas Tacticus 37.6–7.

2 Well-known examples are the introduction to Köchly and Rüstow (1855) 1–101 and Jähns (1889) 3–136. Köchly and Rüstow called their topic: 'Kriegsschriftsteller' (a modern synonym for 'K.' is 'Militärschriftsteller' = 'military writer'). Jähns called his topic 'Kriegswissenschaften' ('sciences of war' and alternatively 'Militärliteratur' = 'military literature' (4). The following is only a small selection of German options: 'Technologische Fachliteratur': Meißner (1999); 'Militärschriftsteller': Le Bohec (2000); 'Fachtexte': Fögen (2005); 'Militärschriftstellerei': Mann (2013); 'Fachliteratur: Poliorketik' / 'Mechanik': Fiorucci (2014). For 'military manuals', cf. Whately (2015). Whitehead (2008) 141 notes the problematic concentration of all these authors in one group. Cf. Wheeler (2010) 434–438 for an introduction with bibliography, and now especially Rance (2017). I avoid the term 'manual' to prevent a connotation with modern 'military field manuals'. Formisiano (2011) 43 criticises the term 'manual'. See also the Introduction to this volume, 4–7.

3 Based on Scholz (2008) and Gerber (2012). The main points of the arguments are to be found in English in Scholz (2014) and Gerber (2014). For an introduction into 'Hermeneutics', see Bühler (2003a), and for the different ways of interpreting a text, see Bühler (2003b). An English introduction, not entirely similar to those works mentioned before, is McCullagh (1998). My understanding of history as a 'science' is based on Schurz (2014).

4 Cf. Schellenberg (2017) 83–85 (self-quotation, modified) with full bibliography. On the topics of 'knowledge', 'expertise', and 'authority', see König and Wolf (2017) and Formisiano and Van der Eijk (2017).

5 These 'challenges' were already noted in *On War* by von Clausewitz (Hahlweg 1980, 281; Howard and Paret 1976, 134 = book 2, 2).

6 This is just an example. For the different intentions in writing such kinds of text, see Wheeler (1980) 74.

7 Military writers of surviving works with personal military experience: Xenophon of Athens, Aeneas 'Tacticus', Sextus Iulius Frontinus, and the emperor Mauricius and Arrianus of Nicomedia; see further later for Arrian. Military writers with personal experience (not necessarily military experience) in what they wrote include: Philo of Byzantium (artillery and maybe siege warfare), Hero of Alexandria (artillery), and perhaps Apollodorus of 'Damascus' (siege engines). The identification of Apollodorus as the famous architect of Trajan is highly insecure).

8　From the surviving works and fragments of the ancient military literature, none seems to be an 'official set of instructions'. The idea of an official 'Roman military manual (Militärhandbuch / Heeresreglement)' of Neumann (1933), repeated in *idem* (1936), (1942), (1946), and (1956), is his assumption and not based on ancient sources. Meredith L. D. Riedel has kindly pointed out to me that this is different in the case of Leo's *Tactica* (*Tactics*).

9　Again, this problem was already noted in 'On war' by von Clausewitz (Hahlweg 1980, 289; Howard and Paret 1976, 140 = book 2, 2), who thought that a 'positive doctrine' could not be achieved at all.

10　For an illuminating and 'deadly' example of personal experience combined with inductive reasoning about the future, I would recommend an internet search for the terms 'Bertrand Russell's chicken' or 'Nassim Taleb's turkey'.

11　Not the best choices of terms, to be sure. There are no direct ancient equivalents.

12　Arrian did write a work about the Hellenistic phalanx and could count as an expert in Roman military matters. In the case of his 'Tactica', he is a layman and antiquarian. His work is in most parts not a useful source for the conduct of war in his own lifetime. The only part that is a source of this kind is the last one, where Arrian switches from the distant Hellenistic past to a description of Roman cavalry exercises. Cf. Arrian, *Tactics* 32–44.

13　This is one of the reasons why, in my view, historical works should not be treated as part of military literature. Jähns (1889) includes such works, as does Campbell (2004); see also the Introduction, 7. For the problems of defining the 'genre' of military literature, see earlier, 39.

14　On readers, see Chapter 3 and the Epilogue.

15　That does not mean that those works could not contain valuable historical information. Hero of Alexandria's 'Belopoeica' is the best source for our understanding of how ancient artillery works. But Hero described the development of the ancient artillery without naming any of his predecessors and sources. Whatever piece of artillery and technical development he describes, it cannot be dated. My rating of him as an expert (artillery) in his lifetime is based on the 'Cheiroballistra' (both texts could be found with text and translation in Marsden (1971), as well as Philo of Byzantium and Biton).

16　This also includes military literature, written with other purposes, as the transmitting of 'actual' knowledge. Cf. the writings of Asclepiodotus and Aelian, which, despite their antiquarian and ideal nature, both ended with the clause: 'These are in brief the principles of the tactician; they mean safety to those who follow them and danger to those who disobey.' Asclepiodotus 12.11); cf. Aelian 42.2 (Devine 1989, 59) with slightly different wording. Arrian omits this phrase in his work.

17　Athenaeus 'On machines' did so and reused the work of Agesistratus, who himself already used the works of earlier experts (?). The ram-tortoise of Hegetor of Byzantium in the transmitted form is non-functioning; cf. Whitehead and Blyth (2004) 120 and 126 (the length of the ram is physically not possible). The personal contribution of Athenaeus is difficult to evaluate.

18　This is only possible for the full works of Biton and Athenaeus: on the former, see Schellenberg (2006) 17, incl. n.15. Cf. Gatto (2010) 67–79 with comparison of Athenaeus and Vitruvius, who both used the lost work of Agesistratus. It is not possible for the work of Vegetius, despite Schenk (1930) or Milner (1991) 253–312 (shortened to Milner (2001) xvi–xxviii), for example.

19　Cf. Campbell (1987). As mentioned earlier, war and warfare are multidimensional processes involving the whole of society. The influence of individual persons ('generals') and their importance therein should not be over-estimated.

20　Cf. now Brodersen (2017) and Pretzler and Barley (2018).

21　There is no other independent literary source or archaeological testimony that one could use to check Herodotus' story. The Persian satrap of Egypt as named by Herodotus is currently not known otherwise. Herodotus' Persians show stereotypical behaviour. This includes helping a woman, resorting to trickery, impaling captives, and enslaving and deporting the surviving citizens. Herodotus' description of the campaign (4.167–205) is mostly about ethnography; crucial military details are missing. Cf. the commentary by Asheri, Lloyd, and Corcella (2007) 693–721. Cf. Schellenberg (2017) for the date and further references. I think that most of the *Libyan logos* is fictitious.

22 ἐνθαῦτα δὴ ἐπολιόρκεον τὴν Βάρκην ἐπὶ μῆνας ἐννέα, ὀρύσσοντές τε ὀρύγματα ὑπόγαια φέροντα ἐς τὸ τεῖχος καὶ προσβολὰς καρτερὰς ποιεύμενοι. τὰ μέν νυν ὀρύγματα ἀνὴρ χαλκεὺς ἀνεῦρε ἐπιχάλκῳ ἀσπίδι, ὧδε ἐπιφρασθείς· περιφέρων αὐτὴν ἐντὸς τοῦ τείχεος προσῖσχε πρὸς τὸ δάπεδον τῆς πόλιος. τὰ μὲν δὴ ἄλλα ἔσκε κωφὰ πρὸς τὰ προσῖσχε, κατὰ δὲ τὰ ὀρυσσόμενα ἠχέεσκε ὁ χαλκὸς τῆς ἀσπίδος. ἀντορύσσοντες δ᾽ ἂν ταύτῃ οἱ Βαρκαῖοι ἔκτεινον τῶν Περσέων τοὺς γεωρυχέοντας. τοῦτο μὲν δὴ οὕτω ἐξευρέθη, τὰς δὲ προσβολὰς ἀπεκρούοντο οἱ Βαρκαῖοι.

23 (Dain 1967): Παλαιὸν δέ τι λέγεται <. . . . .> Ἄμασιν Βαρκαίους πολιορκοῦντα, ἐπεὶ ἐπεχείρει ὀρύσσειν. Οἱ δὲ Βαρκαῖοι αἰσθόμενοι <τὸ> ἐπιχείρημα τοῦ Ἀμάσιδος, ἠπορoῦντο μὴ λάθῃ ἢ φθάσῃ, ἔπειτα ἀνὴρ χαλκεὺς ἀνεῦρεν ἐνθυμήσας· ἀσπίδος χάλκωμα περιφέρων ἐντὸς τοῦ τείχεος ἐπάνω προσίσχεν πρὸς τὸ δάπεδον. Τῇ μὲν δὴ ἄλλῃ κωφὰ ἦν πρὸς ἃ προσίσχοι τὸ χάλκωμα· ἧ δὲ ὑπωρύσσετο, ἀντήχει. Ἀντορύσσοντες οὖν οἱ Βαρκαῖοι ταύτῃ, ἀπέκτειναν πολλοὺς τῶν ὑπορυσσόντων. Ὅθεν καὶ νῦν χρῶνται αὐτῷ ἐν τῇ νυκτὶ γνωρίζοντες ᾗ ὑπορύσσεται.

24 Mörser (2015).

25 Sander / Schellenberg: this question is even more complicated than it seems to be. This result is preliminary and should not be seen as final answer. More research is clearly needed. This would be a good opportunity for a combined attempt by the natural sciences with the support of 'experimental archaeology'. Until more research is done, the statement, that the method mentioned by Herodotus / Aeneas is unlikely to have functioned, is valid. See earlier, 48 n.1.

26 No other ancient source that describes this method in use has survived.

27 So far, there are no other attempts, similar to the previous one, to check the veracity of this story by Herodotus. Asheri, Lloyd, and Corcella (2007) 720 make no statement about it.

28 For Quintus Veranius' governorship in Britannia, see Birley (2005) 37–43 with an addendum by Hassall (2008) 31–32.

29 Cf. Oldfather and Pease in the Loeb edition that contains Onasander, who notes the use of Aeneas 'Tacticus' by Onasander (13); Whitehead (2016) 18–19, who notes the use of Philon of Byzantium; and Chlup (2014) 40 n.15 and 45, who suggests Xenophon's *Agesilaus*. See further Peters (1972); Ambaglio (1981); and Petrocelli (2008) for the sources of Onasander.

30 I focus on Horsmann (1991), whom Rance (2000) follows. Horsmann uses 10.4 as evidence for mock sieges (71 with n.81); 134 n.102 (the 'staves' seen as 'rudis'), 142 (the 'staves' seen as replacement for the 'gladius)', 144 (evidence for Roman mock battles), 150 (the 'shafts of javelins' seen as evidence for 'Roman practice javelins'), 186 (evidence for Roman mock battles and practice weapons). He also thinks (186) that Onasander relied maybe on some kind of Roman 'military field manual' (186). Horsmann follows in this respect Neumann (1933). For the problem of 'Roman field manuals', see 49 earlier.

31 Already noted by Köchly and Rüstow (1855) 85 again by Peters (1972) 154–157 and still highlighted by Petrocelli (2008) 195 with n. 143. On the *Education of Cyrus*, see Chapter 8.

32 Xenophon only mentions 'cudgels' and 'clods' in this passage. They are *not* Greek practice weapons for close combat.

33 ἐκάλεσε δ᾽ ἐπὶ δεῖπνον ὁ Κῦρος καὶ ὅλην ποτὲ τάξιν σὺν τῷ ταξιάρχῳ, ἰδὼν αὐτὸν τοὺς μὲν ἡμίσεις τῶν ἀνδρῶν τῆς τάξεως ἀντιτάξαντα ἑκατέρωθεν εἰς ἐμβολήν, θώρακας μὲν ἀμφοτέρους ἔχοντας καὶ γέρρα ἐν ταῖς ἀριστεραῖς, εἰς δὲ τὰς δεξιὰς νάρθηκας ταχεῖς τοῖς ἡμίσεσιν ἔδωκε, τοῖς δ᾽ ἑτέροις εἶπεν ὅτι βάλλειν δεήσοι ἀναιρουμένους ταῖς βώλοις. ἐπεὶ δὲ παρεσκευασμένοι οὕτως ἔστησαν, ἐσήμηνεν αὐτοῖς μάχεσθαι. ἐνταῦθα

δὴ οἱ μὲν ἔβαλλον ταῖς βώλοις καὶ ἔστιν οἳ ἐτύγχανον καὶ θωράκων καὶ γέρρων, οἱ δὲ καὶ μηροῦ καὶ κνημῖδος. ἐπεὶ δὲ ὁμοῦ ἐγένοντο, οἱ τοὺς νάρθηκας ἔχοντες ἔπαιον τῶν μὲν μηρούς, τῶν δὲ χεῖρας, τῶν δὲ κνήμας, τῶν δὲ καὶ ἐπικυπτόντων ἐπὶ βώλους ἔπαιον τοὺς τραχήλους καὶ τὰ νῶτα. τέλος δὲ τρεψάμενοι ἐδίωκον οἱ ναρθηκοφόροι παίοντες σὺν πολλῷ γέλωτι καὶ παιδιᾷ. ἐν μέρει γε μὴν οἱ ἕτεροι λαβόντες πάλιν τοὺς νάρθηκας ταὐτὰ ἐποίησαν τοὺς ταῖς βώλοις βάλλοντας.

34 (Korzenszky and Vári 1935): Εἶτα διελὼν τὰ στρατεύματα πρὸς ἀλλήλους ἀσιδήρῳ μάχῃ συναγέτω νάρθηκας ἢ στύρακας ἀκοντίων ἀναδιδούς· εἰ δέ τινα καὶ βεβωλασμένα πεδία εἴη, βώλους τε κελεύων αἴροντας βάλλειν· ὄντων δὲ καὶ ἱμάντων ταυρείων χρήσθων ἐπὶ τὴν μάχην· δείξας δὲ αὐτοῖς καὶ λόφους ἢ βουνοὺς ἢ ὀρθίους τόπους κελευέτω σὺν δρόμῳ καταλαμβάνεσθαι· ποτὲ δὲ καὶ ἐπιστήσας ἐπ᾽ αὐτῶν τινας τῶν στρατιωτῶν καὶ ἀναδοὺς ἃ μικρῷ πρόσθεν ἔφην ὅπλα, τούτους ἐκβαλοῦντας ἑτέρους ἐκπεμπέτω· καὶ ἤτοι τοὺς μείναντας ἐπαινείτω καὶ μὴ ἐκπεσόντας ἢ τοὺς ἐκβαλόντας.

35 See Schellenberg (2007) 190 with further references. I do not think that some kind of 'whip' is suggested here. Peters (1972) 156–157 notes the 'leather straps' but has no idea what kind of weapons they should be. Rance (2000) 242 n.43 thinks they are the 'leather straps' mentioned by Josephus in his *Jewish War* 3.95. In Josephus' account, they are part of the normal equipment and not intended as weapons. Petrocelli (2008) 193 n.144 thinks of some kind of horse-gear and is cross-referencing to Xenophon, Cavalry Commander 8.4. This is misleading, because the 'leather straps' have to be weapons, not some sort of equipment. Further, Onasander does not mention horses and cavalry at 10.4. Sestili (2010) makes no special comment.

36 The main point was already made by Köchly and Rüstow (1855) 85, elaborated by Peters (1972) 154–157 and supplemented by Schellenberg (2007). Peters (1972) is seldom mentioned. Without refutation of Köchly and Rüstow and Peters, the argumentation of Horsmann (1991) is pointless.

37 As Horsmann (1991) did with his arguments, relying on 10.4.

38 Cf. Haldon (2014) 207. It is not entirely clear what Leo VI means with this term and how he understands Onasander's 'leather straps'. For Leo's VI *Tactics*, see Chapter 14.

39 Διαμερίσας δὲ τὰ στρατεύματα πρὸς ἀλλήλους ἀσιδήρῳ μάχῃ συμβαλλέτωσαν ἤτοι διὰ κονταρίων ἄνευ ξιφῶν ἢ σαγιττῶν ὁμοίως ἤ, ὡς εἴπομεν, ἀντὶ σπαθίων βεργία ἢ νάρθηκας ἢ καλάμους ἀντὶ κονταρίων ἀναδιδούς. ἐὰν δὲ καὶ βώλους ἔχῃ ἡ γῆ ἐν ᾗ γυμνάζωνται, τούτους βάλλειν κέλευε κατ᾽ ἀλλήλων ἐν τῇ γυμνασίᾳ τῆς συμβολῆς. ποτὲ δὲ καὶ τὰ λεγόμενα χαρζάνια ἢ τούτοις ὅμοιά τινα χρήσθωσαν ἐν τῇ μάχῃ. δείξας δὲ αὐτοῖς καὶ βουνοὺς ὀρθίους κέλευε σὺν δρόμῳ ἀναβαίνειν καὶ καταλαμβάνειν αὐτούς, ἔχοντας δηλονότι τοὺς βουνοὺς ἐκείνους ἑτέρους στρατιώτας ἐφεστῶτας ἐπ᾽ αὐτῶν. (Text and translation Dennis 2014).

40 This point comes from personal discussion with Philip Rance, who is preparing the first full edition of the 'Tactica' of Nikephorus Ouranus, which is only partly edited so far. Cf. McGeer (1995) 79–167 for the chapters 56–65 (text and translation with commentary).

41 On Byzantine warfare and its possible foundation in military texts, see Chapter 13.

42 I would count Onasander's *Strategikos*, Asclepiodotus' *Tactica*, Aelian's *Tactica*, and Arrian's *Tactics* (1–32) as works of this kind. The earlier statement could be applied to one of the most overrated (by laymen) military writer: Sun Tzu. No statements of any military writer should be construed as 'law-like' (see earlier: theories of a timeless and generally valid or universally applicable military knowledge / know-how should be seen as extremely improbable). For Sun Tzu and the Chinese military writers and their content of reality, see also Lewis (2005) 14–15.

43 See Wrightson's chapter in this volume for a problematic approach – in particular, his use of Onasander. Any explanation based on Onasander in regard to Demetrius Poliorcetes should be classified as ahistorical.

44 I would recommend to check, for example, how many times the authors of *The Cambridge History of Greek and Roman Warfare* cite Onasander. How many of these

citations consider the ideal nature of the work and the lifetime of his author (first century CE) and the problems that goes with them? See also the Introduction to this volume, 5.

Bibliography

Ambaglio, D. 1981. 'Il tratto sul commandante di Onasandro'. *Athenaeum* 59. 353–377.

Asheri, A., Lloyd, A., and Corcella, A. 2007. *Commentary on Herodotus Books I–IV*. Oxford: Oxford University Press.

Birley, A. R. 2005. *The Roman Government of Britain*. Oxford: Oxford University Press.

Brodersen, K. 2017. *Aineias/Aeneas Tacticus. Stadtverteidigung – Poliorketika*. Berlin: De Gruyter.

Bühler, A. 2003a. 'Grundprobleme der Hermeneutik', in *idem*, ed., *Hermeneutik. Basistexte zur Einführung in die wissenschaftlichen Grundlagen von Verstehen und Interpretieren*. Heidelberg: Synchron. 3–19.

——— 2003b. 'Die Vielfalt des Interpretierens', in *idem*, ed., *Hermeneutik. Basistexte zur Einführung in die wissenschaftlichen Grundlagen von Verstehen und Interpretieren*. Heidelberg: Synchron. 99–119.

Burliga, B. 2007. *Eneasz Taktyk. Obrona Oblężonego Miasta*. Krakow. Prószyński i S-ka.

Campbell, B. 1987. 'Teach Yourself How to Be a General'. *JRS* 77. 14–28.

——— 2004. *Greek and Roman Military Writers: Selected Readings*. London: Routledge.

Chlup, J. T. 2014. 'Just War in Onasander's ΣΤΡΑΤΗΓΙΚΟΣ'. *JAH* 2. 37–63.

Dain, A. and Bon, A.-M. 1967. *Énée le Tacticien: Poliorcétique*. Paris: Les Belles Lettres.

Dennis, G. T. 2014. *The Taktika of Leo VI: Text, Translation and Commentary*. Revised edition. Washington, DC: Dumbarton Oaks.

Devine, A. M. 1989. 'Aelian's Manual of Hellenistic Military Tactics: A New Translation from the Greek with an Introduction'. *AncW* 19. 31–64.

Fiorucci, F. 2014. 'Poliorketik/Mechanik', in B. Zimmermann and A. Rengakos, eds., *Handbuch der griechischen Literatur. Vol. 2. Die Literatur der klassischen und hellenistischen Zeit*. Munich: C. H. Beck. 591–610.

Fögen, T. 2005. *Antike Fachtexte: Ancient Technical Texts*. Berlin: De Gruyter.

Formisiano, M. 2011. 'The Strategikós of Onasander: Taking Military Texts Seriously'. *Technai* 2. 39–52.

——— and Van der Eijk, P. J., eds. 2017. *Knowledge, Text and Practice in Ancient Technical Writing*. Cambridge: Cambridge University Press.

Gatto, M. 2010. *Il ΠΕΡΙ ΜΗΧΑΝΗΜΑΤΩΝ di Ateneo Meccanico*. Edizione critica, traduzione, commento e note. Rome: Aracne editrice.

Gerber, D. 2012. *Analytische Metaphysik der Geschichte. Handlungen, Geschichten und ihre Erklärungen*. Berlin: Suhrkamp.

——— 2014. 'Causal Explanation and Historical Meaning: How to Solve the Problem of the Specific Historical Relation between Events', in M. I. Kaiser, R. O. Scholz, D. Plenge, and A. Hüttemann, eds., *Explanation in the Special Sciences: The Case of Biology and History*. Heidelberg: Springer. 197–210.

Godley, A. D. 1921. *Herodotus. Vol. 2: Books III–IV*. Cambridge, MA: Harvard University Press.

Hahlweg, W. 1980. *Vom Kriege. Hinterlassenes Werk des Generals Carl von Clausewitz. Vollständige Ausgabe im Urtext*. Nineteenth edition. Bonn: Dümmler.

Haldon, J. 2014. *A Critical Commentary on the Taktika of Leo VI*. Washington, DC: Dumbarton Oaks.

Hassall, M. 2008. 'Footnotes to the Fasti', in H. M. Schellenberg, V. E. Hirschmann, and A. Krieckhaus, eds., *A Roman Miscellany: Essays in Honour of Anthony R. Birley on His Seventieth Birthday*. Gdansk: Foundation for the Development of Gdansk University for the Department of Mediterranean Archeology. 31–41.

Horsmann, G. 1991. *Untersuchungen zur militärischen Ausbildung im republikanischen und kaiserzeitlichen Rom*. Boppard am Rhein: Harald Boldt.

Howard, M. and Paret, P. 1976. *Carl von Clausewitz: On War*. Princeton: Princeton University Press.

Jähns, M. 1889. *Geschichte der Kriegswissenschaften vornehmlich in Deutschland. Erste Abteilung. Altertum, Mittelalter, XV. und XVI. Jahrhundert*. Munich: R. Oldenbourg.

Köchly, H. and Rüstow, W. 1855. *Griechische Kriegsschriftsteller. Griechisch und Deutsch mit kritischen und erklärenden Anmerkungen von H. Köchly und W. Rüstow. Zweiter Theil. Die Taktiker. Erste Abtheilung. Asklepiodotos' Taktik. Aelianus' Theorie der Taktik. Nebst einer Einleitung und zwei Stücken taktischen Inhalts aus Xenophon und Polybios*. Leipzig: Wilhelm Engelmann.

König, J. and Wolf, G., eds. 2017. *Authority and Expertise in Ancient Scientific Culture*. Cambridge: Cambridge University Press.

Korzenszky, E. and Vári, R. 1935. *Onasandri Strategicus*. Budapest: Societas Frankliniana.

Le Bohec, Y. 2000. 'Militärschriftsteller', in *Neuer Pauly*. Vol. 8. Stuttgart. 185–186.

Lewis, M. E. 2005. 'Writings on Warfare Found in Ancient Chinese Tombs'. *Sino-Platonic Papers* 158. 1–15.

Mann, C. 2013. *Militär und Kriegführung in der Antike*. Munich: Oldenbourg.

Marsden, E. W. 1971. *Greek and Roman Artillery: Technical Treatises*. Oxford: Oxford University Press.

McCullagh, C. B. 1998. *The Truth of History*. London: Routledge.

McGeer, E. 1995. *Sowing the Dragon's Teeth: Byzantine Warfare in the Tenth Century*. Washington, DC: Dumbarton Oaks.

Meißner, B. 1999. *Die technologische Fachliteratur der Antike. Struktur, Überlieferung und Wirkung technischen Wissens in der Antike (ca. 400. v. Chr. – ca. 500 n. Chr.)*. Berlin: Akademie Verlag.

Milner, N. P. 1991. *Vegetius and the Anonymus De Rebus Bellicis*. Unpublished DPhil Thesis. Oxford.

——— 2001. *Vegetius: Epitome of Military Science*. Second revised edition. Liverpool: Liverpool University press.

Mörser, M. 2015. *Technische Akustik*. Tenth edition. Berlin: Springer.

Neumann, A. 1933. 'Das Militärhandbuch des Kaisers Augustus'. *Klio* 26. 360–362.

——— 1936. 'Das Augusteische-Hadrianische Armee-Regelement und Vegetius'. *CPh* 31. 1–10.

——— 1942. 'Das römische Heeresreglement'. *HZ* 166. 554–562.

——— 1946. 'Das römische Heeresreglement'. *CPh* 41. 217–225.

——— 1956. 'Militärhandbuch'. *RE Suppl* 8. 357–359.

Peters, W. 1972. *Untersuchungen zu Onasander*. Bonn: Heinrich Trapp.

Petrocelli, C. 2008. *Onasandro: Il generale*. Bari: Edizione Dedalo.

Pretzler, M. and Barley, N., eds. 2018. *Brill's Companion to Aineas Tacticus*. Leiden: Brill.

Rance, P. 2000. 'Simulacra Pugnae: The Literary and Historical Tradition of Mock Battles in the Roman and Early Byzantine Army'. *GRBS* 41. 223–275.

——— 2017. 'Introduction', in N. Sekunda and *idem*, eds., *Greek Taktika: Ancient Military Writing and Its Heritage*. Gdansk: Foundation for the Development of Gdansk University for the Department of Mediterranean Archeology. 9–64.

54 *Hans Michael Schellenberg*

Schellenberg, H. M. 2006. 'Diodor von Sizilien 14,42,1 und die Erfindung der Artillerie im Mittelmeerraum'. *FeRA* 3. 14–23.

——— 2007. 'Einige Bemerkungen zum Strategikos des Onasandros', in L. De Blois and E. Lo Cascio, eds., *The Impact of the Roman Army (200 BC – AD 476)*. Leiden: Brill. 181–191.

——— 2017. 'Reflections on the Military Views of the "Military Writer" Aeneas Tacticus', in N. Sekunda and P. Rance, eds., *The Ancient Taktika: Ancient Military Writing and Its Heritage*. Gdánsk: Foundation for the Development of Gdansk University for the Department of Mediterranean Archeology. 81–93.

Schenk, D. 1930. *Flavius Vegetius Renatus. Die Quellen der Epitoma Rei Militaris*. Leipzig: is Dieterich'sche Verlagsbuchhandlung.

Scholz, O. R. 2008. 'Erkenntnis der Geschichte – eine Skizze', in A. Frings and J. Marx, eds., *Erzählen, Erklären, Verstehen*. Berlin: Akademie Verlag. 111–128.

——— 2014. 'Philosophy of History: Metaphysics and Epistemology', in M. I. Kaiser, idem, D. Plenge, and A. Hüttemann, eds., *Explanation in the Special Sciences: The Case of Biology and History*. Heidelberg: Springer. 245–253.

Schurz, G. 2014. *Philosophy of Science: A Unified Approach*. London: Routledge.

Sestili, A. 2010. *Onasandro: Strategikos. Manuale il Commandante dell'Escrito*. Rome: Aracne.

Whately, C. 2015. 'The Genre and Purpose of Military Manuals in Late Antiquity', in G. Greatrex, H. Elton, and L. McMahon, eds., *Shifting Genres in Late Antiquity*. Farnham: Ashgate. 249–261.

Wheeler, E. L. 1980. 'The Origins of Military Theory in Ancient Greece and China'. *International Commission of Military History: Acta 5*. Bucharest. 74–79.

——— 2010. 'Military Treatises', in M. Gagarin and E. Fantham, eds., *The Oxford Encyclopedia of Ancient Greece and Rome*. Vol. 4. Oxford: Oxford University Press. 434–438.

Whitehead, D. 2001. *Aineas the Tactician: How to Survive under Siege*. Second edition. Bristol: Bristol Classical Press.

——— 2008. 'Fact and Fantasy in Greek Military Writers'. *AAHung* 48. 139–155.

——— 2016. *Philo Mechanicus: On Sieges*. Stuttgart: Franz Steiner.

——— and Blyth, P. H. 2004. *Athenaeus Mechanicus: On Machines (Περὶ μηχανημάτων)*. Stuttgart: Franz Steiner.

3 The blind leading the blind?

Civilian writers and audiences of military manuals in the Roman world

Nadya Williams

In 1616, the otherwise unremarkable Captain John Bingham published what would become for the next four hundred years the definitive translation into English of Aelian's *Tactics* (*Tactica*), originally composed in Greek in the early second century CE.[1] In the preface, Bingham humbly dedicates his translation of the treatise to Charles, Prince of Wales, who would become King Charles I a little less than a decade later, with the hope that Aelian's nuggets of wisdom would serve him as well as they had served Aelian's original addressee, the Emperor Hadrian.[2]

As Bingham himself pointed out in that same preface to his translation of Aelian, the Low Countries had already adopted Aelian as a source of military knowledge, and he did not wish the future King of England to be lacking recourse to such important information himself. Bingham was not exaggerating the perceived value of his subject to his future sovereign. Aelian's work had indeed gathered popularity in early modern Europe, because of the practical descriptions of the Macedonian army's drills and formation orders, which could be adapted wholesale by pike-wielding armies as *exempla* for hand-to-hand combat.

Early modern readers did not seem to appreciate the irony of transmission, for Aelian himself lacked military experience of his own, and his work was largely based on an earlier manual on the same subject by Asclepiodotus. But Aelian was certainly not the only ancient military luddite-turned-writer to be held dear by armies of the Middle Ages and beyond. Among his more popular peers were such writers as Onasander, whose *The General* (*Strategikos*) inspired a variety of spin-offs in the Byzantine Empire and was translated widely in Renaissance Europe and beyond, and Vegetius, whose *Epitome of Military Science* (*Epitoma Rei Militaris*) has been widely read from the day of its publication to the modern times.[3] But while the value of these ancient manuals and their popularity among the military elites throughout the Middle Ages is well established, the same cannot be said of their original audiences. Who read Greek and Roman military manuals originally? A related question is: why did their authors, some of whom (e.g., the philosophers Asclepiodotus and Onasander) lacked military experience themselves, write these military manuals? Did they truly see themselves as teaching real-life generals and soldiers about their craft, or could they have been addressing a less experienced audience? But, if so, *cui bono*? And did their audiences have the same impression of the value of these works as did the authors?[4] The remainder of this paper is dedicated to proposing some answers to these questions, not yet considered in modern scholarship from this angle. Specifically, this paper offers a more nuanced response to Brian Campbell's thoughtful study of the practical uses and influences of military manuals on real generals in the Roman Empire, and to Conor Whately's recent work on the purpose of military manuals in Late Antiquity.[5] While Campbell and Whately have considered military manuals as a single genre *en masse*, I argue for considering those texts that were not written by authors with personal military experience as a related but separate subgenre of sorts, as their readership and influence were more exclusively civilian in nature. Strikingly, no term for 'noncombatant' or 'civilian' existed in the ancient Greek or Latin, but ancient authors highlight as irregular military engagements by individuals who were not members of the sanctioned military forces of their city-state or nation. The use of the anachronistic modern term 'civilian', therefore, seems justified.[6] Ultimately, the identity of both the authors and the audiences of these manuals as civilians without military experience separates these works into a different genre from those manuals written by and for military personnel.

Proceeding chronologically through the works of Asclepiodotus, Onasander, and Vegetius as case studies, this paper suggests that these works, written by authors without military experience of their own, had primarily a civilian audience in mind, and despite some authors' aspiration to practical purpose, these manuals ultimately were received by their original audiences as works of entertainment and personal edification rather than practical training manuals. This ultimately meant a very different impact of these manuals on their audiences than the impact of manuals written by seasoned military generals, such as Frontinus.[7] And while this perhaps detracts from these manuals' worth to military historians of the ancient world, their existence is a valuable cultural artifact that shows the existence in the Roman Empire of armchair aficionados whose participation in war did not extend beyond reading about it for pleasure.[8]

Aeneas Tacticus' military manual on surviving under siege has been described in modern scholarship as the only ancient manual expressly written for civilians.[9] But while it may indeed be the only ancient manual composed with an exclusively civilian audience in mind, the goal of this paper is to show through an analysis of three other Greek and Roman manuals from the Roman world the tempting possibility of a civilian audience being the ultimate audience of their original authors, even if another audience may have been intended at first by the author. Indeed, I would argue, what makes Aeneas even more unique is that he was a trained military general who chose to address a civilian, rather than a military, audience. While Byzantine and Medieval European readers of all ancient manuals viewed them as authentic guidelines for training troops, those authors of military manuals who had no military experience of their own intended them to serve, rather, as works of entertainment and education for an audience equally far removed from the military sphere, but no less fascinated with it as a result of that distance.[10]

I

Before proceeding to the ancient manuals themselves, it is important to establish a methodological approach. What makes scholars so certain that Aeneas Tacticus, assumed to be the Stymphalian general mentioned by Xenophon, wrote to educate civilians, rather than fellow generals?[11] The answer is based on Aeneas' subject matter selection and presentation of topics, in combination with the circumstances of the period during which he composed his work.

In 400 BCE, the *gastraphetes*, a giant belly bow considered to be one of the prototypes of the catapult, entered the world of Greek warfare. Its arrival set a new model for hitting targets from afar. The continuing perfection of such ever-farther-shooting machines over the course of the fourth century BCE revolutionized Greek siege warfare by giving a new advantage to besiegers.[12] While attackers in previous eras largely had to rely on starving out the besieged, the new technology allowed for much more effective (meaning, deadly!) targeting of the people within the besieged city. A side-effect was an involvement of civilians in warfare to a greater degree than ever, at least in times of siege. After all, an arrow or other projectile shot into a besieged city did not discriminate between soldiers and civilians.[13] Such targeting of the besieged aimed to provoke fear in the civilian populations, with the hope of obtaining a swift surrender.

As a career general, Aeneas Tacticus had lived and fought during this crucial period, thus gaining thorough first-hand experience with siege warfare. Presumed to have been written after his retirement from duty, Aeneas' *How to Survive under Siege* (*Poliorketica*) is filled not only with practical advice, but also with concrete examples from a variety of real sieges throughout the Greek world. Strikingly, the work focuses on the perspective of the besieged city and provides advice not for the besiegers, but specifically for those who might find themselves within a city under siege. His methodical and clear yet succinct advice covers such topics as organizing guard for the city walls and preventing perfidy from within.[14] No less memorable are such scare tactics, aimed to frighten away the besieging force, as getting one's cattle drunk and sending them out of the city at night, or arming women with pots and pans and stationing them on the city walls in order to make the city's defense forces look larger than they really are.[15] The practical nature of Aeneas' advice, combined with his perspective from within the city under siege, readily suggests a civilian audience seeking practical advice. The context of Greek warfare in the fourth century BCE suggests, furthermore, a clear need for such advice. To sum up, while the contents of Aeneas' work by no means rule out that some military personnel in the fourth century and beyond read it, a civilian audience, including perhaps even women, appears most plausible.[16] And since Aeneas' work was seen in antiquity to be the beginning of the genre of military didactic literature, its appeal to a particular audience may also have been taken into consideration by subsequent writers in the same genre.[17]

Let us apply now these same criteria of subject matter selection and the context of the period in which a treatise was written, in order to consider likely audiences for the military manuals of Asclepiodotus, Onasander, and Vegetius. In addition, when available, I consider the references to these authors in antiquity by others, as this is evidence for at least some of their readership. The selection of these particular authors' manuals for discussion seems methodologically apt, as they represent several periods of Roman history (Late Republic, Early Empire, and Late Empire). But second, their selection unashamedly privileges manuals that have survived in a more or less complete state and had served as models for later manuals, suggesting a reasonable degree of circulation.[18]

II

Sometime in the mid- to late first century BCE, Asclepiodotus, a student of the famed Stoic philosopher Poseidonius of Rhodes, produced *The Art of* War (*Technê Tactica*), an exceptionally technical treatise on military training, maneuvering, and tactics. Poseidonius was the Aristotle of his age – a philosopher equally versed in all arts and sciences. It is possible, therefore, that Asclepiodotus' treatise was based on his teacher's own manual on the subject, which unfortunately does not survive.[19] At the very least, his interest in the topic of the military, with which he had no personal experience, likely derived from his teacher's encyclopedic approach to knowledge.[20]

One hopes that Asclepiodotus' own teaching style (assuming that he followed in his teacher's steps into the philosophical profession) was less terse and more approachable than his writing style. Not one to mince words, by way of introduction he immediately presents categories of land forces: three types who fight on foot, three types who fight mounted. He provides no description of purpose of his work or any similar address to the reader, until the brief concluding statement at the end of the manual: "These are in brief the principles of the tactician; they mean safety to those who follow them and danger to those who disobey" (XII.11.14).[21] Two hundred years earlier, when the Hellenistic era armies were still using phalanx tactics, this advice would have been apt indeed. By the Late Roman Republic, however, Asclepiodotus' subject has been decisively mothballed. Rome had full control over the Greek world, proof that the legion (never mentioned in this treatise) had surpassed the phalanx.[22] But while the historical context in which Asclepiodotus was writing rules out the usefulness or practical appeal of the work to real military personnel, the content and its presentation suggest a likely different audience: students of philosophy and history in Rhodes and possibly even in Rome. Two particular features stand out in Asclepiodotus' style in the manual. First, the rhetorical structure throughout privileges mathematical equations and visuals, to the point of using tripartite or quadripartite structure for various configurations even at the expense of obfuscating understanding. Second, Asclepiodotus' approach to combining features from all periods and geographical areas of Greek warfare into a single manual results in a chimera-like entity that flamboyantly refuses to be constrained by historicity.

From the opening statement of his work, Asclepiodotus works hard to present all types of categories of military forces, soldiers, and maneuvers in mathematical terms, to the point that mathematical explanations become the defining feature of his rhetoric.[23] For example, in describing the ideal numerical strength of the phalanx, he explains,

> Accordingly you should rather select numbers which are evenly divisible by two down to unity, and you will find that most tacticians have made the

phalanx to consist of 16,384 hoplites, because that number is divisible by two down to unity. . . . Let us also assume that the phalanx will consist of this number of men, and the file of sixteen men.

(II.7)[24]

Not satisfied to stop there, Asclepiodotus provides further mathematical break-down of this structure of the phalanx:

the union of the two wings is called the phalanx, under the command of the general, comprising 2 wings, 4 corps or half-wings, 8 division, 16 brigades, 32 regiments, 64 battalions, 128 companies, 256 platoons, 512 double-files, and 1024 files.

(II.10)[25]

This latter description is redundant from a factual standpoint, as it repeats the information provided earlier in II.7. And yet, it allows Asclepiodotus to showcase his mathematical prowess, as he readily multiplies by two increasingly greater numbers. Variations on these examples repeat throughout the manual, such as when Asclepiodotus discusses the structure and organization of the light infantry in VI.1–3.

In addition to presenting ample arithmetical explanations about the phalanx, Asclepiodotus also resorts to geometry to strengthen his case:

The entire army as well as its units is disposed on the basis of a fourfold division, so that of the four half-wings the bravest holds the right of the right wing, the second and third in point of valour the left and right, respectively, of the left wing, and the fourth the left of the right wing. For with the units ordered in this manner the right wing will have the same strength as the left, since, as the geometricians say, the product of the first and the fourth will equal that of the second and third, if the four be proportionate.

(III.1)[26]

It is significant that Asclepiodotus cites geometricians in this example, rather than military leaders, as the authority on whose word his own argument rests. Geo-metric considerations further come to the fore in his discussion of the various tactical arrangements of cavalry in battle – from rectangle to wedge to rhomboid (VII.2–9). Indeed, the geometrical descriptions of the arrangements of cavalry in particular make little sense without accompanying visuals, and as the text itself makes clear, Asclepiodotus had intended his text to incorporate detailed diagrams of the described formations.[27]

As one might expect from a philosopher, Asclepiodotus' arguments through-out are very logical, albeit dry, and his math impeccable. But while the logic for the arrangement of the phalanx and the number of soldiers therein, to select one example, certainly makes sense on paper, the picture of the phalanx that results would not have been recognisable for any soldier or general of a classical Greek *polis*, although it perhaps comes the closest to a Hellenistic army. The problem is, of course, real-life considerations. As we consider the calculations for the phalanx provided by Asclepiodotus, we have to keep in mind that a real-life phalanx rarely if ever reached the ideal numerical strength, and generals instead fought wars using the men they had, whatever their number. The same is true of the descriptions of cavalry formation and numerical strengths. And yet, the rhetorical impact of Asclepiodotus' readiness with numbers has the same effect on the modern-day reader as it likely would have on ancient lay audiences: such facility with citing numbers impresses the audience and makes the author sound more knowledgeable. The perception is, of course, that true experts cite precise numbers and figures, whereas amateurs deal with vagueness. Thus Asclepiodotus' citation of numbers as facts throughout the manual likely served to reinforce his image as the expert on his topic at least for audiences without much military experience, although it is possible that even readers with experience in the Roman army of Asclepiodotus' day would have been fooled. After all, the Roman army of the Late Republic did not look at all like the Classical or Hellenistic Greek phalanx.[28]

But while Asclepiodotus showcases his mathematical facility to great effect, he shows remarkably little interest in historicity. Already in his opening statement, Asclepiodotus notes that warfare comprises two types of forces (land and naval) and, true to his overall trend of presenting all concepts in mathematical terms, continues to present three types of infantry and three types of cavalry. Yet although all categories listed have indeed existed at one time or another in the ancient world, all of them had never coexisted at the same place and time. A particularly salient example of Asclepiodotus' privileging of rhetoric over historicity is his inclusion of both elephants and chariots in the category of mounted forces: 'In the same way (as with regard to land forces) there are three branches of the mounted force: the first is cavalry, the second is furnished with chariots, and the third with elephants' (I.3).[29] This inclusion of both chariots and elephants did serve a useful rhetorical purpose, as it allowed Asclepiodotus to present a tripartite division of mounted forces, just as for the land ones. And yet, the mere inclusion of these forces in the list is fraught with historical challenges and seems to have been the result of the author's determination to entertain his readers. Chariots, in particular, have had little use in Greek warfare since the Bronze Age and thus have rarely seen the light of day at the same time as the hoplite phalanx.[30] Elephants, likewise, although likely a fascinating novelty for Asclepiodotus' audiences, most of whom had never seen one in person, were rarely used by armies in the Greek world.[31] Asclepiodotus is forced to admit as much himself, when introducing these two types of mounted forces in more detail later in the manual: 'Although we rarely find any use for chariots and elephants, we shall, nevertheless, set forth their *nomenclature* to complete this discussion' (VIII.1).[32] Asclepiodotus' word choice here is apt: it is the terminology related to these categories of mounted forces that interests him, not the tactics proper. And, indeed, having provided nothing more than a list of the names of different subdivisions within units of chariots and elephants, Asclepiodotus rests his case (VIII.1–IX.1). The overall effect is similar to that which the reader gets from the lengthy lists of numbers

and geometrical explanations. Put simply, Asclepiodotus' use of elaborate terminology to describe even the most obscure and rare divisions of military forces
presents a wonderful illusion of his own competence and extraordinary knowledge. Anyone without military experience, and thus without the awareness of how
impractical and anachronistic much of Asclepiodotus' coverage was, would have
likely been wowed by his rhetorical prowess. But who precisely were his audiences, and were they duly impressed?

As it happens, the few references to Asclepiodotus in antiquity provide us with
some insights about his readership. Seneca the Younger, writing in the mid-first
century CE, referenced the works of Asclepiodotus five times, albeit none of
these references was to the present work. In the meanwhile, Aelian relied heavily on Asclepiodotus in writing his own *Tactics* in the early second century CE,
although, in a funny twist, Aelian never mentioned Asclepiodotus by name, but
duly cited Asclepiodotus' teacher Poseidonius.[33]

As my modern historian colleagues would say, this is slim pickings, but it is
enough of a clue, perhaps, to draw some cautious conclusions about Asclepiodotus' readership. First, Asclepiodotus was never able to get out of the shadow of
his teacher Poseidonius. Indeed, Seneca's references to Asclepiodotus suggest that
Seneca (and presumably others by the mid-first century CE) assumed that Asclepiodotus was largely publishing the research of his own teacher. Furthermore, while
Aelian's military manual bears significant enough resemblance to Asclepiodotus'
work to make a relationship between the two works obvious, Aelian mentions
Poseidonius, rather than Asclepiodotus, as his source, suggesting that the two were
easily confused, and the credit could more easily be given to the more famous of
the two philosophers. At any rate, we see that at least a niche audience of Greek
and Roman philosophers were still avidly reading Asclepiodotus throughout the
first and early second centuries CE and using his works as a conduit to the teachings of Poseidonius. This suggests, indeed, that as originally conjectured at the
beginning of this section, Asclepiodotus likely found a place in the curricula of
Greek and Roman philosophical schools. Second, it seems clear that this audience
was not relying on Asclepiodotus for practical advice about the waging of war.
Indeed, much if not all of this audience never saw war first-hand, since by the
first century CE, Rome's army was fully professionalised.[34] Instead, his thoroughly
mathematical approach to the description of the phalanx made the work particularly suitable for use in the teaching of philosophy and rhetoric. These audiences
likely were not looking for a military manual to read, but were looking to acquire
the best education of their day. In a twist of irony, their pursuit of philosophical
education engaged them in the study of military science, albeit via this thoroughly
transhistorical manual. And perhaps, like for many modern armchair military historians, once they caught the military science 'bug', they were eager to read more.

III

Sometime in the mid-first century CE, another little-known career philosopher,
Onasander, noted in the Suidas as the author of a commentary on Plato's *Republic*,

chose to make the same bold disciplinary leap as Asclepiodotus and wrote his own military manual. Onasander's manual has the special distinction of being the only ancient military manual specifically dealing with war from the perspective of the general. Thus, Roman military generals (meaning, largely, Roman consuls and ex-consuls) ostensibly seem to be his target audience. This theory is further strengthened by his dedication of the work to one Quintus Veranius, consul in 49 CE. And yet, the manual's elementary and broad presentation of the role of the general in war would certainly not have served to educate any Roman general in military matters. In fact, one presumes that Quintus Veranius himself would have had plenty of corrections to make to the book, had he been allowed.[35] So who really was the audience of this manual, and why would anyone have been interested in such a book?[36]

A careful reading of Onasander's prologue reveals that while he dedicated the work to Quintus Veranius, the author was hoping for a broader readership among the senatorial class. Since all members of the aristocracy would have hoped to attain the consulship still, and with it the possibility of military leadership, the manual could have served an inspirational role, more than educational. In addition, Onasander's discussion of qualifications for selection as general gives a hint of yet another possible audience. Overall, Onasander's work would have served the purpose of entertaining and encouraging Romans with dreams of military leadership, while ironically providing advice that was minimally substantive, and at times misleading or utterly erroneous.[37]

Unlike Asclepiodotus, who jumps straight into the 'meat' of his subject, Onasander opens his work with a formal prologue. In it, he firmly roots his work in the genre of technical manuals and explains his target audience and the reasons for addressing it:

> It is fitting, I believe, to dedicate monographs on horsemanship, or hunting, or fishing, or farming, to men who are devoted to such pursuits, but a treatise on military science, Quintus Veranius, should be dedicated to Romans, and especially to those of the Romans who have attained senatorial dignity, and who through the wisdom of Augustus Caesar have been raised to the power of consul or general, both by reason of their military training (in which they have had no brief experience) and because of the distinction of their ancestors.
>
> (Prooemium 1)[38]

The prologue's reasoning is fascinating, as it shows a potential purpose in writing technical manuals, other than direct instruction of an audience. And, indeed, ancient manuals typically attempt to present at least the illusion of educating their audience. But Onasander proposes the contrary position by suggesting that each type of work ought to be dedicated to those who are the best authorities on the subject at hand – 'to men who are devoted to such pursuits'. Such reasoning allows him, therefore, to flatter both Quintus Veranius and other potential Roman readers, as he lumps all Romans together into the category of military experts worthy of the dedication of his work.

Onasander's reasoning for directing his work to those who may already be experts in the subject must have indeed struck even the author himself as somewhat unusual. With an eye for further flattering his readers, he clarifies his dedication:

> I have dedicated this treatise primarily to them, not as to men unskilled in generalship, but with especial confidence in this fact, that the ignorant soul is unaware even of that in which another is successful, but knowledge bears additional witness to that which is well done.
>
> (Prooemium 2)[39]

In other words, in a process of expert circular reasoning, Onasander humbly seeks approval and authority for his work through this dedication to those who are best qualified to judge it by virtue of their military heritage.

The remainder of the prologue shows a greater degree of confidence on Onasander's part that although he is addressing an audience that (he claims) already is expert in the subject, he still has something to teach even them: 'It remains for me to say with good courage of my work, that it will be a school for good generals, and an object of delight for retired commanders in these times of holy peace' (Prooemium 4).[40] Finally, Onasander's concluding point, that the Roman successes have been not by chance, but the result of having great generals, who are therefore worthy of study, echoes a more famous prologue, that of Polybius:

> For who is so worthless or indolent as to not want to know by what means and under what system of polity the Romans in less than fifty-three years have succeeded in subjecting nearly the whole inhabited world to their sole government – a thing unique in history?
>
> (Prooemium 1)[41]

The Polybian echo was surely not lost on Onasander's readers, and likely enticed them to expect the manual to use historical *exempla* as an instructional method. But while Polybius argued that the Romans' success was the result of their Republican government, Onasander appropriately updates the explanation to the days of the Empire by attributing Roman military successes to great generals, the chief of them, mentioned in the first sentence of the work, being Augustus Caesar himself. Clearly, despite his interest in Plato's *Republic*, Onasander was not the gadfly type of a philosopher, but one who saw the benefits of being strongly pro-establishment. Moreover, Onasander's use of historical *exempla* turns out to be minimal in the remainder of the manual.

Following the prologue, the structure of Onasander's manual follows roughly in order of events in a general's experience: selection of a general and best qualifications for that role; proper procedure and reasoning for declaring war; the process of training troops; fighting in battle, and various scenarios that might come up as part of leading in a battle; consequences of winning a battle, including how to deal with prisoners and burial of the fallen; how to treat cities that surrender; and how to maintain peace after war. But while each topic is logically presented, a lack of historical examples throughout lends it a vague tone, devoid of anchoring in any one place and time. This approach is in sharp contrast to the *exempla*-based works of writers with military experience who were writing for military audiences, such as Frontinus and Polyaenus.[42] Instead of using *exempla*, however, Onasander privileges logic, an approach that certainly sets his work apart from the military manuals of professionals but also arguably opens the door for more diverse audiences for his work. And, indeed, while certain logically grounded explanations that Onasander provides are contrary to Roman historical practice and show potentially a lack of awareness of how the Roman military actually worked in practice, these explanations use philosophical methods to justify the idealistic picture that he wished to impart to his readers.[43]

An especially salient example of this possible ignorance of Roman practice occurs in the very first section following the prologue – on the selection of a general. Before proceeding to the description of the responsibilities of the general in various circumstances in which he might find himself in war and in peace, Onasander presents an intriguing picture of the ideal qualifications for a general:

> I believe, then, that we must choose a general, not because of noble birth as priests are chosen, nor because of wealth as the superintendents of the gymnasia, but because he is temperate, self-restrained, vigilant, frugal, hardened to labour, alert, free from avarice, neither too young nor too old, indeed a father of children if possible, a ready speaker, and a man with a good reputation.
>
> (I.1)[44]

Onasander elaborates further for why these qualities are indeed ideal for a general, and common sense is certainly on his side. And yet, the inclusion of this section in the manual is surprising, since one would assume that both Onasander, and at the very least his audiences, would have known how Roman generals were appointed. Indeed, if Onasander was hoping that the emperor would read his work, it is possible that he modeled the previous description on Augustus himself, hoping to flatter the princeps, who cultivated a public persona of an ideal general and paragon of familial, civic, and military virtue. But since Roman generals were typically consuls or ex-consuls, they were by default men of a sufficiently noble birth to be members of the senatorial class. In addition, because of the wealth prerequisites for the senatorial class, they literally were millionaires.[45] As for the remainder of the qualifications, they were certainly regularly violated, albeit sometimes with disastrous results. Indeed, Onasander's list here would have seemed prophetic to anyone who might have read it in, say, 69 CE, approximately decade or two after Onasander wrote it. All of the failed generals-turned-emperors in the year 69 CE violated one or more of Onasander's qualifications: Galba was too old, had no children, and was known to be greedy; Otho as well had no children and was in no way to be considered temperate or hardened to labor; and Vitellius had only a very young son and violated every single other qualification on Onasander's list. By contrast to the other three, Vespasian, who emerged as the last man standing, met many (albeit not all) of Onasander's qualifications and was indeed a man of neither noble birth nor significant wealth. And yet, the fact that generals like Galba, Otho, and Vitellius existed in Onasander's day, and were so well placed to launch an imperial bid in 69 CE, is a firm reminder of the ways in which the Roman military appointments were not based on character qualifications, but rather on birth and wealth.

While completely divorced from the reality of how Roman generals were appointed, Onasander's detailed advice on the qualities essential in a good general reveals his main interest throughout the rest of the manual in the character of the general. This interest, rooted in his philosophical background, would likely have intrigued readers of all levels of military experience.[46] In history, on the other hand, Onasander shows about as much interest as Asclepiodotus – meaning, little to none. Sections V–X of the manual, for instance, present tips for the general on how to handle his army during marches, how to set up camp, and how to carry out drills even during peace. All of these topics easily lend themselves to the inclusion of prominent examples from Roman history. Indeed, the approach of Aeneas Tacticus was to include real-life historical examples to accompany each item of advice.[47] Onasander, however, shows no awareness whatsoever of any specific historical events that could have supported the need to heed his advice, and drily narrates his suggestions with no elaborations.

The ultimate message, to which Onasander returns in the concluding chapter, is that of the need for a man to work on improving and maintaining his good character and reputation: 'A good man, then, will be not only a brave defender of his fatherland and a competent leader of an army but also for the permanent protection of his own reputation will be a sagacious strategist' (XLII.26).[48] With this emphasis on the moral character of the general, Onasander has produced a 'feel-good' manual that would have made anyone reading it feel important, and (if he went through the checklist of essential qualifications for a general) could have led many a reader to view himself as a viable potential general, unless he thought hard about how Roman generals actually were appointed. After all, improving one's character would have been much easier than changing one's birth and wealth status. And since the author was a respected philosopher, presumably the same individuals who read his philosophical works – meaning, educated aristocracy, broadly defined – might have given his manual a chance.[49]

IV

Sometime in the late fourth or early fifth century CE, one Publius Flavius Vegetius Renatus, also known to have composed a treatise on veterinary science, a subject in which he had extensive first-hand experience, decided to branch out into military writing and wrote a manual on the ideal practices for recruitment and training of the Roman army.[50] The work, ambitiously dedicated to an unnamed emperor, assumed to be either Theodosius I, Gratian, or Valentinian III, seems to have resonated with its addressee.[51] Vegetius proceeded to write a second volume, on the organization of the legion, a third, on military actions in the field, and a fourth on siege warfare and naval warfare. In the prologues to these subsequent books, he noted that it was the success of his first volume that induced the emperor to order him to keep writing on the subject. Vegetius was delighted to oblige. But there appears to have been a discrepancy, at least initially, between what the author had hoped his audience would gain from the work, and the purpose with which the audience itself endowed his work. While the ostensible purpose of the work was to produce a manual of practical advice for the emperor, and especially to encourage the emperor to use more Romans (rather than barbarians) in the army, no actual changes appear to have resulted from Vegetius' writing. So why did the emperor encourage Vegetius, a man who by his own admission knew much more about horse-breeding than military affairs, so urgently to keep expanding his military treatise?[52] Conor Whately has recently proposed an intriguing argument for the dual purpose of Vegetius' manual to entertain and educate the readers of both civilian and military manuals.[53] Through a detailed reading of Vegetius' prologues, I now add to Whately's argument the possibility that the historical approach of the manual, with its regular contrasts between the army of the Republic and the army of the author's day, reveals its literary purpose in entertaining and encouraging a civilian aristocracy of the new Christian empire with stories of the past glories of Rome at precisely the time when those glories, and Roman identity as a whole, were threatened more than ever by the changing nature of the late Empire.

Vegetius' extensive prologues to each volume of his work are striking, and as seen by the lack of any prologue in Asclepiodotus' manual, for instance, were not necessarily required by the genre. In Vegetius' work, however, the elaborate prologue to each volume serves to emphasise the role of the emperor as both the encourager and the audience of the work, lending authority to the author's voice and thus inviting broader readership. Concurrently, Vegetius emphasises his own role as the researcher of the truths of the past, and thus a conduit of these truths to the Romans of his own day. In the process, his prologues echo a different genre – that of historical writing. Thus, in Book I, Vegetius begins by explaining a long-standing tradition of dedicating such works to the emperor, and the worthiness of his recipient:

> In ancient times it was the custom to commit to writing one's studies in the liberal arts and offer them summarised in books to Emperors. For nothing is begun rightly unless after God the Emperor favours it, nor is it appropriate that anyone should have superior or wider knowledge than the Emperor, whose learning can benefit all his subjects.
>
> (I, Preface)[54]

From the opening sentence, the paradoxical nature of Vegetius' work as rooted in antiquity and yet a product of its own times is clear. Yes, it is a time-honored tradition to dedicate one's literary labors to the emperor. And yet, as the reference to God reminds, this is a different kind of emperor, and a different, now fully Christian, empire. If the recipient is Theodosius, the dismantling of pagan temples and other final vestiges of traditional Roman religion were still underway at the time of writing.[55] And even if the recipient is a slightly later emperor, the Theodosian ban on pagan worship was still a recent phenomenon within living memory. As a result, the emperor is second to God (as Vegetius' statement not so subtly reminds), rather than a god himself. Likewise, Vegetius' note on the responsibility of the emperor to use his knowledge to benefit his subjects acquires a new nuance in the Christian empire, as the emperor now had to answer to God directly for the well-being of his subjects.[56] But can a simple bureaucrat like Vegetius have truly aspired to educate the emperor himself? Careful lest he appear overly presumptuous, as Onasander likewise did in the beginning of his own manual, Vegetius also nuances his writings by noting that he means flattery rather than disrespect:

> We attempt to show then, by a number of stages and headings, the ancient art of levying and training recruits. Not that those things would appear unfamiliar to you, Invincible Emperor, but so that you may recognize in your spontaneous dispositions for the safety of the State the principles which the builders of the Roman Empire long ago observed, and in this little book find whatever you think needful to affairs of State, which are ever pressing.
>
> (I, Preface)[57]

Vegetius misses no opportunity to highlight the antiquity of his subject, proudly calling it an ancient art. Addressing the emperor as 'Invincible Emperor', at the same time, highlights the emperor's continuation of that ancient art, and his ability to apply it in the present time. After all, the emperor's task of preserving the safety of the state, Vegetius notes, is equally old, even if some of the challenges the emperor faces are relatively novel.

Growing bolder after the emperor's kind reception of Book I, in the subsequent two books Vegetius begins by praising a key figure or group from the past, whose military excellence he has researched, and the benefit of whose wisdom he now imparts to his audience. Thus, in his prologue to Book II, Vegetius thanks the emperor for the vote of confidence that his first volume had received, and vows obedience to the emperor's command to keep writing. He also notes the significance of his own role, for '[b]rave deeds belong to a single age; what is written for the state is eternal' (II.3).[58] As a writer, in other words, Vegetius turns singular historical events into eternal *exempla* for edifying his audiences. Then singling out Cato the Elder as a source of key military information for this second book, Vegetius outlines his task:

> These men's recommendations, their precepts, I shall summarise as strictly and faithfully as I am able. For although both a carefully and a neglectfully ordered army cost the same expense, it is to the benefit of not only of present but of future generations also if, thanks to your Majesty's provision, August Emperor, both the very strongest disposition of arms be restored and the neglect of your predecessors amended.
>
> (II.3)[59]

Vegetius is clear about his own position as a researcher, who will combine the best of the wisdom of earlier Roman military leaders and writers (especially Cato).[60] He follows this description with the explanation of the benefit of his work to the emperor: maintaining an army is expensive, so one might as well do it well. Furthermore, a well-managed army will benefit subsequent generations of Romans. Last but not least, in a somber but respectful tone Vegetius hints yet again, as at the end of his prologue to the first volume, about the troubles of the Roman Empire in his day, which are separating this empire from its former glory. This remark, however, is carefully couched not as a criticism to mismanagement of military affairs by the current emperor, but rather as the fault of his predecessors. And, of course, by mentioning that the Roman army's former military strength can be fully restored, Vegetius reminds that military success is indeed part of the identity of the Roman Empire, rather than something new.

In his prologue to Book III, Vegetius goes back yet further in time than the Roman Republic for inspiration, complimenting the Spartans' dedication to military art and (who knew?) military writings. These are important to study, says Vegetius, because they, along with the Athenians, were 'masters of the world before the Macedonians' (III, Preface). The Romans inherited all this knowledge, claims Vegetius, and now he is ensuring that it will continue:

> Following these men's precedents the Romans maintained the principles of warfare in practice and transmitted them in writing. This material, dispersed through various authors and books, Invincible Emperor, you ordered my Mediocrity to summarise, so that neither should boredom arise from excessive detail nor complete confidence be lacking because of brevity.
>
> (III, Preface)[61]

Repeating the same themes as in his other prologues, Vegetius notes that it has been a Roman tradition as well to preserve military achievements in writing, so that achievements of the past can become lessons for all time. Vegetius thus places himself into a tradition of military writing that is distinctly Roman as well. At the same time, however, he highlights the two-fold difficulty of his task, which required him to assemble material 'dispersed through various authors and books'. Research is not easy, after all! To make matters more challenging, he had to present this material in a way most appropriate to his audience, striking just the right balance between boredom from excessive length and confusion from excessive brevity of the work. This final note shows Vegetius' careful awareness of the difficult balance between entertainment and education, while also highlighting his desire to place his own writing in an older tradition.

In the prologue to the fourth and final volume, on siege warfare, Vegetius deems it appropriate to harken back to the first builders of cities, arguing that the art of city-building (and the accompanying art of governing a state) saw its height in none other than his emperor, to whom the treatise is addressed. Vegetius describes the emperor in the most flattering terms, addressing this complimentary statement to him, among many others: 'You surpass all Emperors in felicity, moderation, morality, displays of indulgence and love of studies' (IV, Preface).[62] It is therefore especially apt, Vegetius argues, that

> [t]o complete, then, a work undertaken by command of Your Majesty, I shall summarise in order from various authors the measures by which either our cities should be defended or the enemy's destroyed. Nor shall I regret my labour since foundations are being laid for the benefit of all.
>
> (IV, Preface)[63]

This prologue echoes themes from those to the previous volumes. The reader is reminded that Vegetius' authority as author is based on the emperor's tasking of Vegetius with writing the manual, and his thorough research of earlier authors. Also, Vegetius reiterates the importance of his work as something that will benefit all, at the same time noting again the challenging nature of his task – it truly is a 'labour'.

Throughout the prologues to the four volumes, the repeated emphasis on Roman military excellence as an integral part of Roman tradition sets the stage for the manual to connect the historic past to the present of the audience's day, with the clear aim of using the glorified past to inspire the crumbling present. Vegetius' admiration for the Roman army of the past is clear throughout his work, and it comes to the fore in the contrasts between the Roman army of the nebulous past and that of the present. As a result, certain historical details find their way into the work, despite their irrelevance for purposes other than historical background. These historical examples, however, are mixed together and removed from their original context. What matters, after all, is not history itself, but its value for the present.[64] For example, Vegetius includes a lengthy explanation of the armor of the ancients (I.20), a detailed overview of drills from the days of the early Empire (I.27), a thorough overview of the structure and leadership of the early Roman legion (II.6–8), a list of the titles of officers in the Roman fleet (IV.32), as well as an explanation of how Liburnian warships got their name (IV.33). Perhaps nowhere else in the work are Vegetius' aims in including such historical details as clear as in his conclusion to Book I, where he explains that as the conquerors of the known world, the Romans should study military science as a way to maintain and better understand their own identity.[65] After all, many great civilizations and conquerors have existed, and still exist, throughout the ancient Mediterranean. Although the Romans have conquered them all, they cannot merely rest on their laurels, as their own past history reveals:

> For martial energy has not declined in mankind, nor are the lands exhausted that produced the Spartans, the Athenians, the Marsians, Samnites, Pael-igni, the Romans themselves. Were not the Epirotes once very powerful in arms? Did not the Macedonians and Thessalians overcome the Persians and penetrate as far as India on campaign? Clearly the Dacians, Moesians and Thracians have always been warlike, for fables tell us that Mars himself was born among them. But it is tedious if I attempt to list the strength of all the

provinces, when they all belong under the sway of the Roman Empire. However, a sense of security born of long peace has diverted mankind partly to the enjoyment of private leisure, partly to civilian careers.

(I.28)[66]

Put simply, studying Roman military excellence is the Roman thing to do, and is part and parcel of the 'brand' name of what it means to be Roman. Furthermore, as the darker note on which Vegetius finishes the book hints, the lack of study of military affairs, and the accompanying lack of military training, could only result in disaster for Rome. There is a genuine urgency in maintaining both military knowledge and military training. As Vegetius concludes the book, 'For it costs less to train one's own men in arms than to hire foreign mercenaries'.[67]

Vegetius' extensive prologues and the conclusion to Book I show that he understood a question of relevance that his work might elicit from his audience: why should a Roman emperor and his court decide to devote so much time to reading a military manual that summarizes research on past Roman military operations? After all, the Roman Empire of Vegetius' day was a completely different Rome, not even located in Rome anymore! Also, Theodosius' ban on sacrifices was the end of traditional Roman religion, as seen for instance in the new and fully Christianized military oath (II.5). Finally, regardless of whether Theodosius I or Valentinian III was Vegetius' addressee, the Roman Empire of Vegetius' day was constantly under attack, including such disasters as the Battle of Adrianople in 378 CE and the sack of Rome in 410 CE.[68] But, it appears, this crisis of identity spurred reflective retrospection both on the part of Vegetius and his audiences, emperor and imperial court alike. While the emperor and other readers likely felt rooted in their present troubles and were seeking solutions to them that were current and relevant, Vegetius rather believed that the solutions his audiences actually needed lay in the distant past. Vegetius' work, rooted so deeply in examining the army of the Roman Republic and early Empire, showed a way to reconcile the new Empire to the old 'brand' name of Rome through the study of Roman military excellence as the most valuable part of Roman identity. In Vegetius' treatise on military art, therefore, entertainment and education married to inspire.

IV

In this paper, I proposed that military manuals were a popular reading and entertainment option for Roman audiences who did not have military experience but were fascinated with military history and the concept of military glory in their nation's past either from a philosophical perspective or from seeing a necessity to cultivate such an interest as a matter of solidifying their identity as Romans. This interest on the part of the military luddites-turned-writers and their civilian audiences closely connected the manuals and their readers to the much older genre of epic poetry, which likewise presented war in glorifying and exciting terms, providing military heroes whom audiences from all walks of life could admire. The military manuals of the Late Republic and all periods of the Roman Empire glorified historically successful armies – the Macedonian phalanx and the legion of the Roman Republic, in particular – and presented the military maneuvers of these armies as epic subjects for all times. As a result, the content of these manuals, which at times leaves much to be desired from the perspective of historical accuracy, was irrelevant to their purpose of entertaining and inviting interest, so long as the goal of bringing some epic into the mundane has been fulfilled.

It is impossible to know the authors' motivations, and whether they realized their own inadequacies and lack of informed understanding of military affairs. Indeed, like so many armchair academics of any period, perhaps they believed that they were better informed about the subject of their writing than they actually were. Their adherence to the general rules of the genre of military manuals certainly suggests this possibility. And in this case, it is entirely possible that they were frustrated with the lack of expert readership for their works, especially if, as I argue, the civilian readership relegated their manuals into a different category than the military manuals of the known experts, such as Frontinus. But it is one of the ironies of history that the long-standing popularity of these manuals with medieval and early modern audiences transformed them primarily into works of practical advice. Promoted to the rank of experts by much later readers, Asclepiodotus, Onasander, and Vegetius alike would have been overjoyed with this twist indeed.

Notes

1 This was indeed the translation that I found in the stacks of my university! See Aelian, *The tactiks of Aelian; or, Art of embattailing an army after ye Grecian manner*. Translated by Bingham (1968). This is a reprinting of the original 1616 translation.

2 It is possible that the addressee was Trajan, rather than Hadrian, but the point stands. Charles' subsequent life would prove that he was in need of a different sort of wisdom, but that is not the subject of the present work.

3 See Roueché (2002) 119 on the popularity of Aelian and Onasander in the Byzantine Empire. Onasander's work, for instance, influenced the *Tactica* of Leo VI, Byzantine Emperor from 886 to 912 CE. As for Vegetius, Goffart dubs him "the Bible of warfare throughout the Middle Ages" – cf. Goffart (1977) 65. For a study of Vegetius' influence on military writing after antiquity, see Allmand (2011). For earlier scholarship, see Schrader (1981) 167–172. Vegetius enjoyed quite the popularity in Medieval France – see Knowles (1956) 452–458. For details on the typical reading list of American and French officers in the early modern period, see Powers (2006) 781–814. On the list Powers identified are Aelian and Vegetius, as well as other Classical military (albeit not strictly didactic) writers, such as Polybius, Caesar, and Quintus Curtius Rufus. Finally, see Bulkeley (1983) 233–257, for an overview of the controversial legacy even in the modern world of Vegetius' maxim *si vis pacem, para bellum*.

4 Polybius, at least, was not impressed with historians who wrote generally on military matters, although he does not single out writers of manuals in his disparaging comment (I.4.3).

5 Campbell (1987) 13–29 and Whately (2015) 249–262.

6 For a discussion of definitions and especially how civilians were involved in warfare in the Archaic and Classical Greek world, see Williams (2017) 23–41, and especially 24 on definitions.

7 See Campbell (1987) 20–28 for a summary of ways in which Frontinus' manual and other similarly practical texts influenced real commanders in the Roman Empire. Also

see Chapter 10 for a consideration of how Frontinus viewed the Roman defeat at Cannae as a surprising example of a stratagem in his *Stratagems*.

8 This would seem to allay (at least some of) the concerns about reading military manuals as providing historical examples for a real general to emulate in actual campaigns, against which Schellenberg in his chapter warns.

9 Early scholarship assumed his targeting of a military audience – see especially Handford (1926) 181–184. See also Oldfather (1923) 8–10 for the diplomatic summary of the work as addressing only the defensive side of siege warfare, without stating that this means largely addressing a civilian audience. For a detailed reasoning, however, for assuming a civilian audience under siege, which also may be dealing with enemies from within, see Whitehead (1990) 17–33. Finally, see Williams (2017) 23–41 for an overview of the growing civilian involvement in warfare in the fourth century BCE, based on the evidence of Aeneas' manual.

10 For example, see Chapter 14.

11 For a summary of reasons why Aeneas the writer is likely Aeneas the general, see Whitehead (1990) 12–17.

12 Perhaps the best summary of the changes and the rapid evolution of the catapult in the period can be found in Rihll (2007).

13 For more information on the impact of war on civilians, and especially the existence of any formal or informal laws restricting attacks primarily against civilians in Greek warfare, see the debate between Ober (1998) 57–71 and Peter Krentz (2002) 23–39, as well as more recent works of Lanni (2008) 469–489 and Gaca (2010) 117–161. Ober argues that informal laws of war existed from the Archaic Period until the Peloponnesian War, when they were violated repeatedly with no consequences for the offenders. Krentz agrees with the latter part of Ober's argument but argues that any rules restricting attacks on civilians did not come into existence until 480 BCE. Lanni argues that any laws restricting various aspects of war in Greek warfare, including attacks on civilians, were religious in nature. Gaca, finally, argues that certain aspects of deliberate attacks on civilians, namely andrapodization of women and children, existed in all periods of Greek warfare. Finally, see Garland (2016), for a consideration of the fears of the siege that led the Athenians to abandon their city to the Persians in 480–479 BCE.

14 This last topic has accounted for a significant portion of the scholarship on Aeneas Tacticus. See, for instance, Bengtson (1962) 458–468 and Winterling (1991) 193–229.

15 Aeneas Tacticus 27.14 and 40. For a description of a later cattle-based scare tactic, used by Hannibal against Fabius Cunctator and his army, see Livy 22.17.

16 Literacy rates for women in the Greek world remain a subject of debate, but the portrayal of women engaged in reading in Classical Athenian vase paintings, for instance, shows that at least some of them read. Furthermore, the presence of women in Aeneas' examples, and his emphasis throughout the manual that the defense of the city requires the entire citizen body acting as one, suggests his envisioning the close involvement of women in the process.

17 Cf. Vela Tejada (2004) 141–143, 146.

18 While we know that a number of manuals had been written between the era of Aeneas Tacticus and that of Asclepiodotus, none survive intact.

19 See, however, Kidd (1988) 31 for the argument that, as we know at least based on the relationship of Asclepiodotus' work on meteorological phenomena to that of Poseidonius, 'the Senecan evidence strongly suggests that he wrote his own book(s) on the subject, and did not merely summarize or abridge his master's work'. The same logic can be applied to Asclepiodotus' military manual, argues Wrightson (2015) 79.

20 Cf. Vela Tejada (2004) 144–145 on the Sophists' influence on military writing, and especially the idea that a written book was the best way to preserve thoughts and knowledge permanently. See also Spaulding Jr (1933) 662 for an overview of Asclepiodotus' style and approach as 'lifeless theory, with no historical illustrations and no comment from experience'.

21 αὗται διὰ βραχέων αἱ τοῦ τακτικοῦ καθηγήσεις, τοῖς μὲν χρωμένοις σωτηρίαν πορίζουσαι, τοῖς δ᾿ ἐναντίοις κινδύνους ἐπάγουσαι.

22 For an argument in favor of the continuing use of phalanx-style tactics by the Roman legion in Late Antiquity, however, see Wheeler (2004) 309–358.

23 Spaulding comments on this phenomenon of philosophers' fascination with the military arts, especially from a mathematical perspective: "Socrates, soldier and philosopher, could talk soldier language when he wished; later philosophers, not soldiers, felt that to make good their claim to universality for their doctrines, they must treat the military art. The phalanx lent itself well to logical and mathematical treatment, and so we find a long line of writers who have so treated it" (Spaulding 661–662). Spaulding names Asclepiodotus as the first in this subgenre of technical military manuals.

24 δἰ ὃ τοὺς ἀρτιάκις ἀρτίους μᾶλλον ἐκλεκτέον ὡς μέχρι μονάδος διαιρεῖσθαι δυναμένους: καὶ τούς γε πλείονας τῶν τακτικῶν εὑρήσεις πεποιηκότας τὴν φάλαγγα τῶν ὁπλιτῶν μυρίων ἑξακισχιλίων τριακοσίων ὀγδοήκοντα τεσσάρων, ὡς δίχα διαιρουμένην μέχρι μονάδος, ταύτης δὲ ἡμίσειαν τὴν τῶν ψιλῶν. ὑποκείσθω δ᾿ οὖν καὶ ἡμῖν τοσούτων ἀνδρῶν εἶναι τὴν φάλαγγα, τὸν δὲ λόχον ἑξκαίδεκα. . . . ὑποκείσθω δ᾿ οὖν καὶ ἡμῖν τοσούτων ἀνδρῶν εἶναι τὴν φάλαγγα, τὸν δὲ λόχον ἑξκαίδεκα. The peculiar choice of the translator's words, "divisible by two down to unity", refers to the idea that all numbers are divisible by two with no remainders, down to the last two.

25 ὑποκείσθω δ᾿ οὖν καὶ ἡμῖν τοσούτων ἀνδρῶν εἶναι τὴν φάλαγγα, τὸν δὲ λόχον ἑξκαίδεκα.

26 ἥ τε ὅλη φάλαγξ καὶ τὰ μέρη κατὰ τετράδα, ὥστε τῶν τεσσάρων ἀποτομῶν τὴν μὲν ἀρίστην κατ᾿ ἀρετὴν τοῦ δεξιοῦ κέρατος τετάχθαι δεξιάν, τὴν δὲ δευτέραν ἀριστερὰν τοῦ λαιοῦ καὶ δεξιὰν τὴν τρίτην, τὴν δὲ τετάρτην τοῦ δεξιοῦ λαιάν. οὕτω γὰρ διατεταγμένων ἴσον εἶναι συμβήσεται κατὰ δύναμιν τὸ δεξιὸν κέρας τῷ λαιῷ: τὸ γὰρ ὑπὸ πρώτου καὶ τετάρτου, φασί γεωμέτριοι, ἴσον ἔσται τῷ ὑπὸ δευτέρου καὶ τρίτου, ἐὰν τὰ τέσσαρα ἀνὰ λόγον.

27 E.g., at the end of VII.6 and VII.9.

28 Cf. Lendon (2005) 142 for a discussion of Polybius' astonishment over the formal severity of Roman military discipline, in contrast to the lack of formal discipline in Hellenistic armies.

29 κατὰ τὰ αὐτὰ δὴ καὶ τῆς ὀχηματικῆς δυνάμεως τρεῖς εἰσι διαφοραί: ἡ μὲν γάρ ἐστιν ἱππική, ἡ δὲ δι᾿ ἁρμάτων ἐπιτελεῖται, ἡ τρίτη δὲ δι᾿ ἐλεφάντων.

30 Cf. Lendon (2005) 158–159 on the paradox that despite the continuing fascination with Homeric warfare, the one aspect of epic warfare that did not acquire popularity was the use of chariots.

31 For an overview of the uses of war elephants in the Hellenistic period, see Epplett (2007) 209–232.

32 Lendon (2005) 159 considers it important that Asclepiodotus dismisses both chariots and elephants as 'not naturally suited for fighting'.

33 For possible explanations for Aelian's omission, see Oldfather (1923) 236 and Spaulding (1933) 662.

34 For information on the rise of the professional army in Rome, see Gabba (1949) 173–209, and its sequel Gabba (1951) 171–272. See also de Blois (2011) 164–180.

35 Campbell notes that the claim to provide practical advice is so common in military manuals as to appear to be almost a stock claim common to didactic manuals. "Moreover, that an author claims to be useful does not mean that others found him so, or that he was much consulted" (Campbell, 19). Ultimately, Campbell argues for the usefulness in real life of manuals written by experienced generals, most notably Frontinus and Arrian. Campbell (1987) 26–27.

36 For a highly skeptical answer to this question, cf. Chapter 2.

37 See Chlup (2014) for the intriguing argument that one of Onasander's aims in this manual is actually to present a moral argument in defense of Just War. For a recent edition, translation, and commentary on Onasander (in Italian), see Petrocelli (2008).

38 ἱππικῶν μὲν λόγων ἢ κυνηγετικῶν ἢ ἁλιευτικῶν τε αὖ καὶ γεωργικῶν συνταγμάτων προσφώνησιν ἡγοῦμαι πρέπειν ἀνθρώποις, οἷς πόθος ἔχεσθαι τοιῶνδε ἔργων, στρατηγικῆς

δὲ περὶ θεωρίας, ὦ Κόϊντε Οὐηράνιε, Ῥωμαίοις καὶ μάλιστα Ῥωμαίων τοῖς τὴν συγκλητικὴν ἀριστοκρατίαν λελογχόσι καὶ κατὰ τὴν Σεβαστοῦ Καίσαρος ἐπιφροσύνην ταῖς τε ὑπάτοις καὶ στρατηγικαῖς ἐξουσίαις κοσμουμένοις διά τε παιδείαν, ἧς οὐκ ἐπ᾽ ὀλίγον ἔχουσιν ἐμπειρίαν, καὶ προγόνων ἀξίωσιν.

39   ἀνέθηκα δὲ πρώτοις σφίσι τόνδε τὸν λόγον οὐχ ὡς ἀπείροις στρατηγίας, ἀλλὰ μάλιστα τῇδε θαρρήσας, ἢ τὸ μὲν ἀμαθὲς τῆς ψυχῆς καὶ τὸ παρ᾽ ἄλλῳ κατορθούμενον ἠγνόησεν, τὸ δὲ ἐν ἐπιστήμῃ τῷ καλῶς ἔχοντι προσεμαρτύρησεν.

40   τὸ δὲ σύνταγμα θαρροῦντί μοι λοιπὸν εἰπεῖν ὡς στρατηγῶν τε ἀγαθῶν ἄσκησις ἔσται παλαιῶν τε ἡγεμόνων κατὰ τὴν σεβαστὴν εἰρήνην ἀνάθημα.

41   τίς γὰρ οὕτως ὑπάρχει φαῦλος ἢ ῥάθυμος ἀνθρώπων ὃς οὐκ ἂν βούλοιτο γνῶναι πῶς καὶ τίνι γένει πολιτείας ἐπικρατηθέντα σχεδὸν ἅπαντα τὰ κατὰ τὴν οἰκουμένην οὐχ ὅλοις πεντήκοντα καὶ τρισὶν ἔτεσιν ὑπὸ μίαν ἀρχὴν ἔπεσε τὴν Ῥωμαίων, ὃ πρότερον οὐχ εὑρίσκεται γεγονός.

42   Or, even, Aeneas Tacticus, whose work abounds with historical examples of each issue that he discusses. In Aeneas' case, these *exempla* become cautionary tales for those might be inclined to ignore his advice – see note 47 later.

43   In his analysis of the early portions of the manual, Chlup notes that "[t]he early chapters of the *Strategikos* are superfluous if Onasander's intention is to provide practical advice on generalship only. Rather, their inclusion and placement early in the treatise is clear evidence that the author seeks to impose on the reader the moral frame of war; all subsequent chapters are meant to be read through the moral filter established here". Chlup (2014) 49.

44   φημὶ τοίνυν αἱρεῖσθαι τὸν στρατηγὸν οὐ κατὰ γένη κρίνοντας, ὥσπερ τοὺς ἱερέας, οὐδὲ κατ᾽ οὐσίας, ὡς τοὺς γυμνασιάρχους, ἀλλὰ σώφρονα, ἐγκρατῆ, νήπτην, λιτόν, διάπονον, νοερόν, ἀφιλάργυρον, μήτε νέον μήτε πρεσβύτερον, ἂν τύχῃ καὶ πατέρα παίδων, ἱκανὸν λέγειν, ἔνδοξον.

45   See Cuomo (2007) 71 for the intriguing argument that Onasander's unique criteria for selecting a general add up to his seeing the job of the general as a *techne*. Lendon (2005) 145 likewise describes generalship, especially in the Hellenistic world, as a competitive *techne*. See also the Introduction to this volume, 4.

46   For an overview of Onasander's manual as a work more philosophical than military in nature, see Smith (1998) 151–166.

47   Examples include the practices of recruiting trustworthy gatekeepers by Leucon, tyrant of Bosporus (5.2); the need to watch out for plots from within, based on examples from Argos, Heracleia Pontica, Sparta, and Corcyra (11.3–14); and cautionary tales about getting help from allies and mercenaries, based on examples from Chalcedon and Heracleia Pontica (12.3 and 12.5). Also see section XVII, on precautions to take during celebration of festivals in a city under siege. Aeneas provides an example of a problematic situation that arose during a festival in Argos, and an example of the right way to handle such situations at Chios.

48   ἀνὴρ οὖν ἀγαθὸς οὐ μόνον πατρίδος τε καὶ στρατιωτικοῦ πλήθους ἄριστος ἡγεμών, ἀλλὰ καὶ τῆς περὶ αὐτὸν εἰς αἰεὶ εὐδοξίας ἀκινδύνου οὐκ ἀνόητος στρατηγός.

49   See Wheeler (1988) 18, however, for the intriguing theory that Frontinus may have read Onasander for background material on Greek military theory.

50   Cf. Goffart (1977) 89–91 for a summary of what we know about Vegetius from his own writings.

51   See Goffart (1977) 69–87 for an overview of previous scholarly debate on the date of Vegetius, and an argument that internal evidence within the text rules out Theodosius I, and while Valentinian III does not seem to be a perfect fit, Goffart suggests that he is the only alternative remaining. Charles (2007) presents a more involved argument in favor of Valentinian III. See Milner (1993) xxix–xli, however, for an argument in favor of Theodosius I.

52   See Whately (2015) 253, however, for the caveat that such prologues need not necessarily mean that the emperor really read the books, although they certainly did imbue them with an air of authority.

53 Whately (2015) 260–261.
54 *antiquis temporibus mos fuit bonarum artium studia mandare litteris atque in libros redacta offerre principibus, quia neque recte aliquid inchoatur, nisi post Deum fauerit imperator, neque quemquam magis decet vel meliora scire vel plura quam principem, cuius doctrina omnibus potest prodesse subiectis.*
55 See Alan Cameron's monumental Cameron (2013) for a study of the gradual process of the dissolving of traditional Roman religion over a long period of time, rather than in one fell swoop.
56 See Goffart (1977) 92–93 on the ways in which Vegetius' Christian faith has influenced his manual. The rise of new moral judgments in this newly Christian empire perhaps explains some of the more puzzling items of advice that Vegetius provides – for example, Vegetius' recommendation against recruiting fishermen! Considering the prominence of fishermen among Jesus' disciples, Vegetius' view on the subject arguably makes even less sense in a Christian empire than a pagan one. See Charles (2010) 101–120. Also see Bond (2016) 149–150. Bond argues that the bias against fishermen was evident already in the writing of Cicero and continued into Late Antiquity.
57 *de dilectu igitur atque exercitatione tironum per quosdam gradus et titulos antiquam consuetudinem conamur ostendere; non quo tibi, imperator invicte, ista videantur incognita, sed ut, quae sponte pro reipublicae salute disponis, agnoscas olim custodisse Romani imperii conditores et in hoc parvo libello, quicquid de maximis rebus semperque necessariis requirendum credis, invenias.*
58 This attitude can explain the difficulty that Vegetian scholars have noted with nailing down a precise date for Vegetius' 'antiqua legio' in Book II – cf. Parker (1932) 137–149. If, however, the main point for Vegetius is that any part of the past is valuable for teaching the present, the idea of conflating all of the past into a single "antiqua legio" makes more sense, since the precise date of any one institution is less important for Vegetius than the fact that it existed, and therefore is now valuable for educating the army of the present day.
59 *horum instituta, horum praecepta, in quantum valeo, strictim fideliterque signabo. nam cum easdem expensas faciat et diligenter et neglegenter exercitus ordinatus, non solum praesentibus, sed etiam futuris saeculis proficit, si provisione maiestatis tuae, imperator Auguste, et fortissima dispositio reparetur armorum et emendetur dissimulatio praecedentum.*
60 See Chapters 11 and 12 for case studies on Vegetius' view on rules of war and on naval warfare, respectively.
61 *horum sequentes instituta Romani Martii operis praecepta et usu retinuerunt et litteris prodiderunt. quae per diversos auctores librosque dispersa, imperator invicte, mediocritatem meam abbreviare iussisti, ne vel fastidium nasceretur ex plurimis vel plenitudo fidei deesset in parvis.*
62 *cunctos imperatores felicitate moderatione castimonia, exemplis indulgentiae, studiorum amore praecedis*
63 *ad complementum igitur operis maiestatis vestrae praeceptione suscepti rationes, quibus vel nostrae civitates defendendae sint vel hostium subruendae, ex diversis auctoribus in ordinem digeram, nec laboris pigebit, cum omnibus profutura condantur.*
64 As Goffart aptly summarises Vegetius' approach to historical examples: 'Although teaching by the examples of antiquity, Vegetius was more interested in his lessons than in the accuracy of his examples; he chose his past, and he tailored it to his own ends rather than to the satisfaction of future antiquarians'. Goffart (1977) 92.
65 This attitude is reflected elsewhere throughout Vegetius' text. See, for example, Whately (2015) 251 for an analysis of Vegetius' relationship to research sources other than famous Roman generals and writers.
66 *neque enim degeneravit in hominibus Martius calor nec effetae sunt terrae, quae Lacedaemonios, quae Athenienses, quae Marsos, quae Samnites, quae Pelignos, quae ipsos progenuere Romanos. nonne Epiri armis plurimum aliquando valuerunt? nonne Macedones ac Thessali superatis Persis usque ad Indiam bellando penetrarunt? Dacos*

autem et Moesos et Thracas in tantum bellicosos semper fuisse manifestum est, ut ipsum Martem fabulae apud eos natum esse confirment, longum est, si universarum prouinciarum vires enumerare contendam, cum omnes in Romani imperii dicione consistant. sed longae securitas pacis homines partim ad delectationem otii partim ad ciuilia transduxit officia.

67 *vilius enim constat erudire armis suos quam alienos mercede conducere.*
68 Schrader (1981) 168 assumes that the disaster of Adrianople is what spurred Vegetius to write his treatise.

Bibliography

Allmand, C. 2011. *The De Re Militari of Vegetius: The Reception, Transmission and Legacy of a Roman Text in the Middle Ages*. Cambridge: Cambridge University Press.

Bengtson, H. 1962. 'Die griechische Polis bei Aeneas Tacticus'. *Historia* 11. 458–468.

Bingham, J. 1968. *The Tactiks of Aelian: Or, Art of Embattailing an Army after ye Grecian Manner*. New York: De Capo Press.

Bond, S. 2016. *Trade and Taboo: Disreputable Professions in the Roman Mediterranean*. Ann Arbor: University of Michigan Press.

Bulkeley, R. I. P. 1983. 'Vegetius Vindicatus? Giving an Old Hypothesis a Fair Break'. *Current Research on Peace and Violence* 6. 233–257.

Cameron, A. 2013. *Last Pagans of Rome*. Oxford: Oxford University Press.

Campbell, B. 1987. 'Teach Yourself How to Be a General'. *JRS* 77. 13–29.

Charles, M. 2007. *Vegetius in Context: Establishing the Date of the Epitoma Rei Militaris*. Stuttgart: Franz Steiner.

——— 2010. 'Unseemly Professions and Recruitment in Late Antiquity: *Piscatores* and Vegetius *Epitoma* 1.7.1–2'. *AJP* 131. 101–120.

Chlup, J. T. 2014. 'Just War in Onasander's ΣΤΡΑΤΗΓΙΚΟΣ'. *JAH* 2. 37–63.

Cuomo, S. 2007. *Technology and Culture in Greek and Roman Antiquity*. Cambridge: Cambridge University Press.

de Blois, L. 2011. 'Army and General in the Late Roman Republic', in P. Erdkamp, ed., *A Companion to the Roman Army*. Malden, MA: Wiley Academic. 164–180.

Epplett, C. 2007. 'War Elephants in the Hellenistic World', in W. Heckel, L. Tritle, and P. Wheatley, eds., *Alexander's Empire: Formulation to Decay: A Companion to Crossroads of History*. Claremont: Regina Books. 209–232.

Gabba, E. 1949. 'Le origini dell'esercito professionale in Roma: i proletari e le riforma di Mario'. *Athenaeum* 27. 173–209.

——— 1951. 'Ricerche sull'esercito professionale Romano da Mario ad Augusto'. *Athenaeum* 29. 171–272.

Gaca, K. 2010. 'The Andrapodizing of War Captives in Greek Historical Memory'. *TAPA* 140. 117–161.

Garland, R. 2016. *Athens Burning: The Persian Invasion of Greece and the Evacuation of Attica*. Baltimore: Johns Hopkins University Press.

Goffart, W. 1977. 'The Date and Purpose of Vegetius' De Re Militari'. *Traditio* 33. 65–100.

Handford, S. 1926. 'The Evidence of Aeneas Tacticus on the Balanos and the Balanagra'. *JHS* 46. 181–184.

Kidd, I. G. 1988. *Posidonius. Volume II: The Commentary, (i) Testimonia and Fragments 1–149*. Cambridge: Cambridge University Press.

Knowles, C. 1956. 'A 14th Century Imitator of Jean de Meung: Jean de Vignay's Translation of De Re Militari of Vegetius'. *Studies in Philology* 53. 452–258.

Krentz, P. 2002. 'Fighting by the Rules: The Invention of the Hoplite Agon'. *Hesperia* 71. 23–39.

Lanni, A. 2008. 'The Laws of War in Ancient Greece'. *Law and History Review* 26. 469–489.

Lendon, J. E. 2005. *Soldiers and Ghosts: A History of Battle in Classical Antiquity*. New Haven: Yale University Press.

Milner, N. P. 1993. *Vegetius: Epitome of Military Science*. Liverpool: Liverpool University Press.

Ober, J. 1998. *The Athenian Revolution: Essays on Ancient Greek Democracy and Political Theory*. Princeton: Princeton University Press.

Oldfather, W. A. 1923. *Aeneas Tacticus, Asclepiodotus, Onasander*. Cambridge, MA: Harvard University Press.

Parker, H. M. D. 1932. 'The Antiqua Legio of Vegetius'. *CQ* 26. 137–149.

Petrocelli, C. 2008. *Onasandro. Il generale, manual per l'esercizio del commando*. Bari: Edizioni Dedalo.

Powers, S. 2006. 'Studying the Art of War: Military Books Known to American Officers and Their French Counterparts during the Second Half of the Eighteenth Century'. *Journal of Military History* 70. 781–814.

Rihll, T. 2007. *The Catapult: A History*. Yardley, PA: Westholme.

Roueché, C. 2002. 'The Literary Background of Kekaumenos', in C. Holmes and J. Waring, eds., *Literacy, Education and Manuscript Transmission in Byzantium and Beyond*. Leiden: Brill. 117–123.

Schrader, C. 1981. 'The Influence of Vegetius' De Re Militari'. *Military Affairs* 45. 167–172.

Smith, C. J. 1998. 'Onasander on How to Be a General', in M. M. Austin, J. D. Harries, and *idem*, eds., *Modus Operandi: Essays in Honour of Prof G. E. Rickman: BICS Supplementary Series*. Vol. 71. London: Institute of Classical Studies. 151–166.

Spaulding, O. L., Jr. 1933. 'The Ancient Military Writers'. *CJ* 28. 657–669.

Vela Tejada, J. 2004. 'Warfare, History and Literature in the Archaic and Classical Periods: The Development of Greek Military Treatises'. *Historia* 53. 126–146.

Whately, C. 2015. 'The Genre and Purpose of Military Manuals in Late Antiquity', in G. Greatrex, H. Elton, and L. McMahon, eds., *Shifting Genres in Late Antiquity*. Burlington, VT: Ashgate. 249–262.

Wheeler, E. 1988. *Strategem and the Vocabulary of Military Trickery*. Leiden: Brill.

——— 2004. 'The Legion as Phalanx in the Late Empire, Part I', in Y. Le Bohec and C. Wolff, eds., *L'Armée romaine de Deoclétien à Valentinien I*. Paris: de Boccard. 309–358.

Whitehead, D. 1990. *Aineias the Tactician: How to Survive under Siege*. Oxford: Oxford University Press.

Williams, N. 2017. 'The Evolution of Civilian Participation in Ancient Greek Warfare', in N. Foote and N. Williams, eds., *Civilians and Warfare in World History*. London: Routledge. 23–41.

Winterling, A. 1991. 'Polisbegriff und Stasistheorie des Aeneas Tacticus. Zur Frage der Grenzen der griechischen Polisgesellschaften im 4. Jahrhundert v. Chr'. *Historia* 40. 193–229.

Wrightson, G. 2015. 'To Use or Not to Use: The Practical and Historical Reliability of Asclepiodotus's "Philosophical" Tactical Manual', in G. Lee, H. Whittaker, and G. Wrightson, eds., *Ancient Warfare: Introducing Current Research*. Vol. 1. Newcastle Upon Tyne: Cambridge Scholars Publishing. 65–93.

4   Homeric *Taktika*

Nicholas Sekunda

Poseidonius of Apamea, who produced a *Technē Taktikē* in about 90 BCE, seems to have been the first philosopher to have founded the tradition of writing a new genre which we call 'The Greek *Tactica*' and which seems to be an attempt to describe and divide up contemporary, that is late Hellenistic, armies according to a philosophical and taxonomical system, looking for geometrical symmetry in military organisation.[1]

Although Poseidonius' work is lost, the tactical works of Asclepiodotus, Aelian, and Arrian, all of which were ultimately derived from that of Poseidonius, either directly or through an intermediary, have survived.[2] Of these three works, those of Aelian and Arrian both include an identical list of earlier writers on the military arts. Alphonse Dain thought that the works of Aelian and Arrian were both derived from a common prototype, later than Asclepiodotus, and intermediate between both authors and Poseidonius.[3] The first folio of Arrian's *Tactics* is missing, but the corresponding section of text is preserved in Aelian, and this includes a list of writers of Homeric *Taktika*. These authors are mentioned, to my knowledge uniquely, at the very beginning of the *Tacitics* (*Taktikē Theōria*) of Aelian (1.1–2), immediately following the Preface:

> Homer was, it seems, the first, whom we know of, to discover the theory of arranging troops, and to wonder at those with knowledge of it, like Menestheus
>
>> 'Never on earth before had there been a man born like him For the arrangement in order of horses and shielded fighters',
>
> and with reference to the tactics according to Homer, we have come across those by the prose-writers Stratokles, and by Hermeias, and by Frontinus, the man of our times of consular rank. Aeneas has worked out the theory completely, through putting together in order a great and adequate amount of books on this, and on the art of generalship.[4]

There follows a list of Greek military writers dealing with *Taktika*, beginning with Aeneas Tacticus (included in this translation), and finishing with Poseidonius. The speculation that the tradition of writing 'Greek *Taktika*' was based on a lost work of Poseidonius is largely based on the fact that this list in both authors finishes with the name of Poseidonius himself. The list in Arrian provides more detail on the individual authors mentioned than that in Aelian does, from which it is clear that the list has been drawn up in chronological order. Unfortunately, the beginning of this list is missing from the manuscript of Arrian, which begins abruptly with [Alexander] the son of Pyrrhus of Epirus. Presumably the list in Arrian contained the same names in the same order as the list in Aelian, but with more detail.

The appearance of the three authors of Homeric *Taktika* in Aelian's list before Aeneas Tacticus seems to be determined by the antiquity of Homer himself, not of the three authors. This is confirmed by the fact that one of them, Frontinus, was a contemporary of Aelian himself. There is no way of knowing whether the information about Homer, and the works of Stratokles, Hermias, and Frontinus was repeated in Arrian or not. More broadly, reflection upon the nature of Homeric *Taktika* can provide insight into the emergence of the military manual as a genre, especially as ancient military authors themselves reflected upon the literary-historical tradition in which they sought to work.

I. Homer as a military writer

In the absence of any sacred book, to the ancient Greeks the writings of Homer, and the *Iliad* in particular, fulfilled the role of a *vade mecum* for human conduct.[5] The implication that Homer was the first writer of *Taktika* should, therefore, not be taken too literally.[6] In labelling Homer such, Aelian doubtless had in mind such passages as when an anonymous speaker advises Agamemnon to 'draw up the men by *phylai* and phratries' (2.363).[7] Neither, obviously, should Aelian be taken as saying that Homer wrote a specific work entitled *Taktika*: it seems that Pyrrhus of Epirus should be credited with that literary achievement.[8]

A good example of Homer serving in place of a sacred book comes in Xenophon's *Symposium* (3.5, 4.6) where Nikeratos tells how his father, Nikias the Athenian commander, who, indeed, was profoundly religious in real life, bade him learn by heart the whole of the *Iliad* and *Odyssey*, because, as Nikeratos says (4.6), 'the sage Homer has written about practically everything pertaining to man . . . the art of running a household, or public speaking, or generalship' (4.6).[9] Aristophanes, too, sees Homer as an authority in military matters. For example, in *Frogs* he has Aeschylus say:

> and the divine Homer, what did he get his honour and renown from, if not from the fact that he gave good instruction about the tactics and virtues and arming of soldiers?

> (1034–1036)[10]

Commentators on this passage have pointed to a similar passage in Plato's *Ion* 540 D–541 B, a dialogue between Socrates and Ion the Ephesian rhapsode, a professional performer of epic poetry, Homeric and other, to the accompaniment of a lyre. Egged on to an *argumentum ad absurdum* by Socrates, Ion claims that he is the best rhapsode in Greece and also the best general thanks to his study of Homer. As Alan Sommerstein points out, 'before Socrates sets to work inveigling him into absurdity, he asserts only, reasonably enough, that his knowledge of Homer has taught him "the right things for a general to say when encouraging his troops"'. In *Frogs*, Aeschylus wants to claim that he is a great poet and that all great poets have been teachers, and teaching military virtue is all important. The *Iliad* was an epic of war, and therefore, so Sommerstein's argument runs, 'it is, then, not surprising that Aeschylus is made to present Homer as a teacher of military virtues and the military arts'.[11]

Against this we have Aristotle, *Politics* 4.13.10 (1297b20–21), where it is stated in the context of the drawing up of hoplites in order, 'and the crafts and tactical rules connected with the above-mentioned matters did not exist among the ancients' (αἱ δὲ περὶ τῶν τοιούτων ἐμπειρίαι καὶ τάξεις ἐν τοῖς ἀρχαίοις οὐχ ὑπῆρχον), and therefore that in former times the strength of armies lay with their cavalry.[12] The *Politics* seems to have been composed between the years 335 and 322, the period from when Aristotle was resident in Athens and head of the Lyceum which he founded there, to his death. At any rate, it seems certain that the *Politics* appeared well after the works of Aeneas Tacticus had been published. For the first time, the works of Aeneas met the perceived 'need for a comprehensive handbook on the subject'.[13]

Aeneas was considered by Casaubon to be identical with Aeneas of Stymphalos, whom Xenophon mentions participating in the second battle of Mantineia in 362 BCE (*Hellenica* 7.3.1). Most modern scholars support this identification.[14] The sole surviving work with which he is credited is entitled the *Poliorkētika* (*How to Survive under a Siege*). The majority of the historical incidents in this work date to the two decades between 370 and 360, and a further four incidents to the following decade, during the years 359–355, which suggests the work was composed in the latter half of the 350s.[15]

At this point, it is worth remarking that this does not constitute a date for the composition of the whole range of Aeneas' works. In the *Poliorkētika* at 14.2, he refers the reader to his *Poristikē Biblos* (ἡ ποριστικὴ βίβλος), a book covering military finances, which was therefore earlier, and at 40.8 Aeneas promises us a discussion of naval affairs, which is presumably a book which he has not completed yet, and we do not know whether this did in fact appear or not. It is entirely plausible that Aeneas compiled his works over a decade or more. Aeneas makes no mention of Homer as a military writer, and neither does Aristotle.

II. The nature of the Homeric *Tactica*

At this point it seems appropriate to ask what was the nature of these Homeric *Taktika* mentioned by Aelian? Wheeler has suggested that they constituted 'military instruction through Homeric quotation'.[16] Although he nowhere develops his thoughts further, Wheeler is presumably thinking of examples such as that in Frontinus' *Strategemata* (*Stratagems*), which has Pyrrhus at the battle of Asculum in 279 BCE stationing the Samnites and Epirotes on the right wing, the Bruttians,

Lucanians, and Sallentines on the left, with the Tarentines in the centre (2.3.21). He did this, according to Frontinus, in line with the stricture contained in the *Iliad*:

> First he ranged the mounted men with their horses and chariots, and stationed the brave and numerous foot-soldiers behind them to be a bastion of the battle, and drove the cowards to the centre that a man might be forced to fight even though unwilling.
>
> (4.297–300)[17]

The Homeric allusion does not appear in the corresponding passages in either Plutarch's *Life of Pyrrhus* (21.5–6) or Dionysius of Halicarnassus (20.1–4), which might indicate that Frontinus is using another source tradition, and it is perhaps possible that a parallel passage appeared in Frontinus' *On the Tactics Found in Homer* (see also later, 90).

According to the historian Ammianus Marcellinus, the very same Homeric passage inspired the Emperor Julian in his dispositions for battle (*Homerica dispositio*) before Ctesiphon against the Sasanian Persians in 363 CE, where he allotted the centre space between his two lines to his weakest infantry (24.6.9). Lendon has commented on this passage:[18]

> Pyrrhus had picked it up from Homer and used it, although he interpreted it as a matter of left, right, and center rather than front, middle, and back. (So well known was this ruse that the term was used in rhetoric to denote a speech with a weak middle section.) Julian, the better philologist, corrected Pyrrhus: by Julian's day it was important to get such things right.

Polybius (15.16.3) quotes the same Homeric passage to describe Hannibal's disposition of Carthaginian forces at Zama in 202 BCE.[19]

The use of Homeric quotation was ubiquitous in the Greek and Roman world. As far as military authors are concerned, Onasander quotes from Homer twice (at 1.7; 23.1), Arrian includes an extensive series of Homeric quotations at one point in his *Tactica* (31.5–6), and Polyaenus laces the Preface to the first book of his *Stratagems* (*Strategemata*) with a string of Homeric quotations taken from both the *Iliad* and the *Odyssey* (Preface 3–4).[20] I think this is simply motivated by the desire of contemporary authors to display their high level of culture and, as we have already seen, by the ubiquitous role played by writings of Homer in ancient life. I suspect that the Homeric *Taktika* were something more than a gathering together of Homeric quotations which could be used for military instruction. I suggest that they were commentaries on military passages dealing with tactics in Homer's works.

III. Aristarchus of Samothrace

Given the great interest in the study of Homer in Alexandrian literary and intellectual circles, which, for example, manifested itself in the re-creation of the epic form in Hellenistic literature, it is only natural that the first Homeric commentaries would have been compiled in that city. The easy availability of the resources of the Alexandrian library would have both encouraged scholars already resident in the city to attempt such an undertaking and would have drawn anyone attracted to such work to the city.

Around the second quarter of the second century BCE, Aristarchus of Samothrace held the post of head of the Museion and Library at Alexandria.[21] As the figure of Aristarchus is intimately bound up with the writing of Homeric commentaries, and as I have suggested that the Homeric *Taktika* were commentaries on military passages dealing with tactics in Homer's works, I would like to take this opportunity to establish the career of Aristarchus in some detail. Only when the chronology of his career has been established will it be possible to establish whether the persons either directly named as authors of Homeric *Taktika* by Aelian, or proposed as the same in this text, are disciples of Aristarchus or not. One of our principal sources on the life and achievements of Aristarchus is his entry (A 3892) in the Suda Lexicon:

> Aristarchus, an Alexandrine by adoption, but by birth a Samothracian, his father was Aristarchus. He was active during 156th Olympiad in the reign of Ptolemy Philometor, whose son he taught. It is said that he wrote over 800 books that were commentaries alone. He was a pupil of the grammarian Aristophanes, and he hotly contested very much in Pergamon with the grammarian Krates, the Pergamene. About 40 grammarians were his students. He died in Cyprus, having been overtaken by lack of the means of maintenance, from dropsy. He lived 72 years. And he left behind Aristarchus and Aristagoras as children; both being feeble-minded, and Aristarchus had even been sold: the Athenians redeemed him, when he had come to them.

If Aristarchus died at the age of 72, and if it is correctly surmised that this took place around 144 (see later), then Aristarchus should have been born around 216.[22]

> No information survives regarding his early life, and no indication of when he changed his civil allegiance (if indeed he did, and it is therefore quite uncertain whether he was born in Alexandria of Samothracian parents resident there, or was brought up on this remote and rocky island in the north Aegean (then a Ptolemaic dependency), and later moved to Alexandria,[23]

writes Fraser, who elsewhere suggested that Aristarchus became Librarian about 175 BCE.[24]

Eichgrün argued that Aristarchus was tutor to both the future Ptolemy VI Philometor (180–145 BCE) and Ptolemy VIII Euergetes II Physcon (145–116 BCE).[25] In support of Eichgrun's argument, Physcon is credited with writing a work on Homeric criticism,[26] a work very much in the genre of Aristarchus. Furthermore, according to Athenaeus (2.71b), Ptolemy Euergetes, the King of Egypt (by which we can only understand Ptolemy VIII Euergetes II), was a student of Aristarchus. On this passage, Fraser has commented 'I feel very uncertain that this last phrase indicates that Aristarchus was the formal tutor of Euergetes. It probably means no more than Euergetes as a youth had heard lectures by Aristarchus'.

Aristarchus 'first appears on the scene as tutor to the son of Philometor, the Crown Prince, Ptolemy Eupator, and also probably to his younger brother, who reigned briefly as Neos Philopator'.[27] Ptolemy VI Philometor 'Mother-Loving' had no children until about 165 BCE.[28] His eldest son Eupator was appointed King in Cyprus on the failure of Ptolemy VIII Euergetes II to take the island 153 BCE. He died in Cyprus in 150 BCE. Philometor also had a younger son, born probably in 162/1 BCE.[29]

The 156th Olympiad runs from the years 156 to 153 BCE. It was presumably during this period that Aristarchus was appointed tutor to the royal children, and this is why the date appears in the Suda Lexicon as the date when Aristarchus was 'active'.

Philometor died in the late summer of 145 BCE, having in June the same year made his 17-year-old second son king, with the title of Ptolemy VII Neos Philopator 'Father-Loving'. Euergetes moved from Cyrene and, according to Justin (38.8) as soon as he entered Alexandria, ordered that all the supporters of the young prince be put to death, as well as the young king himself. We next find Aristarchus in Cyprus. As both Pfeiffer and Fraser have pointed out,[30] we should not entertain any idea that Aristarchus 'escaped' there, as the island was loyal to Euergetes at the time. Rather we might think that Aristarchus was transported there for safe-keeping and punishment, dying of disease there, brought on by a reduction into a state of poverty, probably around the year 144 BCE.

To the information contained under the name of Aristarchus in the Suda Lexicon, we can add the evidence supplied by *Oxyrhynchus Papyrus* no. 1241.[31] This is a chrestomathy, in other words a treatise containing lists and catalogues, dating to the second century CE. The second column contains a short account of the Alexandrian librarians. The translation is that supplied by the editors.[32]

> [Apollo]nius son of Silleus, of Alexandria, called the Rhodian, the disciple of Callimachus; he was the teacher of the third king.[33] He was succeeded by Eratosthenes, after whom came Aristophanes son of Apelles of Byzantium, [and Aristarchus][34] then Apollonius of Alexandria the so-called Classifier, and after him Aristarchus son of Aristarchus, of Alexandria, but originally of Samothrace; he became also the teacher of the children of Philopator,[35] He was followed by Cydas, of the spearmen; and under the ninth king there flourished Ammonius, Zenodotus, Diocles, and Apollodorus the grammarians.

Fraser suggested that the text of *Oxyrhynchus Papyrus* no. 1241 should be emended to mean the children of Philometor, and so referring to Eupator and Neos Philopator.[36] It is debatable whether the error occurred through scribal error. It is easy to see that the original compiler of the list might have made a simple mistake, writing Philopator for Philometor.

As well as seemingly writing a new critical edition of the text of the works of Homer, Aristarchus, who was known as ὁ Ὁμηρικός,[37] wrote a number of commentaries on the works of Homer (and of other writers of course): over 800 according the Suda Lexicon. But, in the words of Rudolf Pfeiffer, 'Towards the middle of the second century the imperative demand was not for editing the text anew, but for explaining it in its entirety'.[38] It was in the age of Aristarchus that the genre of Homeric commentary really began. As well as a number of commentaries (ὑπομνήματα) on Homer, Aristarchus seems also to have written a number of treatises (συγγράμματα) clarifying specific problems arising out of the Homeric works.[39] We know that one of them, entitled περὶ τοῦ ναυστάθμου, was composed upon the naval camp of the Greeks.[40]

I have previously suggested that the Homeric *Taktika* were a subgenre of the Homeric commentaries, dealing with the military formations and tactics contained in Homer. It is unlikely that Aristarchus himself produced a work (commentary) on the Homeric *Taktika*. If this had, indeed, been the case, it would be difficult to believe that Aelian made no mention of it. If Aristarchus was not responsible for writing the first work in this subgenre, it is easy to see how pupils of Aristarchus, and according to the Suda Lexicon he had about forty, would be encouraged to produce such works.

IV. Hermolytos the Tactician

I believe a fragment of one of the Homeric *Taktika* has been preserved in Eustathius. The *Commentaries on Homer's Iliad and Odyssey* by Eustathius, who lived in the 12th century, is a vast compilation of earlier commentaries or *scholia* on the two Homeric works by earlier authors. In a passage in the *Iliad*, the poet describes the Greek army thus:

> spear by spear, shield against shield at the base, so buckler leaned on buckler, helmet on helmet, man against man, and the horse-hair crests along the horns of their shining helmets touched as they bent their heads, so dense were they formed on each other.[41]

On this passage Eustathius contains the following comment:[42]

> Hermolytos the Tactician says that Lycurgus later ordained a *synaspismos* of such a type for the Lakedaimonians, while Lysander the Lakonian taught this in his deeds, just as Epaminondas also (taught it) to the Thebans and Charidemos to the both the Arcadians and Macedonians.[43]

The vetera scholia have a slightly different version of the text:[44]

> This *synaspimos*, as Hermolytos the Tactician says, Lycurgus ordained, Lysander the Laconian and Epaminondas taught it, then the Arcadians and Macedonians learned it under Charidemos.[45]

Eustathius calls Hermolytos, who is only known from this single reference, a τακτικός, a term which is used in other authors for a writer of *Taktika*. For example, Josephus describes Philostephanos the general (*Jewish Antiquities* 13.340), whom we know to have also been a military writer from a passage in Plutarch's *Life of Lycurgu*s (23.1), as a military writer (ὁ τακτικός). So the fact that Eustathius calls Hermolytos a τακτικός, and the very fact that he appears in his *Commentary* in the first place, indicate that Hermolytos was the author of a work on the Homeric *Taktika*.

The actual contents of the citation indicate what the nature of these Homeric *Taktika* may have been. They seem to deal with literary passages concerned with military matters in Homer. This fragment constitutes, in effect, a commentary on the word *synaspismos*. Others may have sought to explain military terminology in Homer which had become obsolete. This was, presumably, one of the principal roles of the Homeric *Taktika*: they were in fact commentaries on the military sections of the *Iliad*. Indeed, the elucidation of obsolete or obscure vocabulary was a principal aim of the Homeric commentaries as a whole. In a wider sense, the Homeric *Taktika* may have sought to explain, or deepen the reader's understanding of, military practices or phenomena described in the Homeric works beyond lexical obscurities.

The citation of three historical characters belonging to the Classical period (Lysander, Epaminondas, and Charidemus of Oreus) indicates that Hermolytos must have been writing in the Hellenistic period or later. Perhaps Hermolytos was one of the pupils of Aristarchus.

The statement that Charidemus of Oreus taught the technique of fighting in *synaspismos* may refer to a period in Charidemus' career that has previously been unclear. During the early 360s, he may well have found employment among the Arcadians as a military instructor, and later among the Macedonians. He may have been responsible for teaching these two peoples the fighting methods of the 'Iphicratean peltast', which may have in turn involved fighting with shields locked together in a closer formation than previously.[46]

The name Hermolytos, which means 'freed by Hermes', is otherwise unknown in the ancient Greek onomasticon, though the comparable names Hippolytos,[47] Timolytos,[48] and indeed Theolytos[49] are attested. One is, therefore, tempted to suppose that the name given in the text is a corrupted form of the much more common name Hermolykos. The name must stand however, given that it is cited in a single source, with no possibility of establishing a different reading. Emendation of the text would also be contrary to the normal, though not invariable, preference for *lectio difficultior*. An emendation in this direction would do little to help us identify or date the author however. The name Hermolykos is a very common one, and there would be no reason to prefer an identification of our tactical writer with any particular one of them. It is to be hoped that the name Hermolytos may occur elsewhere in the future. This would not only confirm the form of our author's name, but might also give us some clues as to his date and place of origin.

V. Stratokles

Returning, then, to the preface in Aelian, he lists (*Taktika Theoria* 1.2) first Stratokles, then Hermeias, 'and in our time Frontinus a man of consular dignity' as writers of Homeric *Taktika*. Frontinus is writing shortly before Aelian, but the other two individuals listed may have both been active from any date around the middle of the second century BCE to the first century CE. The sole mention of Stratokles as an author of a book on Homeric *Taktika* comes in the introduction to Aelian, and with this meagre information he is listed as the seventh holder of that name in *Pauly-Wissowa*.[50]

Loreto has suggested that the Stratokles mentioned by Aelian could 'hypothetically' be identified with one Stratokles,[51] the Athenian general who commanded the left of the Greek army at the battle of Chaeroneia in 338 BCE.[52] This would be to place the composition of the Homeric *Tactica* before the earliest composition of the Homeric commentaries, of which I have argued they were a subset. This would be too early in my opinion.

Another possibility to find a match for Stratokles, the author listed by Aelian, would be to search among those who bore the same name and served in the Ptolemaic military. One Stratokles son of Stratokles is attested as serving as a military *grammateus* of the τῶν κατιό[κων ἵππέων] in the Arsinoite nome in papyri dating to the last two decades of the third century BCE.[53] This must be considered an unlikely match, though it is also possible that a descendant of this Stratokles was Stratokles the author of a Homeric *Taktika*.

The personal name Stratokles, a name which has military connotations,[54] is especially popular in Rhodes (and, as we have seen earlier, at Athens as well), where it is borne by a number of individuals prominent in the Rhodian military forces and administration.[55] One of them is the philosopher Stratokles of Rhodes,[56] who is mentioned by Strabo (14.2.13) as among the most eminent Rhodian philosophers, and as pupil of Panaitios of Rhodes in the *Index Stoicorum Herculanensis*.[57] The latter source informs us that Stratokles was the author of 'books' in the plural, and Max Pohlenz apparently believed that Stratokles also wrote a history of the Stoa which was ultimately 'sunteggiata' in the *Index Stoicorum Herculanensis*, which was, in turn, part of the *History of the Philosophers* (Σύνταξις τῶν Φιλόσοφων) of Philodemus.[58]

Panaitios of Rhodes was a pupil of Dionysius 'Thrax', who was, in turn, a pupil of Aristarchus. Dionysius Thrax was an Alexandrian, though as suggested by his soubriquet, probably of Thracian origin. Born in about 160, he must have listened to Aristarchus as a young man. He may have been caught up in the expulsions of 145. He migrated either to Rome or to Rhodes (which is the widely favoured manuscript emendation): 'He was a man of varied culture, whose interests were not wholly linguistic or grammatical; he is known to have been interested in the arts, and he also published a history of Rhodes'.[59]

Panaitios of Rhodes (c. 185–109 BCE), the pupil of Dionysius Thrax, recognized the primacy of Aristarchus as a critic and interpreter, and was so familiar with his work that he called him a 'seer'.[60] This indicates that Panaitios, though primarily a philosopher, also had an interest in the works of the grammarians. He first studied at Pergamon and then at Athens, and moved to Rome in the 140s. He became leader of the Athenian Stoa in 129 BCE on the death of Antipater. He died in Athens in 110/09 BCE.[61] It is probable, then, that Stratokles studied under Panaitios at Athens between these dates.

A section of *Index Stoicorum Herculanensis* has been interpreted as listing the pupils of Stratokles, one son of Stratokles himself, whose name is lost, as well as [–]ων of Alexandria and Antipatros of Tyre.[62] Stratokles of Rhodes is listed as a philosopher, of course, and not as a grammarian, but this would not preclude him from writing a work on the *Taktika* in Homer. Poseidonius of Apamea, who produced a *Technē Taktikē* about 90 BCE, was classified primarily as a Stoic philosopher. Asclepiodotus is designated a philosopher at the title page of his book, although there is no way of knowing whether this designation is by the author himself, or due to a Byzantine editorial intervention. He is generally identified, however, with an Asclepiodotus whom Seneca (*Natural Questions* 2.26.6) mentions as a 'listener' of Poseidonius.[63] A generation later Arrian thought of himself primarily as a Stoic philosopher, and only secondarily as a practical man of affairs – and a writer of *Taktika*.[64]

VI. Hermeias

Likewise, the only mention of Hermeias as an author of περὶ τῆς καθ᾽ Ὅμηρον τακτικῆς comes in the introduction to Aelian's *Taktika Theoria* (1.2), where he is listed between Stratokles and Frontinus. He is simply listed with this information in Pauly-Wissowa's *Real-Encyclopädie*.[65] Nothing more is known of his work than its genre.

The large gap in time which separates Stratokles from Frontinus in Aelian's list does not allow us to 'fix' Hermeias firmly chronologically. Assuming that the list of these three authors is given in chronological order, then it follows that Hermeias is later than Stratokles: at the earliest he may possibly be given a date towards the end of the second century BC. It does not seem to be possible that Hermeias was another of the 'about forty' students of Aristarchus.

Loreto has also made a hypothetical suggestion for the identification of Hermeias,[66] as the Hermeias who was the Seleucid 'chief minister' (ὁ ἐπὶ τῶν πραγμάτων) during the reigns of Seleucus III and Antiochus III.[67] Again, I think this identification is unsuitable for the same reasons as was the case in the identification of Stratokles proposed by Loreto. He is too early and is neither a grammarian nor a philosopher.

There is no grammarian or philosopher of the same name which lends itself to identification with our figure. On the other hand, the name is common in Ptolemaic Egypt. One Ἑρμίας Ἀλεξανδρεύς[68] honours the Egyptian gods Sarapis, Isis, Anoubis-Hermes, and Apollo-Harpokrates in an inscription from Delos erected during the priesthood of Demetrius son of Demetrius.[69] The editors dated the inscription after 166 BCE and thought that perhaps the Demetrius attested holding the priesthood was the son of the Demetrius who is attested as holding the priesthood together with one Telesarchides in two other inscriptions (2116–2117) in the same volume. The latter is given a date of circa 166 BCE by the editors of the *Lexicon of Greek Personal Names*,[70] so we could give a date of the third quarter of the second century to Hermias, but there is still nothing to connect him with Hermeias the author of Homeric *Taktika* mentioned by Aelian other than a coincidence of name.

VII. Neoteles

One Neoteles is cited in the Homeric commentaries to the scene in *Iliad* where, as Teucer is drawing his bow, he is struck in his hand by a rock thrown by Hector:

> and went straight for Teukros, heart urgent to hit him.
> Now Teukros had drawn a bitter arrow out of his quiver,
> and laid it along the bowstring,
> but as he drew the shaft by his shoulder,
> there where between the neck and the chest collar-bone interposes, and this
>   a spot most mortal;
> in this place the shining-helmed Hektor struck him with all his fury with the
>   jagged boulder,
> smashing the sinew, and all his arm with the wrist was deadened.
> He dropped on one knee and stayed, and the bow fell from his hand.[71]

The commentary in Porphyry, writing in the later third century CE, on this passage, runs thus:

> In these lines spoken about Teucer, they inquire in which hand Teucer has been wounded and whether he draws the string towards his shoulder like the Scythians. For Neoteles, who wrote a whole book about archery among the heroes (ὅλην βίβλον γράψας περὶ τῆς κατὰ τοὺς ἥρωας τοξείας), supposed this, claiming that the Cretans draw the bowstring to the breast but make the extension [of the bow] round, whereas the Scythians draw [the bowstring] not to the breast, but to the shoulder, <so that> the right side of the archer does not project beyond the left side.[72]

In the other commentaries on this passage in the *Iliad*, the name of Neoteles is mentioned, once by the *scholia vetera*,[73] in a citation that Erbse apparently thought ultimately went back to Nicanor of Alexandria, a grammarian who was working in the second century CE,[74] and twice by Eustathius,[75] but neither of these sources credits Neoteles with the authorship of a whole book on the subject. Manuel Baumbach suggests that it was not a whole book, but rather a lengthy digression within the commentary.[76]

Didymus, in his commentary on the *Iliad* 24(Ω).110, lists Neoteles and another grammarian Aretades between Apollodorus (born c. 180 BCE) and Dionysius

Thrax (c. 170–90 BCE), both of whom are known to have been pupils of Aristarchus (Ἀπολλόδωρος καὶ Ἀρητάδης καὶ Νεοτέλης καὶ Διονύσιος ὁ Θρᾷξ).[77] This means that Neoteles wrote prior to Didymus, who lived in the first century BCE. In fact, August Blau argued that Neoteles and Aretades were both pupils of Aristarchus.[78] Martin West is more cautious, stating that the 'juxtaposition may perhaps suggest that Aretades and Neoteles were also Aristarcheans of the later second century'.[79]

Blau's suggestion was accepted by Susemihl[80] and by the editors of the *Prosopographica Ptolemaica*, where Neoteles is listed as a 'disciple d'Aristarque', based in Alexandria, and given a date in the second half of the second century BCE.[81] This is the only entry for any person named Neoteles in the whole of the *Prosopographica Ptolemaica*, and so it is not possible to link the individual to any other figure in the Ptolemaic establishment.

The name Neoteles is rare. This Neoteles (a grammarian) is the only entry under that name in Pauly-Wissowa.[82] In the *Lexicon of Greek Personal Names*, the name Neoteles occurs only in two cities. Neoteles son of Eualkides occurs on one side of a military catalogue from Eretria dating to the turn of the fourth and third centuries BCE, and on the other side an individual named Eualkides son of Neoteles is listed.[83] Both are from the deme Styra, and we are probably dealing with a father and son.[84] One Agathon son of Neoteles and his brothers are recorded as being *proxenoi* of the Thourians at Delphi in an inscription dating to the fourth century BCE.[85] In the *Lexicon of Greek Personal Names*, the inscription is given a date of after 373 BCE.[86] It is possible, therefore, that the family of Neoteles, or Neoteles himself, came originally from one of these two cities. If Neoteles was, in fact, a pupil of Aristarchus, then it is not known if Neoteles moved from the city, like his teacher, on the accession of Ptolemy VIII Euergetes II.

Given that Lorimer, who only partially accepted Blau's suggestion that Neoteles was 'possibly' a pupil of Aristarchus,[87] was able to devote some 11 pages (289–300) to the subject of 'Bow in Homer' in her *Homer and the Monuments*, we should not doubt that an erudite Hellenistic scholar was able to write a whole book on the subject. The very title of the work of Neoteles makes it certain that it took the form of a Homeric commentary. In any case, it is extremely doubtful that it would have borne any resemblance to an instructional manual on archery.[88] If Neoteles was, in fact, the author of a commentary on archery in Homer, but, as we have seen, this has been disputed, then it is perfectly possible that he was also the author of a περὶ τῆς καθ' Ὅμηρον τακτικῆς, and hence his inclusion in this article.

VIII. Frontinus

The most authentic recension of Aelian is the *Codex Laurentianus* 55.4 in the Biblioteca Medicea-Laurenziana in Florence. The new critical edition of Kai Brodersen, which is based on this manuscript,[89] has done a great service to scholarship, not least in confirming that the name Frontinus, and not Fronto, should be read among the list of writers of Homeric *Taktika*.

Previous academic debate on the identity of this individual has been based on the edition by Francesco Robortello of the defective text of the *Codex Venetus Marcianus* 516, on which the text and English translation recently published by Christopher Matthew is based.[90] Instead of the Greek Φροντίνῳ, this manuscript in this place reads Φρόντωνι, which should be rendered as 'Fronto' not 'Frontinus'.

Köchly and Rüstow in their 1855 edition stuck to the textual reading of Φρόντωνι, and identified the individual called Fronto with one Ti. Catius Caesius Fronto, attested as *consul suffectus* in 96 CE.[91] This identification was not accepted by Richard Förster, who felt that the attribution of a work dealing with military matters to Ti. Catius Caesius Fronto would be inappropriate, for Fronto as 'Ein Kriegsschriftsteller ist nicht bekannt', and the work could be more appropriately attributed to Frontinus.[92] W. A. Oldfather's and the Illinois Greek Club Loeb Classical Library edition retained the Fronto of the text in their translation of this same passage of Aelian in the *testimonia* of Aeneas Tacticus gathered in the Loeb translation of that author.[93] Dain thought that the Φρόντωνι of the text was a scribal error for Φροντίνῳ.[94] A. M. Devine in his 1989 English translation followed Dain, noting that the manuscript reading was Fronto, but retaining Frontinus in his translation.[95] The reading of Fronto and the identification with Ti. Catius Caesius Fronto was accepted by Luigi Loreto.[96] The 2012 translation of this passage offered by Christopher Matthew retains the Frontinus of John Bingham's translation and omits the author Hermeias from his translation (but includes him, incorrigibly, in the text offered) again copying the translation of Bingham on this point.[97] The new critical edition of Brodersen has finally cleared this problem up.

Frontinus was the author, as well as a book of the *Strategemata* which has survived, of at least one other earlier book upon the military art, which he refers to (without a precise reference to its title) in the beginning of Book 1 of the *Strategemata*. It is presumably this same work to which Vegetius refers at 1.8 of his *Epitome* (*Epitoma Rei Militaris*), which implies that the work is an epitome, and at 2.3 which states it deals with the military science. It is presumably this lost work that is again referred to by Johannes Lydus (*de Magistratibus* 1.47), who renders its title in Greek as περὶ στρατηγίας. Wheeler has emphasised the difference between works bearing this title and *Strategemata*;[98] accordingly, this should be a different work from the *Strategemata*. According to the editor of the text of the *de Magistratibus* of Lydus, Anasasius C. Bandy, this should correspond to a Latin title *de officio militari*.[99] Presumably, the work of Frontinus on *The Tactics Found in Homer* (περὶ τῆς καθ' Ὅμηρον τακτικῆς) was of a completely different nature to his *de officio militari* and was a separate work.[100]

In Section 2 earlier, when writing on 'The Nature of the Homeric *Tactica*', I discussed suggested a passage in Frontinus' *Stratagems* (2.3.21), which has Pyrrhus at the Battle of Asculum in 279 BCE stationing his forces in accordance with the advice of Homer *Iliad* 4.297–300. Whilst denying that the nature of Homeric *Tactica* were collections of Homeric quotations, I accept that this passage may have found a parallel in his work on *The Tactics Found in Homer*. It may, in fact, have inspired it.

IX. Conclusion

As well as the authors named in Aelian – Stratokles, Hermeias, and Frontinus – from whose works no fragments have survived, I believe that the fragment of Hermolytos, who is described as ὁ τακτικός, comes from a commentary of this very type, on the description of *synaspismos* contained in the *Iliad* 13.130–34. I also believe, although with less certainty, that the work of Neoteles entitled *Archery among the Heroes* (περὶ τῆς κατὰ τοὺς ἥρωας τοξείας) could possibly be another work of this type.

The demand for Homeric commentaries was at its height around the second quarter of the second century BCE, when Aristarchus of Samothrace held the post of head of the Museion and Library at Alexandria. I suggest that it was in his wake that most of these commentaries were written. Hermolytos, unfortunately, cannot be further identified, and Hermeias is nothing more than a name, but it has been plausibly suggested that Neoteles was an Aristarchean of the later second century, and if my identification of the Stratokles named in Aelian with Stratokles of Rhodes is accepted, then this Stratokles is also an Aristarchean of the later second century.

Frontinus comes much later, when, as I have argued, the times for Homeric commentaries were well passed. He is best known an author of works of a practical nature. Aelian tells us at the beginning of his *Tactics* that the work arose out of a conversation he had with the Emperor Nerva at Frontinus' house at Formiae. We can but imagine that he was encouraged to try his hand at an obsolete genre after contact with such figures.

In this work, I have tried to gather together all the evidence relevant to a genre of military writing which we call 'The Homeric *Taktika*'. To my knowledge, the only person who has advanced any opinion on the nature of this subgenre of ancient military writing so far is Wheeler, who has suggested that they constituted 'military instruction through Homeric quotation'.[101] Although quotation from Homer would have made up a considerable portion of the texts of these *Taktika*, I have rather suggested that they consisted of commentaries on passages in Homer which dealt with the tactics used by the Heroes.

Notes

1 This chapter includes ideas that I intend to explore further elsewhere. Without the help of my esteemed colleague (and friend) in Gdansk, Bogdan Burliga, the writing of this article would not have been possible. Bogdan's help has not been confined to translating the ancient Greek that had baffled me at many points, or supplying me with articles otherwise unobtainable. His profound knowledge of the literature cited in this article, especially the Greek *Tactica*, has acted as a foil to my less plausible suggestions. Philip Rance has also been of great help, saving me from many mistakes, and guiding me through the world of Homeric Commentaries. I would also like to thank Paul Holder for his help in updating and adding to the materials in this article pertinent to Ti. Catius Caesius Fronto. Translations from the *Iliad* come from Richmond Lattimore's edition.
2 Stadter (1978) 117–128.

3 Dain (1946) 34–40 schematised in the diagram (39). On the relationship between Asclepiodotus, Aelian, Arrian, and Poseidonius, see Rance (2017) 18–19, esp. n.28.

4 πρῶτος μὲν ὧν ἴσμεν δοκεῖ τὴν τακτικὴν θεωρίαν Ὅμηρος ἐπεγνωκέναι θαυμάζειν τε τοὺς ἐπιστήμονας αὐτῆς, ὥσπερ Μενεσθέα· τῷ δ᾽οὔπω τις ὁμοῖος ἐπιχθόνιος γένετ᾽ ἀνὴρ κοσμῆσαι ἵππους τε καὶ ἀνέρας ἀσπιδιώτας. καὶ περὶ τῆς καθ᾽ Ὅμηρον τακτικῆς ἐ<νε>τύχομεν συγγραφεῦσι Στρατοκλεῖ καὶ Ἑρμείᾳ καὶ Φροντίνῳ τῷ καθ᾽ ἡμᾶς ἀνδρὶ ὑπατικῷ. ἐξειργάσαντο δὲ τὴν θεωρίαν Αἰνε<ί>ας τε διὰ πλειόνων <ὁ> καὶ στρατηγικὰ βιβλία ἱκανὰ συνταξάμενος. The Greek text follows that of the new critical edition of Brodersen (2017). The section of text quoted is *Iliad* 2.555–557.

5 Parke (1967) 9; cf. Long (1996) 61.

6 As does, for example, Hornblower (1981) 208 when she writes of Aeneas as 'the first of the Greeks after Homer to be interested in military science'.

7 κρῖν᾽ ἄνδρας κατὰ φῦλα, κατὰ φρήτρας. On this passage, see Hornblower (2007) 29.

8 Apart from the preface of Aelian (1.2), we have the evidence of Plutarch's *Life of Pyrrhus,* which at 8.2 mentions that Pyrrhus left writings (emphasis on the plural) on tactics and generalship (περὶ τάξεις καὶ στρατηγίας). Cicero (*Letters to Friends* 9.25.1) confirms that Pyrrhus wrote 'books' in the plural, without supplying their titles. See Chapter 9.

9 ἴστε γὰρ δήπου ὅτι Ὅμηρος ὁ σοφώτατος πεποίηκε σχεδὸν περὶ πάντων τῶν ἀνθρπίνων . . . ἢ οἰκονομικὸς ἢ δημηγορικὸς ἢ στρατηγικός. See Millett (2013) 59–60.

10 ὁ δὲ θεῖος Ὅμηρος / ἀπὸ τοῦ τιμὴν καὶ κλέος ἔσχεν πλὴν τοῦδ᾽, ὅτι χρήστ᾽ ἐδίδαξεν, / τάξεις, ἀρετάς, ὁπλίσεις ἀνδρῶν; The translation is that found in Sommerstein (1996).

11 Sommerstein (1996) 247–248.

12 The text and translation come from Newman's (1902) Oxford edition. Compare Barker's translation: 'Infantry is useless without a system of tactics; and as the experience and the rules required for such a system did not exist in early times, the strength of armies lay in their cavalry'.

13 Hornblower (1981) 208.

14 Whitehead (2008) 143.

15 Whitehead (1990) 8–9.

16 Wheeler (2016) 581 n.35.

17 ἱππῆας μὲν πρῶτα σὺν ἵπποισιν καὶ ὄχεσφι, / πεζοὺς δ᾽ ἐξόπιθε στῆσεν πολέας τε καὶ ἐσθλούς, / ἕρκος ἔμεν πολέμοιο κακοὺς δ᾽ ἐς μέσσον ἔλασσεν, / ὄφρα καὶ οὐκ ἐθέλων τις ἀναγκαίῃ πολεμίζοι.

18 Lendon (2005) 299. This is not the only occasion in which Ammianus compares Julian's campaign to the Pyrrhic wars; elsewhere, Ammianus (24.4.24) explicitly compares Julian's attack on Maozamalcha to the attack of C. Fabricius Luscinus on a Lucanian camp, Lendon notes (297). So it is possible that the Homeric (or Pyrrhic) inspiration of Julian comes more from the pen of Ammianus than from the learning of Julian.

19 Noted by Burliga (2016) 147. n. 19838–19839 n. 3.

20 On Arrian's use of Homer, see Burliga (forthcoming).

21 Peremans and Van't Dack (1968) 213 no. 16512.

22 Pfeiffer (1968) 210.

23 Fraser (1972) Vol. I, 462.

24 Fraser (1972) Vol. I, 333.

25 Eichgrün (1961) 18–21.

26 *FGrH* 234 F 2; cf. Fraser (1972) Vol. I, 462.

27 Fraser (1972) Vol. I, 462.

28 Fraser (1972) Vol. I, 332.

29 Pfeiffer (1968) 211.

30 Pfeiffer (1968) 211 n.6 and Fraser (1972) Vol IIb, 667 n.133.

31 Grenfall and Hunt (1914) 102, no. 1241.

32 ν[ι]ος Σιλλεως Αλεξανδρευς
[κ]αλοθμενος Ροδιος Καλ
λ.[ι]μαχου γνωριμος· ουτος
εγενετο και διδασκαλος του
πρωτου βασιλεως· τουτον
δ[ι]εδεξατο Ερατοσθενης
μεθ ον Αριστοφανες Απελ
λου Βυζαντιος και Αρισταρ
χος· ειτ Απολλωνιος Αλεξαν
δρευς ο ϊδογραφος καλουμε
νος· μεθ ον Αρισταρχος Αρι
σταρχου Αλεξανρευς ανω
θεν δε Σαμοθραξ· ουτος και
διδ[α]σκαλος [ε]γενε[το] των
του Φιλοπατορος τεκνων·
μεθ ον Κυδας εκ των λογχο
φ[ο]ρωμν· επι δε τωι ενατω
[βα]σιλει ηκμασαν Αμμω
[νι]ος και Ζηνο[δοτος] και Διο
[κλ.]ης και Απολλο[δ]ωρος γραμ
[μα]τικοι.

33 The editors' note that the πρωτου of the text is an obvious mistake for the third Ptolemaic king, i.e. Euergetes I.

34 The editors' note that this first mention of Aristarchus in the list 'is doubtless an interpolation, since Aristarchus recurs with a full description in *ll*. 11–15'; cf. Fraser (1972) I, 332.

35 As noted by the editors (108), the Philopator which is in the text (and underlined in this translation), by which Ptolemy IV Philopator (222–205 BCE) presumably is meant, is probably a mistake for Epiphanes (the father of both Ptolemy VI Philometor and Ptolemy VIII Euergetes II), or Ptolemy VI Philometor. Aristarchus would have been born around the year 216. If he died at the age of 72 in Cyprus around 144, it is highly probable that the Philopator of the script is a mistake for Philometor rather than Epiphanes.

36 Fraser (1972) IIa, 477 n.126–127.

37 He is thus termed in the *Scholia* to the *Cod. Vindob.* at Γ 125: see Bekker (1825) 102.

38 Pfeiffer (1968) 212.

39 On the Homeric output of Aristarchos, see Pfeiffer (1968) 212–219.

40 Cf. 'Aristarchos' (*RE* 22).

41 φράξαντες δόρθ δουρί, σάκος σάκεϊ προθελύμνῳ / ἀσπὶς ἄρ' ἀσπίδ' ἔρειδε, κόρυς κόρυν, ἀνέρα δ' ἀνήρ / ψαῦον δ' ἱππόκομοι κόρυθες λαμπροῖσι φάλοισι / νευόντων, ὡς πυκνοὶ ἐφέστασαν ἀλλήλουσιν (13.130–134).

42 Van der Valk (1979) 449, lines 2–5.

43 Ἑρμόλυτος δὲ ὁ τακτικός φησιν, ὅτι Λυκοῦργος μὲν ἐνομοθέτησε Λακεδαιμονίοις ὕστερον τὸν τοιοῦτον συνασπισμὸν, Λύσανδρος δὲ ὁ Λάκων ἐν ἔργοις αὐτὸν ἐδίδαξε, καθὰ καὶ Ἐπαμινώνδας Θηβαίους καὶ Χαρίδημος Ἀρκάδας τε καὶ Μακεδόνας.

44 Erbse (1974) 425.

45 τοῦτον δὲ τὸν συνασπισμὸν, ὡς ὁ τακτικὸς Ἑρμόλυτος λέγε, ἐνομοθέτησε Λυκοῦργος, ἐδίδαξε Λύσανδρος Λάκων καὶ Ἐπαμινώνδας, εἶτα ὑπὸ Χαριδήμου Ἀρκάδες ἐδιδάχθησαν καὶ Μακεδόνες.

46 For more information on this subject, see Sekunda (2014) 139–141.

47 Peek (1955) no. 1804; Laodikeia on Lykos, first century BCE; *SEG* 24 no. 1197, which dates to Egypt in the fourth century CE.

48 *Palatine Anthology* 7.654, an epigram of Leonidas of Tarentum, who dates to the middle of the third century BCE.

49 See *LGPN* 3.A 204 (Aigina, Akarnania, and Epidauros).

50 'Stratokles' (*RE* 7).

51 Loreto (1995) 568.

52 Polyaenus 4.2; Aeschines, *Ctesias* (3) 143; *Prosopographia Ptolemaica* 12931; 'Stratokles' (*RE* 3); Develin (1989) 343.

53 Peremans and Van't Dack (1952) 74 no. 2482.

54 The name is formed from two particles meaning 'army' and 'glory', thus Pape and Benseler (1884) 1446: 'heere berühmt'. At Rhodes the personal name alternates within families over the generations with other personal names with military connotations such as Straton or Stratarchos.

55 *LGPN* I 414, where 32 examples of the name are listed at Rhodes. Our Stratokles the Rhodian philosopher is listed as no. 20 and given a date of the second half of the second century BCE. The name is so common at Rhodes that there is no particular reason to associate this Stratokles with no. 19, a magistrate whom Polybius (27.7.2, 13) mentions under affairs for 171 BCE.

56 'Stratokles' (*RE* 8); Mygind (1999) 258 no. 13.

57 Traversa (1952) 29–30.

58 Pohlenz (1959) 193 makes a bald suggestion which is further developed in the Italian translation of Vittorio Enzo Alfieri (1967) xxix, 392–393 n. 9.

59 Fraser (1972) Vol. I, 469–470 and IIb, 678, ns. 207–208; Pfeiffer (1968) 266.

60 Pfeiffer (1968) 232.

61 See Inwood (2003) 1104. The life of Panaitios is treated in more detail in Vimercati (2002) 14–41.

62 Dorandi (1994) 128–129 col.79.3 and Mygind (1999) 258 no. 13.

63 Oldfather in the Loeb Classical Library edition of Aeneas, Asclepiodotus, and Onasander, 231; cf. Sekunda (2006) 129–130.

64 Burliga (2013) *passim.*

65 'Hermeias' (*RE* 8).

66 Loreto (1995) 568.

67 'Hermeias' (*RE* 1); Schmitt (1964) 150–158.

68 Peremans and Van't Dack (1968) 157 no. 15859.

69 Roussel and Launey (1937) no. 2135, 3.

70 *LGPN* I 431 no. 22.

71 βῆ δ' ἰθὺς Τεύκρου, βαλέειν δέ ἑ θυμὸς ἀνώγει. / ἤτοι ὁ μὲν φαρέτρης ἐξείλετο πικρὸν ὀϊστόν, / θῆκε δ' ἐπὶ νευρῇ τὸν δ' αὖ κορυθαίολος Ἕκτωρ / αὐερύοντα παρ' ὤμον, ὅθι κληῒς ἀποέργει / αὐχένα τε στῆθός τε, μάλιστα δὲ καίριόν ἐστι, / τῇ ῥ' ἐπὶ οἷ μεμαῶτα βάλεν λίθῳ ὀκριόεντι, / ῥῆξε δέ οἱ νευρήν νάρκησε δὲ χεὶρ ἐπὶ καρπῷ, / στῆ δὲ γνὺξ ἐριπών, τόξον δέ οἱ ἔκπεσε χειρός. (8.322–329)

72 MacPhail (2011) 142–143.

73 Erbse (1971) 359.

74 Erbse (1960) 31.

75 Van der Valk (1976) 588–589.

76 'Neoteles' in *Der Kleine Pauly*.

77 Schmidt (1854) 177 and Erbse (1977) 540.

78 Blau (1883) 78.

79 West (2001) 80.

80 Susemihl (1892) 168 n.136.

81 Peremans and Van't Dack (1968) 223 no. 16875. P. 80 is more cautious.

82 'Neoteles' (*RE*).

83 *IG* XII, 9 no. 246 A, 263; B, 37.

84 *LGPN* 1 p. 326.

85 Bousquet (1988) 20–21.

86 *LGPN* 3.B 297.

87 Lorimer (1950) 297 n. 1.

88 Such as MacLeod (1962) 10–11.
89 Brodersen (2017) 22–24.
90 Matthew (2012) x and Wheeler (2016).
91 Köchly and Rüstow (1853–1855) 90–93. From the text of Köchly (93 n.213), it seems
   that the identification had already been made by Borghesi.
92 Förster (1877) 446–448.
93 Oldfather (1928) 201.
94 Dain (1946) 19 n.3.
95 Devine (1989) 44.
96 Loreto (1995) 568.
97 Matthew (2012) 9, on which see Wheeler (2016). On Bingham, see also Chapter 3 (55).
98 Wheeler (1988) 2.
99 Bandy (1983) 74–75.
100 On this question, see Burliga (2016) 38–39 n.37.
101 Wheeler (2016) 581 n.35.

Abbreviations and bibliography

Abbreviations

FGrHist   Jacoby, F., et al., eds. 1923-. *Die Fragmente der griechischen Historiker.* Leiden: Brill.
IG   1924-. *Inscriptiones Graecae.* Berlin: De Gruyer.
LGPN   Fraser, P. M. and Matthew, E., eds. 1987-. *Lexicon of Greek Personal Names.* Oxford: Oxford University Press.
OCD   Hornblower, S. and Spawforth, A., eds. 2003. *Oxford Classical Dictionary.* Third revised edition. Oxford: Oxford University Press.
RE   Pauly, A., ed. 1894-. *Realencyclopädie der classichen Altertumwissenschaft.* Stuttgart: J. B. Metzler.
SEG   Chaniotis, A., et al., eds. 1923-. *Supplementum Epigraphicum Graecum.* Leiden: Brill.

Bibliography

Adler, A. 1928. *Svidae Lexicon I.* Leipzig: B. G. Teubner.
Alesse, F. 1997. *Panezio di Rodi.* Rome: Bibliopolis.
Alfieri, E. 1967. *La Stoa: storia di un movimento spirituale, Max Pohlenz; presentazione di Vittorio Enzo Alfieri; testo . . . tradotto da Ottone De Gregorio,* Firenze: La Nuova Italia, Vol. 2.
Bandy, A. C. 1983. *Ioannes Lydus on Powers, or the Magistracies of the Roman State.* Philadelphia: American Philosophical Society.
Barker, E. 1946. *The Politics of Aristotle.* Oxford: Oxford University Press.
Bekker, E., ed. 1825. *Scholia in Homeri Iliadem.* Berlin: G. Reimer.
Bingham, J. 1629 [1616]. *The Tactiks of Aelian: Or Art of Embattailing an Army after ye Grecian Manner.* London. = John Bingham, *The Art of Embattailing an Army of the Second Part of Aelians Tacktiks,* Facsimile reprint of the edition London 1631 With an introduction (in German) by Werner Hahlweg, (*Bibiotheca Rerum Militarium, Quellen und Darstellung zur Militärwissenschaft und Militärgeschichte, Herausgeben unter*

Mitwerkung des Militägeschichlichen Forschungsamtes durch W. von Groote und U. von Grsdorff, IV Bingham, Aelians Takticks, Biblio Verlag Osnabrück, 1968.

Blau, A. 1883. *De Aristarchi discipulis*. Diss. Jena.

Bousquet, J. 1988. *Études sur les Comptes de Delphes*. Paris: École française d'Athènes.

Bowen, A. J. 1998. *Xenophon: Symposium*. Warminster: Aris and Phillips.

Brodersen, K. 2017. *Ailianos: Antike Taktiken/ Taktika*. Wiesbaden: Marix.

Burliga, B. 2013. *Arrian's Anabasis: An Intellectual and Cultural Story*. Gdansk: Foundation for the Development of the University of Gdansk.

——— 2016. *Frontyn, Postępy Wojenne*. Warsaw: *wstępem i przepisami*.

——— forthcoming. 'Homerus Tacticus: Arrian's Quotations from the *Iliad* in the "Hellenistic" Section of His *Tactics*'. 31. 5–6.

Cary, E., trans. 1925. *Dio's Roman History*. Cambridge, MA: Harvard University Press.

Cockle, M. J. D. 1900. *A Bibliography of Military Books Up to 1642*. London: Holland Press.

Dain, A. 1946. *Histoire du texte d'Élien le tacticien des origins á la fin du Moyen Age*. Paris: Les Belles Lettres.

Delebecque, É. 1957. *Essai sur la Vie de Xénophon*. Paris: Klincksieck.

Develin, R. 1989. *Athenian Officials 684–321 B.C.* Cambridge: Cambridge University Press.

Devine, A. M. 1989. 'Aelian's *Manual of Hellenistic Military Tactics*: A New Translation from the Greek with an Introduction'. *AW* 19. 31–64.

Dillon, H. A. 1814. *The Tactics of Aelian, Comprising the Military System of the Grecians*. London: Cox and Baylis.

Dorandi, T. 1994. *Storia dei filosofi: la Stoà da Zenone a Panezio (PHerc. 1018)*. Leiden: Brill.

Eck, W. and Pangeri, A. 2005. 'Neue Konsulndaten in Neuen Diplomen'. *ZPE* 152. 229–262.

Eichgrün, E. 1961. *Kallimachos und Apollonios Rhodios*. Diss. Berlin.

Erbse, H. 1960. *Beiträge zur Überlieferung der Iliasscholien*. Munich: C. H. Beck.

——— 1971. *Scholia Graeca in Homeri Iliadem (scholia vetera)*. Vol. 2. Berlin: De Gruyter.

——— 1974. *Scholia Graeca in Homeri Iliadem (scholia vetera)*. Vol. 3. Berlin: De Gruyter.

——— 1977. *Scholia Graeca in Homeri Iliadem (scholia vetera)*. Vol. 5. Berlin: De Gruyter.

Förster, R. 1877. 'Studien zu den griechischen Taktiker'. *Hermes* 12. 426–471.

Fraser, P. M. 1972. *Ptolemaic Alexandria*. 3 vols. Oxford: Oxford University Press.

Gow, A. S. F. and Page, D. L. 1965. *The Greek Anthology: Hellenistic Epigrams, Volume II: Commentary and Indexes*. Cambridge: Cambridge University Press.

Grenfall, B. P. and Hunt, A. S., eds. 1914. *The Oxyrhynchus Papyri, Part X*. London: Egypt Exploration Society.

Groag, E. and Stein, A. 1936. *Prosopographia Imperii Romani Saec. I. II. III Pars II*. Berlin: De Gruyter.

Higgins, W. E. 1977. *Xenophon the Athenian: The Problem of the Individual and the Society of the Polis*. Albany, NY: State University of New York Press.

Hornblower, J. 1981. *Hieronymus of Cardia*. Oxford: Oxford University Press.

Hornblower, S. 2007. 'Warfare in Ancient Literature: The Paradox of War', in P. Sabin, H. van Wees, and M. Whitby, eds., *The Cambridge History of Greek and Roman Warfare*. Vol. 1. Cambridge: Cambridge University Press. 22–53.

Huß, B. 1999. *Xenophons Symposion Ein Kommentar*. Stuttgart: B. G. Teubner.

Inwood, B. 2003. 'Panaitios', in S. Hornblower and A. Spawforth, eds., *The Oxford Classical Dictionary*, Third revised edition. Oxford: Oxford University Press. 1104.

Köchly, H. and Rüstow, W. 1853–1855. *Griechische Kriegsschriftsteller: Griechisch und deutsch, mit kritischen und erklärenden Anmerkungen band 2.1: Asklepiodotos Taktik, Aelianus Theorie der Taktik*. Leipzig: Engelmann.

Lattimore, R. 1951. *The Iliad of Homer*. Chicago: University of Chicago Press.

Launey, M. 1949–50. *Recherches sur les Armées hellénistiques (Bibliothèque des Écoles françaises d'Athènes et de Rome fasc. 169)*. Vols. 1–2. Paris: de Boccard.

Lendon, J. E. 2005. *Soldiers and Ghosts: A History of Battle in Classical Antiquity*. New Haven: Yale University Press.

Lindsay, W. M. 1903. *M. Val. Martialis Epigrammata*. Oxford: Oxford University Press.

Long, A. A. 1996. *Stoic Studies*. Cambridge: Cambridge University Press.

Loreto, L. 1995. 'Il generale e la biblioteca. La trattatistica militare greca da Democrito di Abdera ad Alessio I Comneno', in G. Camiano, L. Canfora, and D. Lanza, eds., *La spazio letterario della Grecia antica*, II. *La ricezione e l'attualizzazione del testo*. Rome: Salerno Editrice. 563–589.

Lorimer, H. 1950. *Homer and the Monuments*. London: Macmillan.

MacLeod, W. E. 1962. 'An Ancient Treatise of Military Archery'. *Journal of the Society of Archer Antiquaries* 5. 10–11.

MacPhail, J. A., Jr. 2011. *Porphyry's Homeric Questions on the Iliad*. Berlin: De Gruyter.

Matthew, C. 2012. *The Tactics of Aelian or on the Military Arrangements of the Greeks: A New Translation of the Manual That Influenced Warfare for Fifteen Centuries*. Barnsley: Pen and Sword.

Millett, P. C. 2013. 'Greece: Winning Ways in Warfare', in B. Campbell and L. A. Tritle, eds., *The Oxford Handbook of Warfare in the Classical World*. Oxford: Oxford University Press. 46–73.

Mygind, B. 1999. 'Intellectuals in Rhodes', in V. Gabrielsen, P. Bilde, T. Engberg-Pedersen, L. Hannestad, and J. Zahle, eds., *Hellenistic Rhodes: Politics, Culture, and Society*. Aarhus: Aarhus Universitetsforlag. 247–293.

Newman, W. L. 1902. *The Politics of Aristotle*. Vol. 4. Oxford: Oxford University Press.

Oldfather, W. A., ed. 1928. *Aeneas Tacticus, Asclepiodotus, Onasander*. Cambridge, MA: Harvard University Press.

Pape, W. and Benseler, G. E. 1884. *Wörterbuch der griechischen Eigennamen*. Third edition. Brunswick.

Parke, H. W. 1967. *Greek Oracles*. London: Hutchinson.

Peek, W. 1955. *Griechische Vers-Inschriften: Grab-Epigramme*. Berlin: Akademie-Verlag.

Peremans, W. and Van't Dack, E. 1952. *Prosopographica Ptolemaica II: L'Armée de Terre et la Police nos. 1825–4983*. Louvain: E. Nauwelaerts.

—— 1968. *Prosopographica Ptolemaica VI: La Cour, les relarions internationals et les possessions extérieures, la Vie culturelle nos. 14479–17250*. Louvain: E. Nauwelaerts.

Pfeiffer, R. 1968. *History of the Classical Scholarship: From the Beginning to the End of the Hellenistic Age*. Oxford: Oxford University Press.

Pohlenz, M. 1959. *Die Stoa. Geschichte einer geistigen Bewegung*. Vol. 1. Göttingen: Vandenhoeck and Ruprecht.

—— 1967. *La Stoa. Storia di un movimento spirituale*. Vol. 2. Florence: La Nuova Italia.

Rance, P. 2017. 'Introduction', in *idem* and N. V. Sekunda, eds., *Greek Taktika: Ancient Military Writing and Its Heritage: Proceedings of the International Conference on Greek Taktika Held at the University of Toruń, 7–11 April 2005*. Gdansk: Akanthina. 9–69.

Roussel, P. and Launey, M. 1937. *Inscriptions de Délos*. Paris: Librairie Ancienne Honoré Champion.

Schmidt, M. 1964 [1854]. *Didymi Chalcenteri Gramaticii Alexandrini Fragmenta quae supersunt omni*. Leipzig: Teubner.

Schmitt, H. H. 1964. *Untersuchungen zur Geschichte Antiochos' des Grossen*. Stuttgart: Franz Steiner.

Sedley, D. 2003. 'The School, from Zeno to Arius Didymus', in B. Inwood, ed., *The Cambridge Companion to the Stoics*. Cambridge: Cambridge University Press. 7–32.

Sekunda, N. V. 2006. *Hellenistic Infantry Reform in the 160's BC*. Gdansk: Oficyna Naukowa.

——— 2014. 'The Chronology of the Iphicratean Peltast Reform', in *idem* and B. Burliga, eds., *Iphicrates, Peltasts and Lechaion*. Gdansk: Foundation for the Development of Gdansk University for the Department of Mediterranean Archeology. 126–144.

Shackleton-Bailey, D. R. 1993. *Martial, Epigrams*. Vol. 3. Cambridge, MA: Harvard University Press.

Sherwin-White, A. N. 1966. *The Letters of Pliny: A Historical and Social Commentary*. Oxford: Oxford University Press.

Sommerstein, A. H. 1996. *Aristophanes: Frogs*. Warminster: Aris and Phillips.

Stadter, P. A. 1978. 'The *Ars Tactica* of Arrian: Tradition and Originality'. *CPh* 73. 117–128.

Susemihl, F. 1892. *Geschichte der griechischen Litteratur in der Alexandrinerzeit, Zweiter Band*. Leipzig: B. G. Teubner.

Thompson, D. J. 2003. 'Ptolemy VI Philometor', in S. Hornblower and A. Spawforth, eds., *The Oxford Classical Dictionary*. Third revised edition. Oxford: Oxford University Press. 1272–1273.

Traversa, A. 1952. *Index Stoicorum Herculanensis*. Genoa: University of Genoa.

Van der Valk, M. 1976. *Eustathii archiepiscopi Thessalonicensis Commentarii ad Homeri Iliadem pertinentes*. Vols. 2 and 3. Leiden: Brill.

Vidman, L. 1982. *Fasti Ostienses. Edendos, illustrandos, restituendos curavit Ladislav Vidman*. Prague: Academia.

Vimercati, E. 2002. *Il Mediostoicismo di Panezio*. Milan: Vita e Pensiero.

West, M. L. 2001. *Studies in the Text and Transmission of the Iliad*. Leipzig: De Gruyter.

Will, É. 2003. *Histoire politique du monde hellénistique (323–30 av. J.-C)*. Paris: Points.

Wheeler, E. L. 1988. *Stratagem and the Vocabulary of Military Trickery*. Leiden: Brill.

——— 2016. 'Aelianus Tacticus: A Phalanx of Problems'. *JRA* 29. 575–583.

Whitehead, D. 1990. *Aineias the Tactician: How to Survive under a Siege*. Oxford: Oxford University Press.

——— 2008. 'Fact and Fantasy in Greek Military Writers'. *AAHung* 48. 139–155.

5 Aeneas Tacticus, Philon of Byzantium, Onasander and the good siege

A case-study of Demetrius at Rhodes[1]

Graham Wrightson

Writing manuals was a common practice of Greek philosophers from the end of the fifth century onwards.[2] Unfortunately, few of those with a military focus survive. Of the four that survive from the Hellenistic period (323–146 BCE), two deal with siege warfare, those of Aeneas Tacticus and Philon of Byzantium. None of the Hellenistic manuals discussing battle tactics rather than sieges survive, even the treatise of the famous general Pyrrhus.[3] The other surviving military treatises on sieges – those by, for instance, Athenaeus Mechanicus, Biton, Heron, and Apollodorus – are much later in date. These works focused on siege machines, with the exception of Apollodorus, and date to the Roman period. It seems that siege machinery at least was considered a different genre – or subgenre – of military writing, and that siegecraft was separate from battle tactics. Onasander's treatise, the sole surviving handbook of all generalship, includes sieges as a separate section from battles, which perhaps is an acknowledgement of its status as a different genre (or 'subgenre').[4]

Whereas military manuals tend to focus on providing generalised guidance, David Whitehead's commentary on Philon's work inserts references to historical sieges as *exempla* of the advice given, including two of the most historically famous in the fourth century: Tyre by Alexander the Great (332 BCE) and Rhodes by Demetrius I Poliorcetes (305–304 BCE), the former a successful, the latter an unsuccessful siege.[5] The former was the most successful of all Alexander's sieges, and the latter was the largest siege in terms of manpower expended by the attack to no avail (that is, it is the largest-scale failure). Heretofore no historian has examined offensive, rather than the defensive, generalship in an entire siege through the lens of a siege manual. This is the case for both Aeneas and Philon, the authors whose siege-manuals survive.[6] The advice given to defenders in these manuals can also in theory inform generals how to *besiege* an enemy city, turning the various tactics of survival as a new avenue for analysis of ancient sieges and generalship. The presumption of the manual writers is that following their advice will lead to a successful conclusion, and the discussion that follows adopts the same methodology here for determining 'the good siege'. The focus here will be on Demetrius' siege of Rhodes, the main surviving source on which is Diodorus Siculus in his *Historical Library* (*Bibliotheke*), where the narrative extends across two not insubstantial sections of the text (20.82.1–88.9 and 92.1–100.6).[7] The expansive nature of this narrative means that one can dissect fully (as much as any ancient battle can be) this event.[8]

This has a potential impact for the wider consideration of military texts. While the surviving siege-specific texts articulate defensive actions that one ought to undertake, there is no explicit prohibition from the author on a reader inferring the appropriate offensive actions to counter the author's prescriptions. That is to say, the reader-general or reader-soldier can choose to utilise the text in a manner that is *not* the author's original intent. There is surely an irony in the fact that in so doing, the reader will prove the utility of said text, counteracting in the field each practice as theorised in the text.[9] The fact that Demetrius' siege was unsuccessful adds a layer of complexity to this approach: the reader must navigate both reading against the grain of the military manual and infer from a historical failure the correct counter-practice to undertake a successful siege.[10]

I

Aeneas' treatise περὶ τοῦ πῶς χρὴ πολιορκουμένους ἀντέχειν (*On the Defence of Fortified Positions* or *How to Survive under Siege*) is the earliest military manual that deals with commanding an army rather than a specific unit (as in the case of Xenophon's *Cavalry Commander*, for example), and the first to deal specifically with sieges.[11] Aeneas' work came from his own experiences and is full of historical examples 'illustrating where the course of action in question had been either adopted with good results or ignored with bad'.[12] He clearly intended his work to serve as a guide for commanders,[13] as Xenophon's were and as Brian Campbell argues for the whole genre of military writing in the Roman period.[14]

In writing his treatise of siegecraft in the century after Aeneas, Philon of Byzantium uses his predecessor's work to describe the means of surviving a siege.[15] Philon adds advice to an attacker on how to conduct successfully a siege. Philon did not have military experience and may have written for, and dedicated his work to, an 'armchair' general.[16] His work is valuable to historians of the Hellenistic period as the fullest surviving analysis of the various siege methods that were in use everywhere at the time.[17] Philon wrote his treatise after Alexander's siege of Tyre, clearly using it as a historical *exemplum* for his advice to generals. In this instance, it adds reliability to the advice Philon gives in suggesting others copy the Tyrians' and Alexander's methods in their own sieges – as defenders or besiegers. In an age where siegecraft was developing apace, Alexander or his engineers came up with significant technological advances to further their cause. The mole he built was the first like it in the Greek world and today forms the promontory connecting the old city with the mainland.[18] His use of ship-borne battering rams and boarding bridges was the first such use in the ancient world. He also employed the two largest siege towers ever built.[19] Alexander's determination and his ability to use every way possible to achieve his goal is what allowed him to capture Tyre, and Philon recognised the importance of including all his innovations in his siege manual and in ensuring that the advice he provides reflects current practice. This fact reveals an important context for the writing and reading of military manuals: the near-continuous occurrence of warfare and the occasional innovations and escalations of scale meant that authors and readers would face a challenge in providing and acquiring the most up-to-date information.

The only other Greek manual that addresses siegecraft is Onasander's *The General*, which has a short section specifically directed towards conducting sieges that is often a summary of Philon's advice.[20] Onasander's manual, though entirely conceptual with no specific historical examples and written by a philosopher for a Roman audience, is the only extant text that provides general advice for a commander on campaign just as the more famous non-Classical manuals such as those of Sun Tzu, Machiavelli, Clausewitz, and modern army handbooks. This is, in fact, the style of military handbook that became and remained popular throughout the medieval, Renaissance, and early modern periods. Although Onasander was not a general, the advice that he gives is relevant and reflective of ancient military practice, and he possibly draws upon Philon.[21] Moreover, it is the only surviving manual that specifically gives advice to a general on how to *conduct* a siege.

II

As topics of analysis, I have extracted from the siege manuals the main areas of siegecraft that are emphasised therein: manpower; technical ability; treachery and secrecy; speed and timing; innovation; combined arms; blockade; and patience and determination. These topics are never specifically spelled out by any of the manuals. Even Onasander's later summary advice to a general in conducting a siege that collates advice from Philon and Aeneas does not present a specific order of when to carry out each action during a siege. But it is clear certain tenets can be grouped together to provide a clearer understanding of the advice given in the manuals. This collating of advice also allows a clearer reading of the manuals as to how they reflect historical reality, especially considering all of these manuals were intended as practical advice for contemporary and future generals. Where the manuals correlate with, or even demonstrate the influence of, actual practice, and vice versa, can perhaps reveal a set of general principles for conducting any siege in the Hellenistic world, principles that may indeed also apply to all siegecraft before the introduction of gunpowder. More precisely, closer study of a historical siege alongside an ancient scholarly discussion of that same siege can yield insights into the possible relationship between historical practice and military theory and points to possible alternate ways that readers can engage with military texts.

Manpower

A besieging army had to have sufficient manpower to prosecute the siege operations. Whether the plan was to surround a city in a blockade and starve it into submission, to undermine the walls, to have the city betrayed, or to assault the walls, a sufficient number of men was required that almost always needed to be substantially more than the defenders. Aeneas summarises the many stratagems that should be used to deter a numerically superior besieging force from landing easily, advice that an attacking general should bear in mind (8.1–5).

Demetrius landed a large army on the island of Rhodes, possibly numbering upwards of 40,000 men, replete with experienced engineers and siege train ready to besiege the city (D.S. 20.82.4–5). He also used around 30,000 workers from the fleet to help build his extensive siege-works (D.S. 20.91.8). Demetrius also employed a navy of 370 ships (200 warships) supplemented by pirates, and as many as 1000 traders, in order to blockade the Rhodian harbours.[22] The Rhodians could muster only a fraction of that number of men and were almost entirely reliant on their excellent navy.[23]

Technical ability

Without the knowledge of how to use catapults, ballistae, ladders, mines and rams, any direct assault in a siege would likely be unsuccessful. As Onasander states, 'A siege demands courage on the part of the soldiers, military science on the part of the general, and equipment of [perhaps better 'preparations of'] machines of war' (40.1).[24]

Demetrius had experts schooled in the art of sieges by Alexander's engineers and who were just as technically gifted as their predecessors.[25] Demetrius was an experienced attacker of cities who was nicknamed 'the Besieger' for his many successes. At Rhodes, Demetrius very much relied for his final overall success on land on the Helepolis, or taker of cities, 'which far surpassed in size those which had been constructed before it' (D. S. 91.2–8).

Diodorus' description of this huge siege tower is very long and shows the impressive size of the structure (D. S. 91.2–8): nine storeys, almost 150 feet high and 75 feet long, it was indeed the largest ever built.[26] The capture of a number of Demetrius' engineers by the Rhodians would have hampered his siege and aided the Rhodians, but not significantly enough for him to be unable to carry on (D. S. 20.93.5). For the siege on land, he built all the machines advised by Philon: 'machines', 'the rock-projectors and the bolt firers', 'wicker-tortoises', and 'rams' (D 17, 19, and 25).[27] Demetrius used the Helepolis carrying many artillery pieces, sappers under tortoises protected also by covered passages, and battering rams,[28] which were, according to Diodorus, 120 cubits long 'sheathed with iron and striking a blow like that of a ship's ram' (D. S. 20.95.1).[29]

Demetrius similarly prepared his fleet for his naval assaults by placing large siege machines on his ships. For the first assault, Demetrius placed two penthouses on two joined cargo ships to protect the artillery, and two siege towers of four storeys, which were taller than the city's towers and also placed on two cargo ships (D. S. 20.85.1). Philon advises, 'if <you are making the> attack from the sea, attack after putting machines on your cargo-vessels and *lemboi*', perhaps with Demetrius' very navy in mind.[30]

Demetrius also prepared 'a floating boom of squared logs studded with spikes', to shield from ramming the ships carrying the engines and towers (D. S. 20.85.2). Then he fortified the light craft to house the three-span catapults and Cretan archers (D. S. 20.85.3).[31] After his initial towers were sunk by a Rhodian sally, he built another tower three times the size in height and width, but this was overthrown by a storm (D. S. 20.88.7).

The Rhodians countered Demetrius' naval preparations well by placing defensive artillery on the mole, on ships near the harbour boom, and on any cargo ships moored in the harbour (D. S. 20.85.4). Their attempt to burn Demetrius' ships as he withdrew failed on account of his defensive boom (D. S. 20.86.2–3). In breaching Demetrius' boom blocking the harbour and sinking some of his engines during the next phase of the battle, the Rhodians displayed their own technical ability at sea (D. S. 20.88.1–6).

Although Rhodian success was centred on their navy, they also had a significant corps of highly trained artillerymen, who probably practised artillery in festivals just as other Greek states practiced the hoplite dash. The Rhodians had busied themselves with readying their defences and siege engines once they realised an attack was imminent (D. S. 20.84.5). Such was the Rhodian ability at artillery that Demetrius, according to Diodorus, was so astonished by the number of missiles fired at the Helepolis by the Rhodians in such a short space of time that he hesitated in his renewed assault (D. S. 20.97.1–2).

Philon advises specifically that '[it is easy] to set alight tortoises and machines when they come near the wall' (C 40–1). Also, to be discharged on them are incendiaries in the greatest possible numbers and burning *triboloi* wrapped in oakum' (Philon means siege towers here). Demetrius was in difficulty throughout his siege at Rhodes because the technical ability of the defenders matched, if not exceeded, his own.

Innovation

Innovation is another thing that was important for besiegers. Defenders just as much as attackers were aware of the many ways to get into a city and were usually ready to counter the normal methods. Philon provides significant detail in all the ways of fortifying a city to withstand a siege, preparations to make before it begins, and specific tactics for countering enemy attacks (D 10–11).[32]

Mining and counter-mining were tactics for both sides, and defenders often had oil and fire at the ready to pour onto rams. Internal walls and ditches could be built to secure vulnerable sections of wall, as Aeneas also advises (32.11–12), and catapults were ready to hurl stones and other missiles at the towers of the attacker. To get the better of the well-prepared defender, the besieger sometimes had to create new and innovative methods of attack, or simply be better at employing the standard methods.

Demetrius built the massive Helepolis, but this was not a new method of attacking a city, just a larger siege tower. The ship-borne towers used against the harbour were also bigger than before, but they were not a new concept and the Rhodians had no problems with countering them. Even before the siege of Rhodes, Demetrius at the siege of Salamis in Cyprus in 307 had also built a Helepolis of great size, 45 cubits long and 90 cubits in height (c. 68 feet by 135 feet). Nine storeys high, it included ballistae that shot a 180-pound ball (D. S. 20.48.2–3). Yet despite its size, the Salaminian defenders burned down this first Helepolis (D. S. 20.48.6–7). At Rhodes a year later, Demetrius, according to Diodorus, reinforced his Helepolis with iron plates, but the Rhodians dislodged one and thus once again set fire to the whole structure.[33] So Demetrius learned from his previous failure but failed to adequately find a definitive solution.

The one successful and important innovation of Demetrius at Rhodes was creating a spiked boom to protect his siege ships from ramming by the Rhodians (D. S. 20.85.2). It was successful in that it held up to the initial Rhodian assault (D. S. 20.86–2–3), and even when the Rhodians finally breached it Demetrius had enough time to save one of his engines (D. S. 20.88.1–6). Protecting attacking naval forces with a boom is not cited in Philon as a method to aid the besieger, but a boom is common advice to protect a harbour (Philon C 52, C 55, D 22, and D 23). Though perhaps entirely innovative, this boom did not really alter the outcome of the siege, as Alexander's mole did at Tyre, and apparently was not a significant enough invention to warrant inclusion in Philon's fairly comprehensive advice for a besieger.

Rather than on innovation, Demetrius relied on the strength and size of his machines to try to force entry into the city.[34] All the Rhodians had to do was use their usual methods of counter-artillery to put the Helepolis out of action, which they did by setting it on fire just as Aeneas (33.1–2) and Philon (C 12 and 40–41) suggest and as done at Salamis (D. S. 20.48.3–7). Demetrius should have known from his experience at Salamis exactly how the Rhodians would counter his Helepolis and should have taken precautions that were more lasting and successful.

Onasander gives general advice for a commander to anticipate the enemy's response and take appropriate measures:

> The general must take no fewer precautions and be no less observant than the enemy; for the army attacked, when it knows just what its danger is, guards especially against the enemy attacking. . . . For whatever the besiegers intend to do can be seen from the walls; but the besieged with the wall as a shield, often without detection pour through the gates and burn the machines or kill the soldiers or do whatever damage comes to their hands.
>
> (40.1–3)[35]

When he had arrived at Rhodes, Demetrius did indeed make necessary precautions. He had his men cut down trees and destroy farm buildings outside the city, make camp outside of missile range, and surround it 'with a triple palisade and with great, close-set stockades' (D. S. 20.83.4) as advised by Philon (D 6, D 10).[36]

Philon gives extensive advice on how to protect machines and tortoises against being set alight (D 34–42). Demetrius tried to fortify the Helepolis with iron plates, but that was not sufficient to prevent it catching fire from the Rhodian incendiaries. That was surely the most important precaution since he relied entirely on the Helepolis for success. He advises that

[s]hould the face (turned) toward the enemy fall off any of your machines, it is necessary to turn the sound one toward the opponents and attend to the damaged one. Against other misfortunes it is necessary, by cogitating on the basis of these same things, always to contrive something intelligently.

(D 57–58)

Rather than find a way to keep it in position and continue the barrage of the city, as advised by Philon, Demetrius simply moved the Helepolis out of range when it was damaged.[37]

Demetrius' siege equipment was not that innovative, just impressive in size. Alexander, along with innovative ship-borne rams, created a mole to turn an island into a peninsula, and the construction was so successful that the peninsula has existed ever since.

Treachery and secrecy

Philon states that '[i]f in your siege you cannot take the city by force because of its being strong (when approached) from every side, an attempt should be made to take it either by <stealth or> treachery or starvation' (D 72).[38] Treachery often plays a role in siege warfare, either genuine or suspected. Aeneas counsels great precautions be taken to prevent treachery from within. In fact, over half of Aeneas' strategies of surviving a siege are countermeasures against one of the besieged betraying the city, from slaves to foreigners or political rivals. Similarly for Onasander, as Smith summarises, 'the world of war for Onasander is as distinctly a world of deceit and false appearances, as for Aeneas Tacticus in his dissertation it was a world of internal treason'.[39] Trickery and deceit were paramount, but another very important method of gaining control of a city was secrecy, or subterfuge. Philon opens his advice for an attacker thus:

He who intends to capture cities must, for preference, make the attack during a festival which they are celebrating outside the gates; otherwise, at (grain-) harvest <or> vintage time; . . . otherwise, at night, when there is a storm or when the enemy are drunk at some public festival, approach the wall secretly with ladders ready and seize some of the towers.

(D 2–4)

Secret mining operations were often discovered through successful countermines, a defence that was again outlined by Aeneas (37.5), though Philon (C 7) and Aeneas (37.1) suggest instead digging open trenches between the wall and an outside rampart in order to expose any tunnel-diggers to detection and attack. Philon suggests a night attack as an option in his advice for taking the city by surprise (D 4) and gives more details of how to accomplish such an attack when he lists it as a method of last resort if other attacks have failed (D 73–75). Onasander also recommends a night attack on a city as the darkness multiplies the defenders' fears and they are more likely to flee (42.1).[40]

Demetrius' very first attack on the harbour was at night, and he succeeded in capturing part of the defensive mole, fortifying it and landing 400 men and artillery (D. S. 20.86.1). Despite attacking the harbour and the land walls for ten days, he was unable to take the city, and the Rhodians made use of the cover provided by a severe storm to oust finally those forces Demetrius had landed on the mole (D. S. 20.88.8). His other attack at night was through the breach made by the Helepolis, and these two were the only occasions that he got men into the city at all (D. S. 20.98.4–6). When Demetrius got men inside the city, Diodorus states that the general population of Rhodes were afraid the city was lost (20.98.8), as Onasander suggests would happen. However, the Rhodians had reserves, as suggested by Aeneas (38.1), and did not panic at the shock of Demetrius' forces gaining entry.

Philon gives advice on how to counter the enemy breaching the city walls (C 30–33). He advises specifically to make advance preparations for night-watches and sentry rounds (C 28), as well as the importance of regularly changing passwords to make night orders more secure (C 34–38). Aeneas also counsels to wait for the opportune moment to respond to a night attack (16.1–7). The magistrates at Rhodes sent calm orders to the defenders to keep to their posts and led the reserves against the intruders.[41]

Moreover, the attackers concentrated in the theatre district, and thus the Rhodians knew where to send their reinforcements to counter the attack (D. S. 20.93.1). Aeneas Tacticus argues that the besieged leave two or three open spaces in a city lest attackers (Aeneas in this passage is actually more concerned with internal conspirators) secure themselves by taking the only one available (2.7–9). It is not clear if the theatre in Rhodes was the only open space left clear, but it was certainly the only space occupied by Demetrius' forces and thus allowed the Rhodians to concentrate their counterattack. The theatre was the nearest large structure to the breach since its outer wall was taken down to furnish building material for the two Rhodian counter-walls. Demetrius eventually withdrew his forces from the city after seeing too many important officers killed (D. S. 20.98.9).

Demetrius did try every method at his disposal. He used sappers to bring down some of the wall and make a tunnel underneath into the city, but a deserter informed the Rhodians, and they were able to use countermines to hinder Demetrius' progress (D. S. 20.94.1–2). Demetrius also thought he had successfully suborned the commander of the Rhodian tunnel guard to let his men into the city. However, this traitor turned out to be a double agent, and a friend of Demetrius sent into the city was captured at the end of the mine (D. S. 20.94.3–5). The guard commander was also commander of the mercenaries sent by Ptolemy, and thus as a foreigner in service to Rhodes he may have been thought by Demetrius to be more susceptible to inducements for betrayal. Aeneas Tacticus spends a significant amount of time talking about avoiding betrayal, but focusing on section 12 focusing on specific precautions to take with allied troops (12). His following section refers specifically to mercenaries hired directly by the city rather than those serving as allied forces (13), such as was the case of the mercenaries at Rhodes who were sent by Ptolemy.

Philon advises that everyone, but particularly foreigners, mercenaries and slaves, be offered great rewards for fighting well, such as impressive funerary monuments (A 86), promotions and crowns (C 46), and benefits for their wives and children (C 47), such that they will want to fight as much as the citizen inhabitants.[42] Philon D 15 advises the attacker to offer rewards to anyone who would betray the city, especially slaves and foreigners, at the very least just to stop the besieged arming them for their defence. It seems from Diodorus' account that the actions of the Rhodians directly influenced Philon's stratagems here. Diodorus suggests that the necessary precaution by the Rhodians of removing all the noncombatants lest they diminish supplies was also done 'partly that there might be no one to become dissatisfied with the situation and try to betray the city' (20.84.2).[43] At the end of the siege, the Rhodians gave great rewards to the mercenary commander who defended the tunnel and whose men also successfully opposed Demetrius' later night attack at the theatre (D. S. 20.94.5).

Speed and timing

Two aspects that go together and were important for any battle, not just for sieges, are speed and timing. For an assault to work, it had to be launched at the right time. Even a small delay could often lead to disaster. Speed of assault in exploiting weaknesses and breaches in the defence were vital in siege warfare. Acting before the defenders had a chance to reorganise themselves was important. To maximise success, Onasander counsels for everything to be geared towards initial surprise:

> He [the general] must fall unexpectedly on an unsuspecting enemy, even if he is not expecting to seize the towns through treachery but to fight openly after a declaration of war, and he must not hesitate but strive in every way to attack fort or camp or town before his advance is known.
>
> (39.5)

Demetrius' failure to act quickly and in a timely fashion after his miners got into the city and to support various breaches during the siege caused all of his gains to be quickly lost. The Rhodian speed and timing in their reaction to Demetrius' advances and in particular to the advance of the Helepolis caused Demetrius to lose the initiative. It is an interesting contrast to Demetrius' first notable success in fighting a siege, at Athens in 307, where Diodorus suggests that Demetrius' army successfully forced a way over the walls and captured Peiraeus without much delay (D. S. 20.45.3). Plutarch, however, states that Demetrius' arrival was so unexpected that initially the Athenians believed his fleet came from Athens' ally Ptolemy, and so he was able to sail into the harbours unopposed and persuade the Athenians that he came to their benefit (*Demetrius* 8.4–5).[44] This suggests that even his most successful siege may have been accomplished more through luck than skill.

Blockade

Arguably the most important way of gaining the submission of a port city, such as Tyre and Rhodes, was by naval blockade. Even if this was not the method intended as the main way of forcing a submission, a complete blockade of the city, if possible, was desirable in order to prevent any outside help getting to the defenders. Without the possibility of any aid, cities could rarely hold out for long. Philip II, for example, failed in taking the cities of Perinthos and Byzantium because he did not fully blockade their ports.[45] As Philon suggests, 'if the city is by the sea, palisade it round from sea to sea, and if you have warships, anchor them at the harbour, in order that nothing can sail in' (D 5).

Demetrius intended to blockade the ports of Rhodes, and his first two large-scale attacks were intended to capture the harbours and secure a blockade (D. S. 20.85–88).[46] These attacks failed through the stout resistance of the Rhodians, who only had to defend that one location rather than all the city walls at once. The Rhodians used some of the methods outlined by Philon C 51–55 to defend their harbours just as the Tyrians. Philon argues for fire ships to be sent out to disrupt a blockade (C 55), something that the Rhodians tried, though unsuccessfully (D. S. 20.86.3).[47] Demetrius instead planned to use his huge pirate fleet to blockade Rhodes. But without safe harbours nearby for the pirates to remain on duty permanently and without any organized rotation of sentry ships, it was very easy for the Rhodian fleet to escape and gather supplies.

Demetrius' pirates were more concerned with raiding nearby islands than maintaining a blockade, and Demetrius never enforced it. Consequently, the Rhodians were allowed relative freedom of their harbours, and their allies even sent in significant amounts of supplies, troops and ammunition, rendering the blockade pointless. Demetrius was completely unable to prevent Rhodian ships from sallying out and sinking the pirates' ships and capturing valuable supplies, including a number of renowned engineers and catapult bolts as well as dresses intended for Demetrius' wife. Plutarch even suggests that it was the capture of his wife's belongings that provoked a rage in Demetrius that caused him to prosecute the siege even though he was not getting anywhere (*Demetrius* 22.1).

Demetrius had a huge navy and could have continued his blockade on the harbours rather than on land had he not used so many workers selected from the ships' crews to build the giant Helepolis and the associated siege-works. So had Demetrius continually pressed his attacks on the city, not necessarily even just at the harbours, the Rhodians would have been unable to sally out and cause trouble. He had the manpower to blockade fully the harbours if that remained his sole focus, but because he switched solely to land attacks, after the failure of his repeated attacks from land and sea, the Rhodians never felt the requisite pressure to be compelled to surrender the city.

Demetrius had agreed a price for exchanging prisoners, suggesting that he was not concerned with fully preventing the Rhodians from sallying out to sea (D. S. 20.84.6). However, Diodorus suggests that Demetrius' second large attack on the harbour 'centred upon capturing this and shutting off the people of the city from their grain supplies' (D. S. 20.88.1). Obviously, it was on his mind to try to keep the Rhodians shut in completely, but he was simply unable to accomplish the task, or lacked the determination to continue this necessary strategy despite a few initial setbacks. He also failed to prevent support entering the city from Cassander, Lysimachus and Ptolemy, who sent such large amounts of grain that the plentiful supplies encouraged the Rhodians to resist with renewed vigour. When Demetrius switched his attack to a land assault, he could not even prevent the Rhodians from receiving military support from the other Successors, and 1500 mercenary reinforcements from Ptolemy were instrumental in defeating Demetrius' only foray into the city itself. Philon is very clear about the danger posed by reinforcements arriving by sea and stresses the need to face them before they land (D 101–110).

Even at his successful siege of Salamis in 307, Demetrius did not prevent messengers getting into the city from Ptolemy (D. S. 20.49.3) and in the end could not prevent the defenders launching their navy of 60 ships to assist the relieving Egyptian fleet (D. S. 20.52.5). This is despite Diodorus' assertion that Demetrius pressed the siege from land and sea (D. S. 20.48.8). By contrast, Antigonus Monophthalmus, Demetrius' father, had successfully captured Tyre in 315 largely through being able to employ a successful blockade for a year and three months and forcing the inhabitants to surrender for want of supplies (D. S. 19.61.5).

Demetrius did not learn from these past examples, and the main reason for his failure to capture the city was the failure to block the Rhodian harbours completely. Demetrius never at any point, save perhaps his initial forays against the harbours, fully committed every means at his disposal to maintain a complete blockade and so was unable to achieve one. The extra resources that went into the city as a result allowed the Rhodians to maintain, and even increase, their resistance.

Patience and determination

Perhaps the most important human factors in winning a siege were patience and determination. Sieges could be long affairs, but any cessation by the besieger provided the defenders with the opportunity to relax and refresh themselves for the next assault. Keeping the defenders constantly engaged with new tactics and methods of attack wore them down and could lead to errors through fatigue. Usually the besieged were more determined than the besieger at the outset as they were fighting for their lives, and so the besieger had to wait until the defenders' resolve sufficiently waned. Onasander counsels that 'after making a truce he [the general] should neither make an attack nor himself remain unguarded', and that 'he must be suspicious enough to watch for festering deceit on the part of the enemy, for the intentions of those with whom the treaty has been concluded are uncertain' (37.1–3).[48]

Throughout the siege of Rhodes, Demetrius showed that he was open to peace negotiations from Rhodes' allies who sought to mediate a cessation of hostilities, none of which were successful and all of which allowed the Rhodians breathing space to refortify and resupply. However, it is difficult to tell who was more desirous of peace, Demetrius or the Rhodians. Demetrius was right to continually allow peace negotiations to continue as his overall strategy was subduing Rhodes, something an advantageous peace treaty would accomplish and make the continuation of the siege unnecessary. Unfortunately, breaking off attacks for negotiations worked more to Rhodes' advantage than Demetrius', since he could not be seen to be preparing for future attacks while engaged in negotiations, yet the defenders would have welcomed and perhaps needed such time even if just to recover and get some rest and nourishment.

For Demetrius, perhaps the most damaging of these peace negotiations was the arrival of Cnidian envoys promising to persuade the Rhodians to surrender that caused Demetrius to cease his initial attack with the Helepolis and bought the Rhodians more time to prepare for what they could see was coming (D. S. 20.95.4). During Demetrius' second assault with the Helepolis, envoys from over 50 Greek states came to negotiate a peace, and Demetrius again broke off his attacks for fruitless talks (D. S. 20.98.2–3). Demetrius' continual cessation of hostilities for unsuccessful peace negotiations, or to prepare for his own large-scale assaults, gave the Rhodians numerous breaks throughout the siege where they were able to mend their defences, resupply, rest the defenders and prepare for the next assault.

To aid in the wearing down of the enemy during a siege, Onasander suggests that

> the women and children and feeble men and old people he [the general] should send of his own accord into the city. These will be useless in action but will consume more quickly the supplies of the besieged and will serve the purpose of enemies rather than friends.
>
> (42.23)[49]

Demetrius by contrast had allowed the noncombatants in Rhodes to leave the city (D. S. 20.84.2). This was another way in which the expected hardship of the Rhodians in the siege was not as great as it should have been.

Demetrius' willingness to break off the siege in order to negotiate afforded his enemy rest, preventing him from fully exploiting the advances that he did make. He also rushed into assaulting the harbours in a naval assault before he could coordinate a simultaneous land assault because he was too impatient to wait for the land assault engines to be prepared. Demetrius' prolongation of the siege neutralised his superiority in numbers. Eventually, Antigonus ordered his son to make peace on whatever terms he could obtain. Plutarch suggests that by then Demetrius was looking for an excuse to end the war and an Athenian disputation induced both sides to agree to a peace (*Demetrius* 22.4).

Combined arms

Arguably the most important tactic for a successful siege was the constant use of combined arms in order to maintain unceasing pressure on the enemy. It is this method of attack coupled with the determination to maintain a permanent blockade that usually ensured a siege's success. Without using multiple tactics, armaments, and locations to attack, the defenders could always tell where an assault was coming and prepare their defences and resolve accordingly.

Onasander states that 'just as a good wrestler, the general must make feints and threats at many points, worrying and deceiving his opponents, here and there, at many places, striving, by securing a firm hold upon one part, to overturn the whole structure of the city' (42.6). Both Philon (D 26) and Onasander argue for fighting in relays (42.7–13).[50] Onasander even suggests splitting the army into five or six equal divisions so that one can fight and the others rest so as to maintain a constant assault. The use of relays was common in Hellenistic and later siege warfare, again requiring sufficient men to attack in force all around the city uninterruptedly.

To capture any city, it was important to use every method of attack at one's disposal. For coastal cities such as Rhodes and Tyre, the use of the navy in assisting any successful assault was vital. Moreover, the navy and army should attack all around the city simultaneously and constantly:

> When you have brought up the machines (and) exhorted the troops and made the same proclamation as before, conduct the assault on the city from all sides, both by land and by sea, if any of the wall is sea-fronting, in order to create as much fear as possible and to split those fighting inside into many (parts).[51] Let all your artillery be in action, and the rams and the drills and the ravens and the assault-bridges, both by land and by sea, into the appropriate places; and conduct the assault with successions of troops, exempting no-one, in order that the combatants are always vigorous – the result will be (an assault) always strong and continuous. Also, make a great hubbub and sound trumpets at the strongest (parts) of the city, in order that (those inside), supposing that the wall is being captured there, would flee from the curtains with the others; as for you, when you have drawn away most of those inside who are there, you may subdue the city by capture.
>
> (D 24–27)

Land and naval artillery, sea-borne infantry, missile troops, miners, heavy infantry and light infantry were all-important in a successful assault, and the general should use them in combined and coordinated attacks. On the specific use of siege engines, Onasander advises thus:

> dividing his army into many parts, he should [station his engines at certain points and should] command his men to bring forward their ladders against the other parts of the wall, since in this manner the besieged are rendered helpless. For if the besieged disregard the other parts of the wall and make a defence against the attacks of the engines, all the besiegers who attack with ladders will easily climb over the wall without opposition, but if the defenders divide their forces . . . [they] will be unable to repel the advancing menace.
>
> (42.4–5)[52]

Demetrius used all the troops at his disposal at different times during the siege of Rhodes, and often together, but he did not coordinate multiple assaults all around the city using them all together *every time*. Though his initial assaults involved attacks by land and sea, after a while Demetrius primarily concentrated on land assaults only. Demetrius' first attack was aimed at the main harbour but initially delayed by a storm (D. S. 20.86.1). The Rhodians were able to counter his attack because they knew where it was coming and turned all their efforts into counter-measures (D. S. 20.85.4). Philon provides a very useful account of the precautions to be taken against assaults on harbours. He states that stone throwers should be placed at either side of the entry to the harbour (C 56). This is precisely what the Rhodians did using five penthouses in total housing an unknown, but likely significant, number of machines. Despite gaining a foothold at the harbour through a night attack, after a whole day's fighting, Demetrius withdrew his troops at nightfall (D. S. 20.86.1).[53]

After this initial confrontation, Demetrius did launch attacks on the city from both land and sea for eight days but eventually withdrew in order to repair his ships and siege engines after suffering significant losses of men and commanders (D. S. 20.86.4–87.4). What is interesting is that Demetrius apparently made no efforts to prevent the Rhodians repairing the damage caused by over a week of intense fighting and gave them as much time to recover as his own army (D. S. 20.88.1). The Rhodians had time to bury their dead and rebuild the wall where it had been damaged (D. S. 20.87.4).

After the cessation, he launched a huge attack solely on the harbour again, rather than on both land and sea as before. After many sallies from the Rhodians, another storm forced Demetrius' forces to withdraw, after destroying a number of his ships and sinking his siege tower (D. S. 20.88.7). The Rhodians used the confusion caused by the storm to dislodge Demetrius' troops, who had made an initial breach, and regained the harbour, in part by using attacks in relays as advised by Philon and Onasander for attackers (D. S. 20.88.8).

After this, Demetrius switched to land assaults only focusing on the Helepolis (D. S. 20.91.1 cf. 20.91.8). Though he cleared a huge space to assault the walls, between seven towers, it was still only one point of the wall that the Rhodians had to defend rather than the whole perimeter of the city. Philon's advice here runs thus: '[d]o not make it obvious where you will make the attack; rather, feint in some places and deploy the machines in others, in order that the besieged will make mistakes in their preparations' (D 18).

The large scale of Demetrius' preparations took time, and this allowed the Rhodians to build a second defensive wall behind the main rampart using stones from the theatre, nearby houses and temples (D. S. 20.93.1).[54] They also dug a ditch behind the wall. Aeneas advises this exact stratagem in defending the walls, building a ditch and a new reinforcing wall behind the breach (32.12); likewise Philon (A 10 and C 7). The Rhodians also sent out tunnels to counter Demetrius' efforts at undermining the wall (D. S. 20.94.2). Aeneas gives methods on detecting and preventing tunnels and (37.9) to shield one's own tunnelers (37.1–4, 9). Philon gives advice to attackers on how to deal with the defenders' countermines, something that Demetrius did not do (D 31–32). After enduring Demetrius' bombardments of the wall and gaining respite by using counter-artillery to force him to withdraw the Helepolis, the Rhodians further reinforced the breach by building a third wall and another internal ditch (D. S. 20.97.4).

In his final night attack on the breaches created in the land walls, while also launching diversionary attacks from both land and sea, Demetrius finally got men into the city. Philon suggests that

> During captures it is particularly necessary to gain possession of the walls and the acropolis and the agora and the generals' office and any other strong place there may be; if (your) force is less (than is needed to do so, at least gain possession of) the towers and the most important place, in order that you are not expelled from the city again.
>
> (D 71)

Demetrius managed to get a force into the city but he only sent 1500 men, who occupied the theatre (D. S. 20.98.4). Though they were chosen from among his best men, this is still a very small force to take a city considering the total size of the army Demetrius had at his disposal. It proved to be an insufficient number of troops, and the Rhodians, thanks largely to the Egyptian forces sent by Ptolemy, were able to cut off the enemy inside. Plutarch writes that at this battle around the theatre Demetrius' best officer was killed, and this may have induced Demetrius into calling off the assault (*Demetrius* 21.4). Onasander advises that the attacker should attack in relays to maintain a constant attack by night and day to wear out the defenders (42.7–13). Demetrius only maintained his assault for one day. It was not a long enough timeframe for the Rhodians to feel the effects of constant battle (D. S. 20.98.4–9).

Diodorus' account does not suggest that in the morning Demetrius tried to reinforce his elite troops through the breach, so Demetrius perhaps believed that only 1500 men was enough to capture a city the size of Rhodes. This again demonstrates Demetrius' overconfidence, or his lack of astute, tactical thought. In most cases, capturing a city requires the entry of a large force in order to overwhelm defenders and take as many strongpoints as possible, as advised by Philon previously.

Demetrius continually assaulted the harbour and the city walls but rarely together, constantly, and in enough force. His assaults on the harbour were not supported so that the breach was not followed up, albeit because of bad weather, and the breach near the Helepolis was penetrated by his elite troops, but he did not follow up their advance or use missile troops to clear the walls of defenders to shelter them from missile attack. If Demetrius, after maintaining a secure blockade, had assaulted the harbours at the same time as he assaulted the breaches in the main wall, and in full force, there was probably little chance for the dwindling Rhodian defenders to ward off the attack. Demetrius' narrowly focused assaults allowed the limited Rhodians to defend the breaches he made. Although the breach created by the rams near the Helepolis was very wide, it was still the only place that the Rhodians had to defend against a strong attacking force. Demetrius' failure to follow up his momentary gains with renewed assaults also cost the lives of many of his men as well as victory. So although Demetrius had the technical skills again, he failed to implement the tactical knowledge to best use the expertise at his disposal.

By contrast, in Demetrius' successful attack on Munychia in 307, Diodorus states that Demetrius assaulted by land and sea using many engines of war and that his men fought in relays, as advised by Onasander discussed earlier, maintaining an attack for two days (D. S. 20.45.5–7). Likewise at the siege of Salamis in the same year, Demetrius 'pushed the siege persistently by both land and sea' (D. S. 20.48.8). Clearly Demetrius had the knowledge of how to attack successfully using combined arms, but for whatever reason he failed to do so *every time* at Rhodes.

III

This analysis of Demetrius' siege of Rhodes demonstrates that military manuals concerned with defensive tactics during a siege can also provide insight on how to conduct a siege of an enemy state, and that a focused examination of both defensive and offensive aspects of a historical siege can serve to highlight the thoroughness and generalised utility of siege manuals. One might humorously speculate that the Rhodians were better readers of siege manuals than Demetrius. Perhaps ironically, future city-besiegers could benefit from reading siege-focused military manuals *and* historical narratives such as Diodorus, mining the former for information on how best to execute a siege and learning what not to do from Demetrius' failure at Rhodes. In this case, Demetrius' nickname 'city-besieger' (πολιορκητής) proves misleading or ironic, as Heckel has argued, but other generals following him may have benefitted from such accounts of his experiences both successful and not.[55]

More generally, this analysis of Demetrius' siege through the filter of military manuals alongside the historical narrative establishes that perhaps one ought not to rely upon one genre of source only in order to understand ancient warfare. Each genre can serve to complement the other, and when considered side-by-side, the specifics of historical narrative can enhance understanding and validate the utility of the generalised advice of the military manual. And, equally important, consideration of the advice of military manuals with respect to a specific historical event and narrative can provide clarity to what the author describes – or raise important questions about the historian's representation of that particular event or warfare more generally. In this case, one must acknowledge that the interpretation of a military manual can change over time, as recent battles and wars provide new examples and counterexamples that can support, or possibly contradict, the advice given by the writer. Most importantly, this chapter raises the possibility that one can read a military text in a manner that the author does not perhaps intend, to discover the kinds of stratagems that the enemy may undertake and thus be able to anticipate and counteract them.

Notes

1 This chapter comprises material that I intend to explore in more detail elsewhere. I warmly thank the editors of this volume for their suggestions.

2 On this subject, see in particular Campbell (1987) and Vela Tejada (2004), and for the philosophical context of Aeneas, see in particular Whitehead (1990) 34–35 and Burliga (2009) 94–98. All references to Aeneas are to Aeneas Tacticus.

3 Cicero mentions Pyrrhus at *Letters to Friends* 9.25.1; cf. Plutarch, *Pyrrhus* 8.2. See also Chapter 9. Hellenistic siege manuals generally focused upon more with siege machines than with generalship, and so are of less concern here.

4 See also the Introduction (4–7) and Chapter 7.

5 Whitehead mentions Alexander's siege of Tyre 21 times and Demetrius' siege of Rhodes over 50 times.

6 Bettalli (2017) 117–121 gives a brief but succinct analysis of the relevance of Aeneas' section on siege warfare (32–40) to fourth-century siegecraft, Philip of Macedon, and sieges in the ancient world in general. He argues, as I do here, that Aeneas' evidence is far from out-of-date, but useful in presenting practical countermeasures to all the common methods of besieging a city: ramming, ladders, tunnels, artillery, siege towers, and maintaining reserves.

Whitehead (2016) 272–273 begins his commentary on Philon's section concerning defensive measures with an abridged but lengthy quotation from Diodorus Siculus of the Tyrians' defence against Alexander (17.41.2–45.5), stating that 'the siege is noteworthy for the ingenuity, versatility and determination of the defenders *just as much as the attackers*' (272; italics mine). This is the closest to a full analysis of the siege of Tyre through the lens of tactical manuals concluding that 'overall it remains the case that we have here a graphic vignette of the array of defensive tactics available to a city under attack a century before Ph. framed his recommendations on the matter' (273).

7 In the discussion that begins in the next section, references to Diodorus appear *passim*. While the approach of this paper is to focus on specific actions undertaken during a siege in turn, the reader may benefit also from experiencing the siege in a continuous narrative. On Diodorus as a military historian and theorist, see Roisman (2018) and Williams (2018). Plutarch provides a more succinct narrative of events at Rhodes in *Demetrius* 21.1–22.4.

8 Diodorus also narrates in a not brief narrative Alexander's siege of Tyre: 17.40.2–47.6. Tarn (1948) 2.120–121 concluded that Diodorus' description of the measures used by the Tyrians was so inventive that it was excerpted from a lost Hellenistic manual on siege warfare; Whitehead (2016) 273 rightly dismisses this suggestion.

9 I here acknowledge the concern expressed by Hans Michael Schellenberg in his chapter (see specifically 51 n.43) about utilising military manuals in the way I do in this chapter as evidence of prior conflicts. In addition to providing additional insight into ancient siege warfare, my approach seeks to demonstrate the potential utility in military manuals in acquiring a better understanding of ancient warfare in general. Manuals were usually based on intimate knowledge of previous conflicts, and thus, often for lack of other evidence, analyses of them can highlight what any prior influences may be on later manuals, why, and to what extent. The reader will hopefully appreciate the different scholarly opinions that exist about the place of the genre in ancient historical study. In general, the reader hopefully will also appreciate the ongoing influence of warfare and military writing in a continuous timeline with everything influenced by what came before and connected to what came afterwards. Though it may be an anachronistic method, it is also foolhardy not to posit some influence of past events on later writings.

10 The subject of failure (that is, defeat) is becoming increasingly important in ancient military studies: see Chapter 10 with additional bibliography.

11 See especially Whitehead (1990); Vela Tejada (1993); and recently Pretzler and Barley (2017).

12 Whitehead 2016: 25. On his use of examples, see most recently Pretzler (2017a). See also Burliga 2009 for Aeneas' use of examples as rhetorical devices. And see Wrightson (2015) for how even simple references to historical practices lend reliability to the philosophical tactical manual of Asclepiodotus.

13 Whitehead (1990) 34–42, esp. 34: 'the vast majority of scholars who have studied Aineias' treatise have judged it the work of a practical soldier seeking to distil and systematize his experience and transmit it to others'.

14 Campbell (1987). See also Vela Tejada (2004).

15 On just how much Philon relies on Aeneas' advice, see most recently Whitehead (2016) 17. Whitehead (2017) states that the works of these two should often be read together in order to appreciate better the content of each. He particularly notes that '[a]n in-depth Aineias/Philon comparison would be greatly to the benefit of our appreciation of both of them'.

16 On Philon having little military experience, see Whitehead (2016) 23–25. See also Chapter 3 on amateur authors and readers of military texts.

17 Though Philon's work is let down by not making use of historical exempla as Aeneas. Whitehead (2016) 25: 'If Ph.'s original had, in reality, included historical material of this kind, the summarizer has done us a signal disservice by jettisoning it'.

18 A simple glance at Google Maps or Google Earth can demonstrate this.

19 Arrian, *Anabasis* 2.18–24; D. S. 17.40–47; Curtius 4.2–4.

20 On Onasander, see Smith (1998); Petrocelli (2008); and Formisiano (2011).

21 On Onasander's probable amateur status as an author, see Chapter 3 (specifically 61–65).

22 Diodorus discusses Demetrius' navy at 20.82.4.

23 Cf. D. S. 20.81.2.

24 See Petrocelli (2008) 258.

25 Athenaeus Mechanicus (27.2) and Vitruvius (10.16.4) are the only sources that provide the name of Demetrius' main engineer who built the Helepolis, Epimachus the Athenian. Little is known about him other than this stated connection to the Helepolis. Billows (1990) App. 3 Nos. 37, 47, 124 lists the three engineers of Demetrius known by name, but he had at least 11 others as that many were captured by the Rhodians (D. S. 20.93.5).

26 See also Plutarch, *Demetrius* 21.1–2; Athenaeus Mechanicus 27.4–5; Vitruvius 10.16.4; Ammianus Marcellinus 23.4.1ff. Other scholars have suggested that Diodorus excerpted this description from a now-lost military manual (Marsden 1971; Hornblower 1981; Lendle 1983; Wiemer 2001; Pimouget-Pédarros 2011; Wheatley 2016). But since Diodorus does not state his sources for book 20, it is very difficult, if not impossible, to formulate any reliable conclusions about the origins of this or any other passage in his overall description. As such, out of a lack of evidence to the contrary it is necessary to take Diodorus at his word.

27 Philon usually uses the term machines to refer to siege towers exemplified by the Helepolis: see Whitehead (2016) 32, 279.

28 See Philon D 40.

29 It is highly unlikely that the ram was this large as its structural integrity at that length is questionable (see Schramm 1928; Geer 1954; Meiggs 1982). Nevertheless, however large, Demetrius likely did house battering rams in the Helepolis. The same goes for doubts about the size of the catapults on the Helepolis (Schramm 1928). Even if they were not as big as Diodorus describes, there were certainly catapults in significant number on various levels of the siege tower.

30 Whitehead (2016) 354 cites Demetrius' naval forces as the only historical example of Philon's advice and quotes this same passage of Diodorus.

31 Philon suggests that three-span catapults are suitable weapons to use in the limited space of a tunnel (D 31), and two of them should be issued to each *amphodon* in a besieged city (C 26). Clearly they were small antipersonnel artillery pieces.

32 See in particular (cf. Lawrence (1979) 89–99, parts 2 and 3) parts A–C of Philon's text: Whitehead (2016) 66–109 text, 133–332 commentary.

33 Vitruvius (10.16.4) contradicts Diodorus that there was iron plating. Diodorus' account is fuller and so perhaps more probable, but regardless at each siege defenders likely succeeded in setting the Helepolis on fire whether covered in iron or not.

34 Demetrius' enthusiasm for building huge machines also stretched to ships, and Plutarch recounts how even his enemies would stop and wonder at the size of his vessels or siege engines (*Dem.* 20.4).

35 See Petrocelli (2008) 258–259.

36 Philon states, 'as regards anything round the city that furnishes either security or benefit <to those> inside, eradicate some items and cut out others' (D 8).

37 If the Helepolis was indeed sheathed with iron plates all over, it would have been very heavy and so difficult to move (Schramm 1928). But if there was no iron, or other, sheathing at all then it would have been very exposed to incendiaries. It is also unlikely that Demetrius' soldiers could construct such a huge siege tower in situ under fire of the defenders. So it must have been movable towards the city and thus out of range too. Perhaps the covering was not iron plating all over but only on the side facing the city and perhaps not from top to bottom. This may explain why Demetrius did not choose to turn the Helepolis and continue using it once one side was damaged. Of course, we will never know the nature of the tower or if Diodorus' account is accurate. But we can conclude that there was a large tower that had some protection against fire arrows and yet was still set aflame, which is the main point made here. Demetrius' intended precautions against incendiaries were insufficient and costly.

38 Starvation I discuss under 'Blockade', since the one stems from the other.

39 Smith (1998) 161.

40 Petrocelli (2008) 261.

41 All translations of Aeneas are from Whitehead (1990).

42 The arming of slaves was somewhat normal in sieges and the Rhodians did so, as D. S. 20.84.3 and 20.100.1 indicate. See Whitehead (2016) 216 (on the Rhodians' inducements to fight; Philon repeats this at C 47, 297 and 307.

43 See Whitehead (2016) 350 for Rhodes expelling the foreigners who did not want to fight in reference to this advice of Philon.

44 Polyaenus 4.7.6 also records the success of this attack as gaining complete surprise.

45 Perinthos: D. S. 16.74.2–76.4; Byzantium: D. S. 16.77.2–3.

46 Philon suggests anchoring warships across the harbour to enforce a blockade (D 5), as Alexander did at Tyre.

47 Cf. Whitehead (2016) 317.

48 Petrocelli (2008) 253–254 and Chlup (2014) 56–57.

49 Petrocelli (2008) 270 and Chlup (2014) 53.

50 Petrocelli (2008) 266–267.

51 This division of enemy forces is of crucial importance as outlined by a number of stratagems in Polyaenus (for example, 4.7.3.) and Frontinus (for example, *Stratagems* 3.9.5, 6, 8, 10).

52 See Petrocelli (2008) 265–266.

53 Onasander suggests a night attack as the darkness multiplies the defenders' fears and they are more likely to flee (42.1).

54 Cf. Philon's advice for building inner walls and defences (C 18–27).

55 Heckel (1984).

Bibliography

Adcock, F. E. 1957. *The Greek and Macedonian Art of War*. Berkeley: University of California Press.

Berthold, R. M. 1984. *Rhodes in the Hellenistic Age*. Ithaca, NY: Cornell University Press.

Bettalli, M. 1990. *Enea Tattico, La difesa di una città assediata (Poliorketika). Introduzione, traduzione e commento*. Pisa: ETS.

———— 2017. 'Greek Poleis and Warfare in the IVth Century BC: Aineias' *Poliorketika*', in M. Pretzler and N. Barley, eds., *Brill's Companion to Aeneias Tacticus*. Leiden: Brill. 166–181.

Billows, R. A. 1990. *Antigonos the One-Eyed and the Creation of the Hellenistic State*. Berkeley: University of California Press.

Bloedow, E. F. 1998. 'The Siege of Tyre in 332 BC'. *La parola del passato* 53301. 255–293.

Bosworth, A. B. 1976. 'Arrian and the Alexander Vulgate'. *Entretiens Hardt* 22. 1–46.

———— 1988. *Conquest and Empire: The Reign of Alexander the Great*. Cambridge: Cambridge University Press.

Burliga, B. 2009. 'Aeneas Tacticus between History and Sophistry: The Emergence of the Military Handbook', in J. Pigoń, ed., *The Children of Herodotus: Greek and Roman Historiography and Related Genres*. Newcastle-upon-Tyne: Cambridge Scholars Publishing. 92–101.

Campbell, D. B. 1987. 'Teach Yourself How to Be a General'. *JRS* 77. 13–29.

———— 2005. *Ancient Siege Warfare: Persians, Greeks, Carthaginians and Romans 546–146 BC*. Oxford: Osprey.

———— 2006. *Besieged: Siege Warfare in the Ancient World*. Oxford: Osprey.

Champion, J. 2014. *Antigonus the One-Eyed: Greatest of the Successors*. Barnsley: Pen and Sword.

Chaniotis, A. 2013. 'Greeks under Siege: Challenges, Experiences, and Emotions', in B. Campbell and L. A. Tritle, eds., *The Oxford Handbook of Warfare in the Classical World*. Oxford: Oxford University Press. 438–456.

Chlup, J. T. 2014. 'Just War in Onasander's ΣΤΡΑΤΗΓΙΚΟΣ'. *JAH* 2. 37–63.

Clausewitz von, C. 1968 [1832]. *On War*. Translated into English by C. F. N. Maude. 1908. ed. A. Rapoport. London: Penguin.

De Souza, P. 1999. *Piracy in the Greco-Roman World*. Cambridge: Cambridge University Press.

De Voto, J. G. 1996. *Philon & Heron: Artillery and Siegecraft in Antiquity*. Chicago: Ares Publishers.

Formisiano, M. 2011. 'The Strategikós of Onasander: Taking Military Texts Seriously'. *Technai* 2. 39–52.

Fuller, J. F. C. 1958. *The Generalship of Alexander the Great*. London: Eyre and Spottiswoode.

Garlan, Y. 1973. 'Cités, armées et stratégie à l'époque hellénistique d'après l'œuvre de Philon de Byzance'. *Historia* 22. 16–33.

———— 1974. *Recherches de poliorcetiques grecques*. Athens: École Française d'Athènes.

Geer, R. M. 1954. *Diodorus of Sicily with an English Translation by Russel M. Geer. Bd. X: Books XIX, 66–110 and XX*. Cambridge, MA: Harvard University Press.

Gilliver, C. 1996. 'The Roman Army and Morality in War', in A. B. Lloyd, ed., *Battle in Antiquity*. Swansea: Classical Press of Wales. 219–238.

Griffith, S. B. 2005. *Sun Tzu: The Illustrated Art of War*. Oxford: Oxford University Press.

Gruen, E. S. 1985. 'The Coronation of the Diadochi', in J. W. Eadie and J. Ober, eds., *The Craft of the Ancient Historian: Essays in Honor of Chester G. Starr*. Lanham, MD: University Press of America. 253–271.

Hansen, M. H. 2007. 'Aeneas Tacticus', in M. H. Hansen, ed., *The Return of the Polis*. Stuttgart: Franz Steiner. 243–245.

Hauben, H. 1974. 'A Royal Toast in 302'. *AncSoc* 5. 105–117.

Heckel, W. 1984. 'Demetrios Poliorketes and the Diadochoi'. *La parola del passato* 39. 438–440.

Hornblower, J. 1981. *Hieronymus of Cardia*. Oxford: Oxford University Press.

Kern, P. B. 1999. *Ancient Siege Warfare*. Bloomington: Indiana University Press.

Keyser, P. T. 1994. 'The Use of Artillery by Philip II and Alexander the Great'. *AW* 25. 27–59.

Lawrence, A. W. 1979. *Greek Aims in Fortification*. Oxford: Oxford University Press.

Lendle, O. 1983. *Texte und Untersuchungen zum technischen Bereich der antiken Poliorketik*. Wiesbaden: Franz Steiner.

Lund, H. S. 1992. *Lysimachus: A Study in Hellenistic Kingship*. London: Routledge.

Marsden, E. W. 1969. *Greek and Roman Artillery: Historical Development*. Oxford: Oxford University Press.

——— 1971. *Greek and Roman Artillery: Technical Treatises*. Oxford: Oxford University Press.

McNicoll, A. 1986. 'Developments in Techniques of Siegecraft and Fortification in the Greek World CA. 400–100 B C', in P. Leriche and H. Treziny, eds., *La Fortification dans l'Histoire du Monde Grec*. Paris: CNRS. 305–313.

Meiggs, R. 1982. *Trees and Timber in the Ancient Mediterranean World*. Oxford: Oxford University Press.

Murray, W. M. 2012. *The Age of Tyrants: The Rise and Fall of the Great Hellenistic Navies*. Oxford: Oxford University Press.

Petrocelli, C. 2008. *Onasandro, Il Generale*. Bari: Edizioni Dedalo.

Pimouget-Pédarros, I. 2011. *La Cité à l'épreuve des rois. Le siège de Rhodes par Démétrios Poliorcète (305–304 av. J.-C.)*. Rennes: Presses Universitaires de Rennes.

Pretzler, M. 2017a. 'Aineias and History: The Purpose and Context of Historical Narrative in the *Poliorketika*', in M. Pretzler and N. Barley, eds., *Brill's Companion to Aeneias Tacticus*. Leiden: Brill. 68–95.

——— 2017b. 'The *Polis* Falling Apart: Aineias Tacticus and *Stasis*', in M. Pretzler and N. Barley, eds., *Brill's Companion to Aeneias Tacticus*. Leiden: Brill. 146–165.

Pretzler, M. and Barley, N., eds. 2017. *Brill's Companion to Aeneas Tacticus*. Leiden: Brill.

Roisman, J. 2018. 'Fate and Valour in Three Battle Descriptions of Diodorus', in L. I. Hau, A. Meeus, and B. Sheridan, eds., *Diodorus of Sicily: Historiographical Theory and Practice in the Bibliotheke*. Leuven: Peeters. 507–518.

Schramm, E. A. 1928. 'Poliorketik', in J. Kromayer and G. Veith, eds., *Heerwesen und Kriegführung der Griechen und Römer*. HdAW 4.3.2. Munich: C. H. Beck. 209–247.

Shipley, G. 2017. 'Aineias Tacticus in His Intellectual Context', in M. Pretzler and N. Barley, eds., *Brill's Companion to Aeneias Tacticus*. Leiden: Brill. 49–67.

Sinclair, R. K. 1966. 'Diodorus Siculus and Fighting in Relays'. *CQ* 16. 249–255.

Smith, C. J. 1998. 'Onasander on How to Be a General', in M. Austin, J. Harries, and *idem*, eds., *Modus Operandi: Essays in Honour of Geoffrey Rickman*. London: Institute of Classical Studies. 151–166.

Stewart, A. 1987. 'Diodorus, Curtius, and Arrian on Alexander's Mole at Tyre'. *Berytus* 35. 97–99.

Tarn, W. W. 1930. *Hellenistic Military and Naval Developments*. Cambridge: Cambridge University Press.

——— 1948. *Alexander the Great*. Cambridge: Cambridge University Press.

Vela Tejada, J. 1993. 'Tradicion y originalidad en la obra de Eneas el Táctico: la génesis de la historiografía militar'. *Minerva* 7. 79–92.

——— 2004. 'Warfare, History and Literature in the Archaic and Classical Periods: The Development of Greek Military Treatises'. *Historia* 53. 129–146.

Wiemer, H.-U. 2001. *Rhodische Traditionen in der hellenistischen Historiographie*. Frankfurt: Marthe Clauss.

Wheatley, P. 2016. 'A floruit of Poliorcetics: The Siege of Rhodes, 305/04 BC'. *Anabasis* 7. 43–70.

Whitehead, D. 2001 [1990]. *Aineias the Tactician: How to Survive under Siege*. Oxford: Oxford University Press.

——— 2008. 'Fact and Fantasy in Greek Military Writers'. *AAHung* 48. 139–155.

——— 2015. 'Alexander the Great and the Mechanici', in P. Wheatley and E. Baynham, eds., *East and West in the World Empire of Alexander: Essays in Honour of Brian Bosworth*. Oxford: Oxford University Press.

——— 2016. *Philo Mechanicus: On Sieges, Translated with Introduction and Commentary*. Stuttgart: Franz Steiner.

——— 2017. 'The Other Aineias', in M. Pretzler and N. Barley, eds., *Brill's Companion to Aeneias Tacticus*. Leiden: Brill. 14–33.

——— and Blyth, P. H. 2004. *Athenaeus Mechanicus, On Machines*. Stuttgart: Franz Steiner.

Williams, N. 2018. 'The Moral Dimension of Military History in Diodorus of Siculus', in L. I. Hau, A. Meeus, and B. Sheridan, eds., *Diodorus of Sicily: Historiographical Theory and Practice in the Bibliotheke*. Leuven: Peeters. 519–540.

Wrightson, G. 2015. 'To Use or Not to Use: The Practical and Historical Reliability of Asclepiodotus' "Philosophical" Tactical Manual', in G. Lee, H. Whittaker, and G. Wrightson, eds., *Ancient Warfare: Introducing Current Research (IAWC vol. 1)*. Newcastle-upon-Tyne: Cambridge Scholars. 65–93.

6   Mercenaries and moral concerns

Aaron L. Beek

Many ancient military manuals rely heavily on historic *exempla* to make or support their points. This chapter discusses the didactic purposes of early military manuals and argues that, in their selections of *exempla*, some authors of military manuals subordinated military advice to political advice. Moreover, it argues that the historicity of the past examples was of secondary importance in securing good future outcomes. Writers of ancient military manuals often held the mercenary (e.g. *mercennarius*, *misthophoroi*) in disdain and advocated against the employment of mercenaries – advice that flew in the face of both the historical trend of an increasingly professional and wage-earning army, and more importantly, evidence of mercenary armies outperforming their volunteer counterparts on the battlefield.[1] What is their rationale? This situation is best explained by the realisation that our tacticians betray vested motives in preventing the professional mercenary soldier from becoming the standard of the Roman army. Moreover, in Polyaenus in particular, the tacticians' genre ceases to be either a miscellany of historic anecdotes or a simple didactic explanation of tactics, but rather mixes these together along with political philosophy to make broader claims about the role of the army and its leaders. In keeping with the imperial setting and the emperors' combined roles as monarch and general, he adds practical though indirect advice for ruling the state to that for leading armies.[2] Accordingly, it is worthwhile to examine Polyaenus here alongside some *comparanda* from the *Stratagems* of Frontinus and from Xenophon (mainly found in *Hiero* and the *Cavalry Commander*).[3]

I

Polyaenus, writing in the 160s CE, wrote an eight-book compilation of approximately 900 military stratagems, which he dedicated to the emperors Marcus Aurelius and Lucius Verus for use in the war against Parthia. He organised them mainly chronologically but also broadly geographically, while excerpting the strategies of Romans and women to the final book – a decision that seems to be an afterthought upon reading the *Virtues of Women* of Plutarch.[4] The resulting text is thus not quite the practical and pragmatic work that Frontinus penned a century before, and quite probably Polyaenus, shaped by the traditions of the Second Sophistic, was trying to write something fairly distinct, eschewing the direct and explicit political advice popular in previous centuries in favour of something simultaneously didactic and philosophical.[5] Polyaenus tries to shift the genre by teaching the craft of war through examples, rather than precepts (that is, generalised discussion of tactics), while also resurrecting the combination of warcraft and statecraft in his Hellenistic models. For we see that not all the stratagems, strictly speaking, are set in a military context; several of the examples are more stratagems of rule rather than of war. This sort of non-explicit political advice is another feature of writing in the atmosphere of the Second Sophistic, and this, along with the frequent references to Plutarch (or Plutarch's sources), reveals the impact of the Second Sophistic upon Polyaenus. The *Stratagems* (*Strategamata*) is overtly a rhetorical work designed to focus on the historic exploits of the Greeks since other military writings tended to focus on the Romans' wars. Polyaenus desired not to write a text too similar to Frontinus, but to advance the genre of military writing by creating a new form of literature, blending historical, philosophical, anecdotal, and didactic forms together.[6] His intended reader, Marcus Aurelius, likely did make use of the book. Even Polyaenus suggests this at the beginning of his fifth book, an assertion dangerous to make if untrue.[7] Moreover, Polyaenus produced this work quite rapidly, probably having the entire work complete by 163, less than two years after publishing the first book of the collection. It is probable that Polyaenus wrote this work hoping to gain a court position, perhaps to be asked to write a history of the emperors. In writing, Polyaenus asserts himself unable to wage war in person, but also that Greek tactics have been overlooked in favour of Roman tactics.[8]

Scholars have not considered Polyaenus as a particularly careful or reliable source for historical events.[9] Without attempting to attest to his reliability, it is argued here that Polyaenus has not sought to include every possible Greek military strategy he could for his compendium, but that he curated these *Strategemata* carefully to include only certain strategies (or perhaps to *exclude* certain strategies).[10] Compared with other writers, Polyaenus seems to have particular interest in the financial aspects of generalship and in the employment of appropriate soldiers. It is not possible to know for certain whether these interests are personal or instead targeted to the concerns of the day, but in all likelihood, it is primarily the latter – that Polyaenus collected a selection of anecdotes that are meant to have an impact on the Antonine emperors. This is relevant because it suggests that this didactic work on strategy also had a rhetorical purpose with contemporary political concerns.[11]

This curation becomes clear when we examine the way Polyaenus discusses a commander's mercenaries – more as a potential foe to be dealt with than an ally. Several passages are devoted to the methods other leaders had used in dealing with mercenaries. Of particular interest appear to be the Greeks of Sicily, such as Dionysius, Hippocrates, or Agathocles, and for these extracts, he seems to be drawing heavily on Diodorus Siculus. Polyaenus presents not so much a negative view of mercenaries as he does an assumption that the reader also holds this view. Consider the following passage:

> Hippocrates very much wanted to get possession of the city of the Ergetini. He always gave his Ergetine mercenaries the largest share of the booty and

greater pay; excessively praised them as the bravest men in his army, and especially indulged them, so that he might have more allies from this city. All this was reported to the men at home. Envious of the benefits the soldiers were receiving, they all came out of the city voluntarily, leaving it deserted. Hippocrates welcomed the men in a very friendly way, and that same night led his forces through the Laestrygonian plain, stationing the Ergetini by the sea and the rest of his army inland. When the Ergetini were hemmed in by the sea, Hippocrates captured their deserted city by sending the cavalry ahead, ordered the herald to declare war on them, and gave the signal to the men of Gelo and Camarina to kill all the Ergetini with impunity.

(5.6; cf. 5.2.1)[12]

Hippocrates was able to capture a desired city by drawing all the men out in hopes of employment in his army. This passage, one may argue, concerns itself more with betraying an ally rather than preventing mercenaries from being a problem. But the mercenary aspect became particularly relevant for Hippocrates, who treated his conquests so ruthlessly that he could not recruit from them.[13] To Polyaenus, one can manipulate mercenaries by playing upon their stereotypical greed. Other passages describe how a general eliminates defecting mercenaries or deserters or convinces them not to leave (for example, Eumenes at 4.8.2). Still other passages, such as the following example, offer suggestions on how one can effectively cheat the men of their pay:

Dionysius wanted to stop paying his older soldiers. They were unhappy with him, and the younger ones agreed with them that it was unjust to be driven out because of age. Learning of the uproar, he called an assembly and spoke as follows: 'I am assigning the younger men to the danger of battle and the older men to garrison duty, for which they will receive equal pay, since having proven their reliability, they will be diligent guards and they will suffer less hardship.' All the soldiers were pleased and separated from one another in good spirits. After the army had broken up and was divided into small groups for many garrisons, then he dismissed the older men without pay individually, for they no longer had the crowd's support.

(5.2.11)[14]

The mercenaries' numerical strength posed a problem to Dionysius, and thus they are offered an incentive to split up, whereupon they could be defeated in detail. Consider also here the example of Andron, the mercenary pirate captain hired by Lycus to take Ephesus.[15] Andron, originally in the employ of Demetrius, switched sides, let the forces of Lysimachus into Ephesus, and then was dismissed immediately afterward. For Polyaenus, this is the fundamental risk of hiring mercenaries to fight for you: that they will switch sides.

While the individual tactics employed to get rid of mercenaries differed, one implication in the aggregate is that these mercenary forces were strong enough to pose a threat to their employers. Polyaenus includes these anti-mercenary tactics in his work because he considers them good strategy. The overarching portrayal from these passages and others is that mercenaries cannot be considered trust-worthy and need to be eliminated before they can harm one's goals. Effectively, Polyaenus urges the reader to betray their mercenary soldiers for financial, tacti-cal, and *moral* gain. The mercenaries will betray you, and you need to betray them first, which one can do easily, by manipulating them through false promises of higher pay and benefits. The overall objective is to move to a military system in which the soldiers are not motivated *principally* by pay.

By contrast, consider the example of the Athenian Lachares, whose mercenar-ies stopped the hoplite general Charias from betraying Athens to the Macedonians (which would thus also prevent Lachares from establishing a tyranny in Athens).[16] This example, recorded in earlier second-century Greek writers, would seem to fit right in with the others posed by Polyaenus on treachery, yet it is omitted. Instead, Polyaenus relates Lachares later escaping from Athens through bribing the mer-cenary cavalry (3.7.1).[17] Indeed the only stratagems employed by Lachares that Polyaenus mentions are his devious ways of escaping after losing a city.[18] Other clever acts, such as raising money to pay for his troops or his thwarting of the ini-tial attacks of Demetrios Poliorcetes on Athens go unmentioned. It is possible that Polyaenus, a self-proclaimed Macedonian, had little sympathy for these Athenian attempts to regain independence from various Diadochi overlords, but a clearer explanation at hand is that Lachares relied heavily on mercenaries for his success-ful actions and these mercenaries stayed loyal to Lachares. Polyaenus therefore decided against presenting these stratagems in his work, where they would work against his implications that the pay-motivated soldier is always inferior.

Ancient writers often provide advice or commentary concerning shortages of funds in the field.[19] Frequently enough, it is not the commanders' finances in short supply as much as it is their liquidity. Polyaenus is particularly interested in financial stratagems. At least 3% – a notable quantity in a work that attempts to cover a wide range of military strategy – of the examples in Polyaenus concern such battlefield funding issues, and these shortages are often explicitly described as caused by the employment of mercenary soldiers. This is not a unique fea-ture of Polyaenus. Xenophon details similar strategies for funding mercenaries in the *Hiero*. Additionally, to Polyaenus, mutinies are never to be negotiated with, except through deception. Here we have Polyaenus on one of the most famously merciful commanders, Julius Caesar: 'If, however, someone mutinied or deserted his post, he [Caesar] would not let him go unpunished'.[20] This is a stark and didac-tic statement that ignores other accounts of Caesar (see later, 127). Moreover, for Polyaenus, this is a moral injunction for the present day – commanders should never show the weakness of negotiation.

Not only that, society's distrust of mercenaries is so great that even if the spe-cific force of mercenaries *is* trustworthy, a general can play upon the negative reputation of mercenaries to get his opponent to dismiss his own mercenaries. The best example here is the instance of the Athenian commander Iphicrates, who is reasonably famous for employing and commanding mercenary soldiers:[21]

> When 2,000 mercenaries deserted to the Laconians, Iphicrates sent a secret
> letter to the deserters' leaders, exhorting them to remember the time agreed

upon, when he expected help from Athens. Iphicrates knew the letter would be intercepted by the troops guarding the roads. When the guards brought the letter to the Lacedaemonians, they hurried to arrest the deserters. They were glad to escape, having become untrustworthy to the Athenians and seeming untrustworthy to the Lacedaemonians.

(3.9.57)[22]

None of the sixty-three stratagems attributed to Iphicrates praise his frequent use of mercenaries; besides this passage, Polyaenus praises how Iphicrates withholds a quarter of their pay (3.9.51), marches them around the countryside to prevent mutiny (3.9.35), and tricked them into stopping a mutiny (3.9.59, which likely includes Athenians as well). This trick of getting the enemy to dismiss their own troops is known elsewhere by other commanders and ancient writers.[23] Moreover, this evidence also serves as a mild attestation of the battlefield effectiveness of mercenaries – many commanders, including Alexander the Great, took measures to avoid fighting enemy mercenaries.[24]

II

So far it would appear that Polyaenus opposed the employ of mercenaries. Poly-aenus is consistently more negative on employing mercenaries than many of his sources, who were somewhat more even-handed about it. Polybius records Scipio as saying that the soldiers must be well paid to prevent mutiny and betrayal.[25] Xenophon repeatedly makes this point as well, arguing that states relying on mer-cenary forces must, of necessity, maintain higher incomes, because 'their greatest expenses are those necessary to pay and maintain their mercenary force'.[26] Xeno-phon is less criticising the expense as cautioning the reader that it is an expense that must be accounted for. Xenophon also marks the success of the Spartan Teleutias as being at least partially because of his generosity and fair dealing with his paid soldiers and rowers.[27] Elsewhere, Polybius advises against employing a mercenary force larger than the citizen force.[28] For these earlier writers, mutiny is a situation avoidable with proper precautions. For Polyaenus, however, these precautions lead to bad, even immoral, domestic policy.

The modern reader might conclude that this is an unfair weighting of the evidence, but there are conspicuous absences. There is no evidence of the urg-ing of Xenophon or Scipio (by way of Polybius) to keep the soldiers well paid, only methods to trick the soldiers into compliance. Xenophon seems a crucially important comparison as an author who also describes many of the fifth- and early-fourth-century events Polyaenus uses as examples, but he may not be the source for the material Polyaenus uses. H. W. Parke notes that Polyaenus fre-quently diverges from Xenophon in describing the same event and does not appear to be using Xenophon as his source.[29] If Polyaenus *is* relying on Xeno-phon's accounts here, he appears to be retrojecting some corrections into the past. It remains possible that he is simply relying on a source lost to us.

Nevertheless, examples of commanders having positive relationships with mercenaries remain absent. Iphicrates, probably the most salient example, was discussed earlier. In Polyaenus, Philopoemon features in a mere three stratagems, most notably the adoption of Macedonian equipment by Achaean forces, but absent are his employment of mercenaries or as a mercenary, such as are found in Plutarch.[30] Alexander the Great, when marching through Anatolia, made deals with Greek mercenary garrisons that did not involve them breaking their word to their employers.[31] The Roman orator Cicero, when governing Cilicia, claimed credit for quelling a mutiny in Cilicia caused by his predecessor's failure to pay the local recruits fairly.[32] Diodorus' *Library of History* and (Pseudo-)Aristotle's *Oeconomica* also detail several successful negotiations between mercenary and employer, episodes that Polyaenus seems to ignore.[33] Successful commanders of the Hellenistic period often relied on mercenaries, sometimes with a majority of the troops being foreign mercenaries, and several are said to have treated their hired soldiers quite well. Pyrrhus and Hannibal, renowned tricksters (and employers of mercenaries), receive far less attention than we should expect.[34] If we expect Polyaenus to be trying to provide a range of stratagems, we should expect Polyaenus to record some of these situations and negotiations.

Polyaenus was quite thorough in his own research. Even if there may be some glaring omissions in the works he read, and even if we acknowledge that he is trying to limit his use of Roman *exempla*, these cannot explain the gap we find.[35] His relative lack of interest in Latin sources like Livy cannot explain the omission of other common sources used by Frontinus or ignoring sections of the writings of major historians like Polybius and Xenophon.[36] These must be conscious and intentional omissions.

What, then, can be the purpose of such omissions? By excluding examples of successful negotiation with mercenaries, and retaining examples of tricking and betraying them, Polyaenus conveys the message that the only successful tactics for dealing with mercenaries are overwhelmingly ruthless. Where other historians discuss commanders short of funds who pay off their hired soldiers with land or citizenship instead of coin, Polyaenus does not mention this technique, quite possibly because he does not consider that a good idea.[37] While attempts to bribe guards are noted, acts of hiring away enemies' mercenaries or local bandits are not (tactics which are noted in Livy, Diodorus, and Plutarch, among others).[38] Few examples of loyal mercenaries are brought before the reader, and it is implied that these loyal mercenaries are used by the leader to suppress the citizenry – such as in the case of Lachares at Athens or with Sicilian tyrants like Dionysius.

The economic aspect of mercenaries remains important. As David Whitehead points out, some generals were noted for economic stratagems.[39] While these may seem out of place for the overtly military context, such military stratagems could be key to a general's remaining supplied and effective in the field. These logistical considerations were important, especially because a mercenary force would be quicker to abandon a general failing to pay them.

Indeed the evidence that Polyaenus was interested in economic stratagems goes far beyond the five noted by Whitehead: Polyaenus has at least thirty examples that solve what we might call 'field cash-flow problems'.[40] The concern that

Polyaenus has for economic aspects of generalship elsewhere means that he is not ignoring the economic aspects of hiring mercenaries. But his solutions to the problem are overwhelmingly how to save money on employing mercenaries, which is to say by not paying them, dismissing them, or even killing them. Alternate methods of financing mercenaries (such as offering them land from the conquered territory or citizenship) are absent, even though these are well-attested methods in other authors.

It is difficult to know whether Polyaenus displays a personal concern or a situational concern in his discussion of economic matters. The funding system for armies on campaign was markedly different in the second century CE, and the Greek examples would seem to have little potential bearing on Roman armies in the field. His suggesting that the emperors would need to have recourse to tricks to solve field cash-flow problems in turn suggests a more tenuous hold over the empire and the army than we are generally led to believe for this period, anticipating the power indicated in the famous deathbed advice of Septimius Severus.[41] Moreover, the attention that Polyaenus pays to economics has further impact on how we interpret his work.

Namely, the concern Polyaenus seems to show about the financial aspects of waging war suggests that the imperial treasuries may have already been showing strain at the advent of the Parthian War. Fully 33 of his 900 examples of stratagems concern commanders using tricks to get money (see earlier, 126 n.40) to further their campaigns; we can add to this number if we broaden our definitions (for example, to include grain or other supplies). While it is well known that the treasury was empty at the time of the later wars with the Marcomanni, which had Marcus Aurelius selling the palace silver and jewellery, it is usually believed that Antoninus Pius, despite dealing with large rebellions, had left Lucius Verus and Marcus Aurelius a full treasury of 2700 million sesterces.[42] This reading of Polyaenus may suggest that that assumption is a more optimistic reading than warranted, if Polyaenus regards a mutiny driven by lack of funding to be a real danger.

III

At this point, let us compare Polyaenus to another writer of military tricks. Frontinus, who wrote a similar work in the previous century, both omits many of the attacks on mercenary forces and also includes a few examples of generals quelling a mutiny of the soldiers. He has four examples as their own section and others beside.[43] Pertinently, Frontinus includes an example that contradicts Polyaenus on Caesar (see earlier, 123):

> When certain legions of Gaius Caesar mutinied, and in such a way as to seem to threaten even the life of their commander, he concealed his fear, and, advancing straight to the soldiers, with grim visage, readily granted discharge to those asking it.

$$(1.9.4)^{44}$$

While Polyaenus insists that a great general like Caesar would not deign to nego-tiate, Frontinus allows that it might be necessary and there is a right way to do it. Though these examples in Frontinus are often described simply as 'soldiers', the line between paid foreign soldier and paid citizen soldier had become unclear even before the first century BCE, let alone the second century CE.

In Frontinus, the closest example of Roman attacks on one's own mercenary forces is Sulla's alleged claim to have done this to conceal his own strategic mistakes:

> The same Sulla, when certain auxiliary troops dispatched by him had been surrounded and cut to pieces by the enemy, fearing that his entire army would be in a panic on account of this disaster, announced that he had purposely placed the auxiliaries in a place of danger, since they had plotted to desert. In this way he veiled a very palpable reverse under the guise of discipline, and encouraged his soldiers by convincing them that he had done this.
>
> $(2.7.3)$[45]

It is extremely interesting that one might prefer to be regarded as intentionally sending a unit of men to their deaths rather than making a strategic error. Fron-tinus also provides a single example of the Carthaginians sacrificing their Gal-lic mercenaries for financial benefit (3.16.3). These examples do corroborate our earlier accounts of writers' low opinion of mercenaries. Overall, however, we find nothing like the volume of such suggestions we find in the corpus of Polyaenus, who has around fifteen examples of suppressing mercenaries by force or trick.

Already we have a definite shift in how these writers consider mercenary sol-diers. Frontinus also seems less concerned with the financial aspects of keeping an army in the field. None of his stratagems seem particularly concerned with money, with one possible exception that a general could choose to hire men with-out the money to pay them, since, if they were victorious, they could be paid out of the plunder. If not, they would be dead and therefore not need to be paid (4.7.35). The few examples of dealing with desertion tend to be placed with non-Roman commanders (half of them are Carthaginian generals stopping mercenaries from deserting to Rome, for example). Frontinus has a different didactic approach, per-haps driven by his greater use of later sources and events.

What, then, are we to make of these examples from Polyaenus? Of advice that lists several ways to murder one's own troops and get away with it? Of advice on how to cheat soldiers of their pay? Of advice on handling military mutinies ruth-lessly, whatever the cause? These surprising anecdotes, seen within their second-century context, suggest a few conclusions, seen in the following.

First, we must acknowledge that despite the fact that Polyaenus is writing in the second century CE, his examples dwell heavily on the events of the fifth–third centuries BCE, including the infighting in Greece and Sicily after the Pelopon-nesian War, the rise of Macedon, the wars of the Diadochi, and the Punic Wars.[46] This particular period saw a lot of mercenary activity and also saw many states straining to control and retain a sufficient supply of military manpower. In other words, his source material may drive him to draw conclusions less applicable to contemporary warfare. On the other hand, Polyaenus does assert that these stratagems are relevant to warfare in the east in the second century CE, and it may be reasonable to consider that the relationship between soldier and general and their motivations were of particular interest.

Examining the Hellenistic period, we see that the professional mercenary soldier appears to have been outperforming the traditional drafted militia soldier, and accordingly, the most successful states move towards permanent standing armies, just as the Romans would do in the first century; however, states and generals had to come up with the wherewithal to pay such forces, and this appeared to be a problem for many, as noted earlier. Mercenaries who became unemployed might resort to brigandage as did the many bands of mercenaries fired by the satraps of Persia on Alexander's orders.[47] The in-lieu-of-pay grants of land and citizenship by various Greek states appear to be forerunners of the Romans' land-grants that we see in the first century BCE and the granting of citizenship to auxiliaries that we see in the late republic and empire. Despite this trend, we also see no shortage of voices clamouring for the good old days.[48]

A noted trend later in the Roman Empire was the 'barbarization' of the military, often argued to have accelerated under Septimius Severus.[49] Leaving aside the modern debates of what exactly was going on and whether or not this was a bad thing, many ancient commentators, such as Cassius Dio or later, Ammianus Marcellinus, were thoroughly opposed to what they saw. For many years the Roman military had continually sought additional sources of men, extending recruitment to more and more groups (whether ethnic groups or social classes). The extension of military service to new groups generally also brought with it the extension of citizenship and land grants.[50] Offers of citizenship and land had long proven to be effective methods of attracting men. At the same time, however, these offers were frequently resented by those who already possessed these things.[51] The social benefit of citizenship could be seen in the differing organisation (and treatment) of the legions and the *auxilia*.[52] In particular, the 140 CE rescindment of citizenship grants to the children of veterans theoretically functioned to make those children more likely to enlist in the *auxilia*.[53]

Arguably, the emperors were supposed to rethink their position on the employment of a professional mercenary army instead of a drafted volunteer force. In this interpretation, Polyaenus warns of a social danger – namely, that it is immoral and dangerous to employ mercenaries for too long, and this internal danger should trump practical military considerations. The wars in the east, to Polyaenus, offer Rome the opportunity to rid themselves of multiple dangers simultaneously. With the mercenaries sent away and somehow tricked or eliminated, the state can be stronger and wealthier.

Polyaenus, if this is his motive, would not be alone in this sort of line of thought. Joseph Roisman argues that this sort of manipulation was indeed considered praiseworthy.[54] Many Greek and Roman historians voice similar sentiments in dealing with mercenary forces: preferably do not deal with them at all, otherwise dismiss them as soon as possible. We always must face *a priori* assumptions by the historians about the mercenaries when talking about them. So, while Polyaenus is interested in avoiding the intrinsic problem of treachery, Xenophon is concerned about using mercenaries while avoiding citizen resentment, and Polybius, in some prescriptive statements about mercenaries in books 1–6, voices various concerns about maintaining a suitably strong citizenry.[55] Some authors will advise retaining a few specialist mercenaries, such as Xenophon's advice to maintain 20% of the army as mercenaries in *Cavalry Commander* (9.3),[56] but most appear to advocate for an unpaid, drafted militia, something that had not been the case at Rome for over two centuries.

Advocating this was not an *overt* motive on the part of Polyaenus. Fewer than 10% of the anecdotes concern mercenaries *per se*, and that is simply not enough in the work to try to make a compelling argument about mercenary service to the emperors. This is instead indicative of a philosophical view of the time, which Polyaenus shares. Effectively, Polyaenus does not try to make the argument that mercenaries are undesirable, but accepts it as a given. With this *a priori* assumption in place, this work becomes not a curation of anecdotes that attempts to convince directly, but a curation that follows and accepts this train of thought, excluding scenarios that go against accepted wisdom. This remains fundamentally the rationale behind his curatorial decisions, but the resulting selection lacks a certain amount of didactic force by being unconscious.

IV

The simple existence of collections of stratagems also underlines the importance of the commander, an early form of Carlyle's 'Great Man Theory'.[57] Military success in these anecdotes is not determined by the quality of the soldier as much as the quality of the leader. By contrast, later military writers, such as Vegetius, do not echo this ideal. The idea of the well-trained professional soldier being the linchpin of the army gives less credit to the general's genius. This would generally correspond well to the apparent hostility towards mercenary soldiers in a work that is trying to extol the abilities of famous commanders.

The advice to exclude mercenaries, and by fairly drastic measures at that, attests to the power of such forces, even in the second century CE. These anecdotes hide an ulterior motive – to stop common mercenary soldiers from attaining the social power or privilege that belongs to the aristocratic class. In some authors, this is a clear motivation; in others, it is less overt, and potentially unrealised. Whether the bias is intentional is not necessarily a question that needs to be answered to assert that it exists. The Antonine army was increasingly made up of immigrants and non-citizens. For Polyaenus, I believe it is clear that he does not want these soldiers of the contemporary army to gain further power in society. Whether he overtly intends this work to convey such a message is nowhere near as clear-cut.

Notes

1 To address this point (of mercenaries' effective performance) in any detail is beyond the scope of this paper (and addressed in Beek (2015) 205–218, or Trundle (2004)

27–39). The most famous is likely the Ten Thousand of Xenophon's *Anabasis,* but other examples of mercenaries outperforming citizen-soldiers on the battlefield, from Athenian mercenaries at Lechaeum to Hannibal's mercenaries in the Second Punic War, exist in some abundance.

2 This perhaps leads Polyaenus to have more stratagems by tyrants and kings such as Dionysius and Agesilaus, than favourites of Frontinus, like Hannibal, Pompey, and Epaminondas (Caesar and Alexander appear prominently in both). Cf. Chapter 10 *passim,* esp. 191–192.

3 Unless otherwise indicated, references are to Polyaenus' *Stratagems* (*Strategemata*). All quotations from Polyaenus come from Krentz and Wheeler's edition. Translations from Xenophon, Plutarch, and Polybius (except where noted) are my own.

4 A dozen of Plutarch's examples (16–20, 253F–258C) very closely resemble their counterpart accounts in Polyaenus, particularly the stories of Pieria, Polycrita, Lampsace, Aretaphila, and Camma, which Polyaenus records in the same order (8.35–39).

5 See Wheeler (1993) 23 for one assertion that Polyaenus and Frontinus wanted to be seen as writing literature.

6 I am inclined to avoid the classification of all military texts as a single 'genre'. I perceive major distinctions between the different manual writers depending on whether their prospective audience achieves position through election, appointment, or inheritance. For example, Polyaenus includes much material about ruling that Xenophon includes in the *Hiero* but does not in the *Cavalry Commander.* If it is a single genre, then it is one that evolved over time, as this volume as a whole indicates.

7 See, for example, book 5's preface (addressing the emperors): 'I think I do not deserve so much praise for the writing as you do for the diligent reading of such large works' ~ προσφέρω τῶν Στρατηγημάτων οὐχ οὕτως ἐμαυτὸν ἄξιον ἐπαίνου ἡγούμενος ἐπὶ τῷ συγγράφειν, ὡς ὑμᾶς ἐπὶ τῷ σπουδάζειν ἀναγιγνώσκειν τοσαῦτα συγγράμματα.

8 Polyaenus makes this assertion in the preface of book 1.

9 See Wheeler (2010) 7 for modern assessments and further bibliography.

10 Excluding strategies only makes sense if the author disagrees with them either in their validity or in their longer-term ramifications. In this paper, I focus only on the careful curation of stratagems dealing with mercenaries in order to exclude regular mercenary employment as a matter of public policy. Polyaenus may well intentionally exclude other strategies promoted in manual authors like Frontinus or historians like Polybius as well.

11 See Chapter 3 for arguments on the impact of these manuals on 'civilian' readers.

12 Ἱπποκράτης κρατῆσαι τῆς Ἐργετίνων πόλεως ἐσπουδακώς, ὅσους Ἐπεγετίνους εἶχε μισθοφόρους, τούτοις ἔνεμεν ἀεὶ τῆς λείας τὸ πλεῖον μέρος καὶ μισθοὺς μείζονας ὑπερεπαινῶν αὐτοὺς ὡς πλείονας ἐκ τῆς πόλεως ταύτης ἔχοι συμμάχους. ταῦτα ἠγγέλλετο τοῖς ἐν τῇ πόλει · οἱ δὲ ζηλώσαντες τὴν ὠφέλιαν τῶν στρατευομένων ἐθέλοντες πάντες ἐξῆλθον καταλιπόντες ἔρημον τὴν πόλιν. Ἱπποκράτης φιλοφρόνως δεξάμενος τοὺς ἄνδρας, αὐτῆς νυκτὸς ἀναλαβὼν τὴν δύναμιν διὰ τοῦ Λαιστρυγονίου πεδίου προῆγε, τοὺς μὲν Ἐργετίνους τάξας πρὸς τὴν θάλατταν, τὴν δὲ ἄλλην στρατιὰν πρὸς τὴν ἤπειρον. ἐπεὶ δὲ ἀπεφράχθησαν πρὸς ταῖς ῥαχίαις τῶν κυμάτων οἱ Ἐργετῖνοι, τοὺς ἱππεῖς προπέμψας Ἱπποκράτης τὴν πόλιν αὐτῶν ἔρημον οὖσαν κατελάβετο καὶ τὸν κήρυκα πόλεμον αὐτοῖς προειπεῖν ἐκέλευσε καὶ σύνθημα Γελῴοις καὶ Καμαριναίοις ἔδωκε κτείνειν ἀδεῶς Ἐργετίνους ἅπαντας.

13 This became particularly problematic in Zancle, where Hippocrates wanted to build and recruit a fleet. See Dunbabin (1948) 404–405 and Asheri (1963) 759–766. After Hippocrates mistreated his mercenaries so, he was unable to recruit further men from there.

14 Διονύσιος τοὺς πρεσβυτέρους τῶν στρατιωτῶν ἀπομίσθους ἐβούλετο ποιῆσαι· οἱ δὲ ἠγανάκτουν πρὸς αὐτὸν καὶ οἱ νέοι ὡς ἄδικον, εἰ γηρῶντες ἐκβάλλοιντο. ὁ δὲ, τὸν θόρυβον μαθὼν συναγαγὼν ἐκκλησίαν ἀναγορεύει τάδε· 'τοὺς μὲν νεωτέρους ἐς τὸν κίνδυνον τῆς μάχης τάττω, τοὺς δὲ πρεσβυτέρους ἐς τὰς φυλακὰς τῶν χωρίων ἴσην σύνταξιν λαμβάνοντας· πεῖραν γὰρ πίστεως δεδωκότες ἐπιμελῶς τὰ χωρία φυλάξουσι καὶ πονήσουσιν ἔλαττον.' ἥσθησαν οἱ στρατιῶται πάντες καὶ διῃρέθησαν ἀπ᾽ ἀλλήλων

ἄσμενοι. ἐπειδὴ δὲ διελύθη τὸ πλῆθος καὶ κατ᾽ ὀλίγους εἰς πολλὰς φυλακὰς ἐμερίσθη, καὶ δὴ τότε ἑκάστους τῶν πρεσβυτέρων ἀπομίσθους ἐποίησεν οὐκέτι τὴν παρὰ τοῦ πλήθους ἐπικουρίαν ἔχοντας. Krentz and Wheeler translate the verb phrasing at the beginning of this section (ἀπομίσθους ἐβούλετο ποιῆσαι) as 'retire', which fails to get across entirely the financial motive.

15 5.19.1; cf. Frontinus 3.3.7. Inscriptions from the same period (*SIG* 363–364) note that Ephesus was in severe financial trouble and also sold citizenship to some of their allied mercenaries (who were exiles from Priene) for cash. See also Laale (2011) 112–117 for the taking and retaking of Ephesus.

16 In this example, Lachares himself is only referred to as the general of the mercenary troops, not necessarily a mercenary himself. See *FrGH* 257a (Phlegon?); cf. Pausanius 1.25.7, Hammond and Walbank (1988) 206–207.

17 The 'Tarentine' cavalry here probably indicates mercenary cavalrymen outfitted like Tarentines, not that the mercenaries were necessarily Tarentine in origin (though they easily may have been). See Eckstein (2006) 147–158. For further discussion on whether or not 'Tarentines' came from Tarentum, see Bugh (2011).

18 Lachares is admittedly not a well-known figure: Ferguson (1929).

19 These could be quite substantial: Aeneas Tacticus claims to have written a separate treatise on military finance (mentioned at 14).

20 8.23.21; cf. 8.6.5, 8.23.15.

21 For how Polyaenus treats Iphicrates more generally, see Whitehead (2003).

22 Ἰφικράτης δισχιλίων μισθοφόρων πρὸς τοὺς Λάκωνας αὐτομολησάντων ἀπόρρητα γράμματα ἔπεμψε πρὸς τοὺς ἡγεμόνας ἀποστάντων παρακαλῶν μεμνῆσθαι τοῦ συγκειμένου καιροῦ, καθ᾽ ὃν καὶ τὴν ἐξ Ἀθηνῶν συμμαχίαν προσδέχοιτο εἰδὼς τὴν ἐπιστολὴν ἐμπεσουμένην τοῖς φύλαξι τῶν ὁδῶν. ἐπεὶ δὲ οἱ φύλακες ἐκόμισαν τὰ γράμματα τοῖς Λακεδαιμονίοις, οἱ μὲν ὥρμησαν συλλαμβάνειν τοὺς αὐτομόλους· οἱ δὲ ἀγαπητῶς ἐξέφυγον Ἀθηναίοις μὲν ἄπιστοι γενόμενοι, Λακεδαιμονίοις δὲ δόξαντες.

23 For these false betrayals, see Polyaenus 5.2.17, Xenophon *Cavalry Commander* 4.7, Diodorus Siculus 19.26, and Livy 24.31. Cf. Polyaenus 4.3.15, for Alexander's pretence to have suborned the mercenary general Memnon.

24 For more hints at the effectiveness of Alexander-era mercenaries, see Roisman (2012) 175, 220, who indicates that the Successors' actions denote higher reliance on the mercenaries than our sources would prefer to admit.

25 See Polybius 11.25–30 on Scipio's words at Sucro (cf. Livy 28.24–32; Cassius Dio 16 Fragments (217 in Cary's Loeb). Cf. Polybius 10.16–17 on equitable distribution of booty in a more general sense.

26 Xenophon, *Hiero* 4.9 (cf. 6.11); *Cavalry Commander* 9.3.

27 See Xenophon, *Hellenica* 5.1.13–24.

28 The most notable of these is his commentary on the Celtic mercenaries in Epirus; this advice mirrors Aeneas Tacticus (12–13), whom we know Polybius to have read.

29 Parke (1931) 33–34.

30 E.g. *Philopoemon* 7, 8, 20. For discussion of this Life, see Swain (1988). These details probably came from the earlier biography of Philopoemon by Polybius, which is no longer extant. For likely differences between Plutarch and Polybius on Philopoemon, see Beek (2015) 212; Pelling (1997) 135–139 and 291–309 and (2002) 246, 250 n.22, 350 pointing at Polybius 24.12 in particular; Errington (1969) 27–28.

31 One example of this occurs at Arrian 1.29, where Alexander agrees to a truce that lets the mercenaries leave after the Persians fail to relieve them.

32 Cicero, *Letters to Friends* 15.4 (to Cato): 'When, **before my arrival**, the army was broken up by **something like a mutiny**, five cohorts were lacking a legate, a military tribune, or even a single centurion' ~ *cumque ante adventum meum seditione quadam exercitus esset dissipatus, quinque cohortes sine legato, sine tribuno militum, denique etiam sine centurione ullo*. On the other hand, see *Att.* 5.4 for a rather different account of the same event (giving credit to Appius).

33 [Aristotle] *Oec*.1347b–1353b, including some eight examples of placating mercenaries demanding funds.

34 See Wheeler (2010) 37f. on this topic. For Hannibal in Polyaenus and Frontinus, see Chapter 10.

35 This gap has been noted for other authors. For example, Polybius also has very little to say about his hero Philopoemon's travels as a mercenary commander in Greece, though the relevant sections (for example, 10.21–24) are admittedly fragmentary. See Errington (1969) 27–45.

36 Overall, the case that Polyaenus went to substantial effort in assembling his examples in a short time is credibly defended in a short note by Dorjahn (1929). See also Wheeler (2010) 31.

37 This was fairly common in Sicily (e.g. Gelon and Theron as well as under Dionysius and later Hieron), Anatolia (Attalus, in particular), and Ptolemaic Egypt, where cleruchies were offered to would-be phalangite settlers.

38 Plutarch, *Cato Maior* 10.2; Livy 26.40, 27.11, 34.17–23; Diodorus Siculus 16.47–52, 37.18; Josephus, *Life of Apion* 14. Cf. the later accounts in Cassius Dio (2.10, 12 *passim*) and the life of *Marcus Aurelius* in the *Augustan History*. 21.7. See also the reverse threat of mercenaries to switch sides if not paid: for instance, Diodorus Siculus 13.88.

39 Whitehead (2003) 615.

40 Leaving aside a few less clear examples, Polyaenus displays such financial preoccupation at: 1.34.2; 3.9.30, 35–36, 51; 3.10.1, 9–10; 3.11.5, 4.2.6; 4.6.17; 4.10.1–2; 5.1.1, 5.2.11, 19, 21; 6.1.1–7, 6.2.2; 6.9.1–2; 6.51.1; 7.21.1; 7.23.1; 7.27.3; 7.31.1; 7.32.1.

41 Cassius Dio 77.15.2: 'Get along with each other, enrich the soldiers, and despise everyone else' ~ ὁμονοεῖτε, τοὺς στρατιώτας πλουτίζετε, τῶν ἄλλων πάντων καταφρονεῖτε. Almost needless to say, this is not advice with which Polyaenus would agree. For the general stability and peacefulness of the empire under Antoninus Pius, see the life of Pius in the *Augustan History*, and also Heller (2016) 358.

42 The *Augustan History* also speaks frequently of his frugality, though still more often of his not making the exactions of earlier emperors (for example, 7.5–8.5, 11.1). For a study of what the author(s) of the *Augustan History* thought about prosperity in the Roman Empire, see Reekmans (1979) 240–246. See also the assessment of Heller (2016) *passim* that Pius engaged in more warfare then generally thought.

43 This is section 1.9. Other examples are to be found: for example, 1.11.1.

44 *C. Caesar, cum quaedam legiones eius seditionem movissent, adeo ut in perniciem quoque ducis viderentur consurrecturae, dissimulato metu processit ad milites postulantibusque missionem ultro minaci vultu dedit.*

45 *Idem, cum auxiliares eius missi ab ipso circumventi ab hostibus et interfecti essent verereturque, ne propter hoc damnum universus trepidaret exercitus, pronuntiavit auxiliaris, qui ad defectionem conspirassent, consilio a se in loca iniqua deductos. Ita manifestissimam cladem ultionis simulatione velavit et militum animos hac persuasione confirmavit.*

46 For a discussion of anecdotes in education, see Gibson (2004) 110–111.

47 Diodorus Siculus 17.111.1.

48 These tensions probably resulted in the creation of the *civitas sine suffragio* status common in the Republic. See Mouritsen (2007) for several examples to senate opposition to such grants. In the principate, Juvenal and Tacitus stand out as obvious purveyors of nostalgia. For opposition to particular events, see Plutarch, *Lucullus* 42, disputing the *acta* of Pompey; see Haynes (2016) 31–34 for citizenship grants to auxiliaries and allies before and after the Social War. See Cassius Dio 78.9 for condemnation of Caracalla's later extension of citizenship.

49 Cassius Dio makes much of this (for example, 75.2.5). See Haynes (2016) 75–92. On the role of foreign soldiers in the high empire, see Southern and Dixon (1996) 46–53 for the fourth and fifth centuries.

50 For the trend, see Haynes (2016) 31–34, 63 (in turn following Keppie (1984) 79). Famous examples include the enfranchisement of the *turma Salluitana* in 89 BCE (*CIL*

1.709), which was unusual but not anomalous, and the claims of Augustus to grant land to veterans in *Res Gestae* 28.
51 Diodorus Siculus 11.72 recounts an early civil war in Syracuse over land and citizenship. Consider also Juvenal, *Satire* 16, for evidence of general resentment of soldiers' privileges.
52 For some of the citizenship distinctions, see Haynes (2016) 98–99.
53 Luttwak (1976) 122 makes this point, but it appears elsewhere as well.
54 Roisman (2012) 239, attributes the following sentiment, *inter alia*, to Hieronymus of Cardia: 'Tricking troops out of their due payments is considered creditable in a commander'.
55 For Xenophon, see *Hiero* 8.10, 10.3 and *Cavalry Commander* 9.3. See also Polybius 1.65, 4.74, 6.52 for some prescriptive statements about mercenaries, some of which may be rooted in Aeneas 12–14.
56 'Further, I am of the opinion that the full complement of a thousand cavalry would be raised much more quickly and in a manner much less burdensome to the citizens if they established a force of two hundred foreign cavalry' (Loeb translation).
57 For examinations of ancient prototypes of this idea, see Ferrario (2014) and Grethlein (2015). Ferrario identifies a growing importance of individual agency in ancient writing. Several authors take Suetonius and Plutarch to indicate an early Great Man Theory: for example, Benario (2003) 402 and so also Tritle (1995) 110 describe early German scholarship and poetry on Plutarch.

Bibliography

Anson, E. M. 1985. 'The Hypaspists: Macedonia's Professional Citizen-Soldiers'. *Historia* 34. 246–248.
Asheri, D. 1963. 'Laws of Inheritance, Distribution of Land and Political Constitutions in Ancient Greece'. *Historia* 12. 1–21.
———— 1988. 'Carthaginians and Greeks', in J. Boardman, N. G. L. Hammond, D. M. Lewis, and M. Ostwald, eds., *Cambridge Ancient History Volume IV: Persia, Greece and the Western Mediterranean, c. 525–479 B.C.* Cambridge: Cambridge University Press. 739–790.
Beek, A. L. 2015. *Freelance Warfare and Illegitimacy: The Historians' Portrayal of Bandits, Pirates, Mercenaries, and Politicians*. PhD Dissertation. University of Minnesota.
Benario, H. W. 2003. 'Teutoburg'. *CW* 96. 397–406.
Bugh, G. H. 2011. 'The Tarentine Cavalry in the Hellenistic Period: Ethnic or Technic', in J.-C. Couvenhes, S. Crouzet, and S. Péré-Noguès, eds., *Pratiques et identités culturelles des armées hellénistiques du monde méditerranéen*. Bordeaux: Ausonius. 283–292.
Dorjahn, A. P. 1929. 'Polyaenus and the Cycle'. *CJ* 24. 530.
Dunbabin, T. J. 1948. *The Western Greeks*. Oxford: Oxford University Press.
Eckstein, A. M. 2006. *Mediterranean Anarchy, Interstate War, and the Rise of Rome*. Berkeley: University of California Press.
Errington, R. M. 1969. *Philopoemen*. Oxford: Oxford University Press.
———— 1989. 'Rome and Greece to 205 B.C.', in A. E. Astin, F. W. Walbank, M. W. Frederiksen, and R. M. Ogilvie, eds., *Cambridge Ancient History Volume VIII: Rome and the Mediterranean World to 133 B.C.* Cambridge: Cambridge University Press. 81–106.
Ferguson, W. S. 1929. 'Lachares and Demetrios Poliorcetes'. *CPh* 24. 1–31.
Ferrario, S. B. 2014. *Historical Agency and the 'Great Man' in Classical Greece*. Cambridge: Cambridge University Press.
Gabbert, J. J. 1986. 'Piracy in the Early Hellenistic Period: A Career Open to Talents'. *Greece & Rome* 33. 156–163.

Gibson, C. A. 2004. 'Learning Greek History in the Ancient Classroom: The Evidence of the Treatises on Progymnasmata'. *CPh* 99. 103–129.

Grethlein, J. 2015. 'Review of Ferrario 2014'. *Histos* 9. 56–60.

Griffith, G. T. 1935. *The Mercenaries of the Hellenistic World*. Cambridge: Cambridge University Press.

Hammond, N. G. and Walbank, F. W. 1988. *A History of Macedonia Volume III: 336–167*. Oxford: Oxford University Press.

Haynes, I. 2016. *Blood of the Provinces*. Oxford: Oxford University Press.

Heller, A. 2016. 'The Religious Legitimation of War in the Reign of Antoninus Pius', in K. Ulanowski, ed., *The Religious Aspects of War in the Ancient Near East, Greece, and Rome*. Leiden: Brill. 358–375.

Keppie, L. 1984. *The Making of the Roman Army*. London: Batsford.

Krentz, P. and Wheeler, E. L. 1994. *Polyaenus: Stratagems of War*. Chicago: Ares Publishers.

Laale, H. W. 2011. *Ephesos: An Abbreviated History from Androclus to Constantine XI*. Bloomington, IN: WestBow Press.

Loman, P. 2004. 'No Woman No War: Women's Participation in Ancient Greek Warfare'. *Greece & Rome* 51. 34–54.

Luttwak, E. N. 1976. *The Grand Strategy of the Roman Empire from the First Century A.D. to the Third*. Baltimore: Johns Hopkins University Press.

Mouritsen, H. 2007. 'The *Civitas Sine Suffragio*: Ancient Concepts and Modern Ideology'. *Historia* 56. 141–158.

Parke, H. W. 1931. 'The Evidence for Harmosts in Laconia'. *Hermathena* 46. 31–38.

———— 1933. *Greek Mercenary Soldiers: From the Earliest Times to the Battle of Ipsus*. Oxford: Oxford University Press.

Pelling, C. B. R. 1997. *Plutarco: Vite Parallele: Filopemene-Tite Flaminino*. Milan: Biblioteca Universale Rizzoli.

———— 2002. *Plutarch and History: Eighteen Studies*. Swansea: Classical Press of Wales.

Reekmans, T. 1979. 'Prosperity and Security in the "Historia Augusta"'. *AncSoc* 10. 239–270.

Roisman, J. 2012. *Alexander's Veterans and the Early Wars of the Successors*. Austin: University of Texas Press.

Southern, P. 2007. *The Roman Army: A Social and Institutional History*. Oxford: Oxford University Press.

———— and Dixon, K. R. 1996. *The Late Roman Army*. New Haven: Yale University Press.

Swain, S. 1988. 'Plutarch's "Philopoemen and Flamininus"'. *ICS* 13. 335–347.

Tritle, L. A. 1995. 'Plutarch in Germany: The Stefan George "Kreis"'. *IJCT* 1. 109–121.

Trundle, M. 2004. *Greek Mercenaries from the Late Archaic Period to Alexander*. London: Routledge.

Wheeler, E. L. 1993. 'Methodological Limits and the Mirage of Roman Strategy: Part I'. *Journal of Military History* 57. 7–41.

———— 2010. 'Polyaenus: Scriptor Militaris', in K. Brodersen, ed., *Polyainos: Neue Studien—Polyaenus: New Studies*. Berlin: Verlag Antike. 7–54.

Whitehead, D. 2003. 'Polyaenus on Iphicrates'. *CQ* 53. 613–616.

7 Xenophon's *On Horsemanship*

The equestrian military manual

Lucy Felmingham-Cockburn

Xenophon's *On Horsemanship* (Περὶ ἱππικῆς or *Peri Hippikes*) generally dates to the late 360s or early 350s and ostensibly provides instruction on the selection, training, and maintenance of the cavalry's defining piece of military equipment: the horse.[1] The text is one of the earliest extant examples of a treatise on horses and horsemanship, and is preceded (as far as we know) only by the Hittite Kikkuli text (c. 1400 BCE) and a fifth-century-BCE text by Simon of Athens.[2] After Xenophon, the subgenre extends through the Roman world mainly in the form of military training exercises (*hippika gymnasia*).[3] It continued into the Byzantine World, in the form of equine veterinary texts such as the *Hippiatrica* (which translates loosely as 'matters concerning equine veterinary care').[4] The genre, and particularly Xenophon's *On Horsemanship*, are then revived again in Renaissance texts and the practice of the art of horsemanship among the nobility of Europe.[5] It is largely thanks to this revival that modern equestrian communities still frequently refer to Xenophon's text.

It is unfortunate that scholars consider *On Horsemanship* to be a minor text of the Xenophonic corpus and it thus has received minimal scholarly attention, which may itself be a symptom of the author's varying popularity across time.[6] However, scholars do use the treatise as a source of evidence in the study of the cavalry in ancient Athens and the wider Greek world, a topic that itself is now under revision.[7] One aspect of Xenophon scholarship in general over the last fifty years has been linguistic analysis and the study of metaphor.[8] Application of this approach to *On Horsemanship* can provide insight as to whether the agenda of the text is limited to communication of (perfectly sound) horsemanship knowledge or whether the author also uses the horse as a metaphor for sociopolitical issues. Closer analysis of the language of *On Horsemanship* reflects upon how and why Xenophon makes word and imagery choices, which results in intertextual links to other literary texts (including other military literature), which in turn exposes an additional level of meaning of *On Horsemanship*. One ought not to underestimate the possible significance of this: it tethers a military manual to ancient literature more broadly (and ancient military thought to broader intellectual enquiry generally), which frees military texts from being a recluse genre. That is, military manuals are not an isolated genre, but arguably its authors seek to provide an alternate angle of enquiry into, and to engage with, broader public debate on important military and political topics.

I

There are some important obstacles that one ought to acknowledge when considering *On Horsemanship*. Most significantly, the ambiguity as to exactly when in his life and career Xenophon wrote the treatise means that his precise motivations are unclear: does this text represent realities of the classical Athenian cavalry, theoretical musings, or both? Does Xenophon write as a member or supporter of the classical Athenian cavalry or at least the Athenian 'cavalry class'?[9] This in turn raises the question of the intended addressees of the text. The opening sentence provides no clear indication to the target audience nor, in fact, to whom the authoritative voice of the work belongs:

> Since on account of our association with the equitation activity we consider ourselves experienced in the art of horsemanship, we wish also to demonstrate to the younger of our friends that which we consider to be the most correct way to manage horses.

$$(1.1)[10]$$

One cannot be sure that this is simply Xenophon, as a seasoned cavalryman, outlining advice for new recruits. The use of the first-person plural verbs, the phrase τοῖς νεωτέροις τῶν φίλων, and the consequential notion of an older generation handing down information to a younger one evoke the sentiment of sociopolitical education. Indeed, Xenophon's equestrian 'writings' are not dissimilar to many of his works in that they have the potential to be interpreted as metaphorical discussions of wider societal issues such as leadership, education, and government. This applies not only to *On Horsemanship* and the *Cavalry Commander*, but also to the *Memorabilia* and *Oeconomicus*.[11] This analysis is not lost on recent scholarship.[12] Developing on from such investigations, a focus specifically on the symbol of the horse in the representation of the cavalry in classical Athens – through linguistic and thematic commentary of *On Horsemanship* – is compelling. Xenophon's personal history with the Athenian cavalry, from his own service before his exile to the service of his sons around the time to which *On Horsemanship* is dated would explain Xenophon's interest in the social-integration of the classical Athenian cavalry. At this time, the classical Athenian cavalry was generally relegated to the auxiliary part of the Athenian war machine, which either led to or was because of its rather unflattering aristocratic identity in an Athens of developing democracy.[13] So at least one of his motivations for *On Horsemanship* is to address an existing need to promote the cavalry on both a military and social level.[14]

With this in mind, exploration of two connected aspects of *On Horsemanship* may assist in understanding Xenophon's efforts to maximise the utility of the text:

the first is the author's comparing the horse to the hoplite, chiefly through body imagery. The second aspect is the author's possible comparison of the horse to Athenian maritime war culture through naval imagery. In addition, although to some extent it will apply to both working examples, naval imagery exemplifies how Xenophon can avail himself of a well-established motif in Greek literature, namely the 'ship-of-state'.

II

Regardless of military realities, contrast between cavalry and hoplites is certainly a theme which runs through contemporary literature. In his seminal work on the cavalry both on the battlefield and in society, Iain Spence documents the traditional accusations against the cavalry – wealth, aristocratic connotations, lack of bravery – and charts how these accusations find their way, with varying degrees of subtlety, into contemporary philosophical discourse, forensic speeches, drama, and material culture.[15] Despite such repeated portrayals of the cavalry (and therefore reasonably established characterisation in the public consciousness), there are admittedly counterarguments for these accusations, which Spence himself examines before concluding that the only irrefutable claim is that the cavalry are not hoplites: the cavalry and its developmental history to the point of its identity in classical Athens does not represent the new radical democracy with which the hoplite becomes associated, where the emphasis is on the community as a whole.[16] Most crudely, the cavalry represents an older, aristocratic world of individualistic interest, and the hoplites are the embodiment of a dominating sociopolitical ideal in the developing democracy of classical Athens. The result is that many literary references can be categorised either as accusations against the classical Athenian cavalry (and Spence's 'cavalry class'), or as (self-) justification by the cavalry and their supporters.[17] Indeed, this is not only seen in literature but also in public art – the most compelling example of which is the Dexileos Memorial.[18] One way in which supporters of the cavalry could counter such accusations would logically be to find a way to equate the cavalry with hoplites.

The most simplistic, physical difference between the cavalryman (that is, the rider) and the hoplite is that the former is mounted on a horse and the latter is a foot soldier – standing on his own two feet. Xenophon contrasts συμβαίνω with διαβαίνω in order to describe how the horse of ideal conformation should stand:

> If he [the horse] has thighs which are divided under the tail by a wide line, so he will place under his legs at the back [i.e., the hind legs] much apart. And in doing this he will be both more fierce and more strong with regard both to standing under [i.e., pushing power through from the haunches] and to being ridden, and in all respects he will be the better of himself. One judges this also from the example of men. For when men wish to lift something from

the ground, they endeavour to lift it up, **standing with their legs completely apart** rather than **standing with the feet close together**.

(1.14)[19]

The verb διαβαίνω, translated more frequently as 'to stand with feet planted firmly apart', is particularly evocative of the hoplites because of their battle stance. The Homeric poet uses it to describe soldiers, as does Aristophanes.[20] Given his importance in ancient Greek military literature, one should not be surprised to find it also in Tyrtaeus' verse.[21] It appears in the poet's description of the ideal hoplite:

> Come, **one should plant oneself firmly fixing both feet on the ground**, biting his lip with his teeth, covering thighs, shins below, chest and shoulders with the broad belly of his shield; and let him brandish a mighty spear in his right hand, and shake the fearsome crest over his head; by doing mighty deeds let him learn to make war, and make him not stand holding his shield beyond the missiles, but coming with a long spear or sword wounding the enemy take the man, and placing foot alongside foot and having pressed shield against shield, crest on crest and helmet to helmet and chest to chest having drawn near, let him fight a man, holding sword blade or long spear. And you, light-armed men, crouching beneath a shield in one place or another throw large stones, hurling smooth spears at them, standing close to those in heavy armour.
> (Fr. 11, ll. 21–38)[22]

βαίνω-compounds do not frequently appear in Homer, Thucydides, Herodotus, and other fourth- and fifth-century authors to indicate the horse moving.[23] Indeed, when authors describe horses as standing, they tend to use the verbs εἰμί and ἵστημι.[24] So, through his language choice, Xenophon is able to construct textual interaction that sees the horse equated with the hoplite. Other equine character traits of the horse contribute to the portrayal of the horse as hoplite: the herd mentality of the horse also lends itself well to a representation of the hoplite force, for instance.[25] Xenophon himself emphasises this connection in the *Cavalry Commander* when he describes how in a battle situation horses are better in a group (5.5–6). Such comparison between the horse and the human hoplite comes into clearer focus when Xenophon explicitly likens the horse to a human:

> The forelegs below the shoulder blades are both stronger and have a better appearance when they are stout, just as in a man. Likewise, chests that are broader are well-suited to attractiveness, strength and for supporting the legs without them crossing over.

(1.7)[26]

Although it is not surprising that advice on the selection and training of a horse should contain extensive reference to the correct anatomy of the horse, in describing the horse's ideal conformation, Xenophon chooses many words used for the human anatomy. For example, the use of μηρός (thigh), γαστήρ (belly), and πούς (foot) to describe parts of the horse (1.7, 1.12, and 1.2, respectively) evokes the anatomical descriptions in *On Horsemanship* with the prior Tyrtaeus passage:

> Come, one should plant oneself firmly fixing both feet on the ground, biting his lip with his teeth, **covering thighs**, shins below, chest and shoulders with **the broad belly of his shield**; . . . and placing **foot alongside foot** and having pressed shield against shield, crest on crest and helmet to helmet and **chest to chest** having drawn near, let him fight a man.
>
> (Fr. 11 (Stob. 4.9.16) 21–38)[27]

Tyrtaeus describes the protection of the firmly standing hoplite as provided by the belly of the shield, just as Xenophon (at 1.14 quoted earlier) describes the horse standing with his legs placed out underneath him, which would seem to mean under the belly.[28] This is strengthened by the use of γαστήρ in another image of the horse planted with feet apart – although this time the horse is rearing, providing a physical block between the rider and anyone standing in front of the horse and rider team, which means here that the horse is the shield of the rider, and therefore it is the rider (the cavalryman) who is the hoplite:

> Rather, One who has loins (and we do not mean the loins down from the tail, but rather those that come from the middle of both the ribs and the hip down to the flanks) that are supple, and also short and strong will be able to place the legs from behind [that is, hind legs] far under the fore ones [that is, front legs]. Therefore, when one checks [the horse] who is placing [his legs] under himself with the bit, he will sink onto Ones from the back [that is, hind legs] on the hocks and fetlocks, and he will raise up the body in front, so that to those who are opposite it displays **the belly** and the sheath.
>
> (11.2–3)[29]

One also sees the rider of the horse (the cavalryman) equated to the hoplite thanks to βαίνω-compounds again. Xenophon uses διαβαίνω to describe the rider's position once he is mounted on the horse but not as if sitting on a chair or on a Homeric chariot-board.[30] He means standing upright with his legs astride:

> When he [the rider] is seated, be it on the bare back [of the horse] or on a saddle-cloth, we recommend he does not sit as if on a chair, but rather as if standing upright with the legs planted firmly. For as such he will grip the horse more with his thighs, and, being upright, he will also, if the need arises, be able more powerfully to both hurl a javelin and to strike.
>
> (7.5)[31]

Furthermore, various forms of the verb ἀναβαίνω appear frequently in *On Horsemanship* – not most crudely as 'to go up', but rather 'to mount a horse' as also Xenophon's other works and, for example, Theopompus.[32] The rider, one who mounts the horse, is the ἀναβάτης;[33] the horse, One whom the rider mounts, is ἀναβαινόμενος.[34] In the *Anabasis*, Xenophon introduces another aspect of the theme of standing: not so much *how* one is standing, but rather *on what* someone is standing. Xenophon describes the hoplites in the *Anabasis* on a 'stronger' and 'safer' ὄχημα (foundation). In this instance, Xenophon is potentially not expounding his true opinions, but rather seeking to boost the morale of stranded foot soldiers by comparing their tactical advantages to those of cavalry forces:

> But if any one of you is discouraged that we do not have cavalry/ horsemen, whereas the enemy has many cavalrymen, be encouraged that ten thousand cavalrymen are nothing but ten thousand men: For no one has ever in battle been killed by being bitten or kicked by a horse, but rather men are the perpetrators of the things which happen in battle. Furthermore, we are on a much safer **vantage point/ foundation** than the cavalry: For those who are hanging from horses fear not only us but also that they might fall off: but we, standing on much firmer ground, will strike, if someone comes towards us, and we will shoot more accurately wherever we wish [aim]. The cavalrymen have advantage over us in one way only: to flee is safer for them than for us.
>
> (3.2.18–19)[35]

In addition to such translations as 'foundation' in this passage, ὄχημα can signify anything that supports something, referring therefore also to vehicles, both metaphorical and literal, including chariots and ships.[36] It also includes animals that one rides, such as Trygaeus' riding beetle substitute for Pegasus in Aristophanes' *Peace*.[37] Although ὄχημα does not appear ostensibly in *On Horsemanship*, Xenophon does allude to the concept of a firm base when he describes the most important part of the horse's body to examine: the feet. He elaborates by using the analogy of a house with bad foundations, which will be unsound even if it is aesthetically pleasing (1.2).[38] In an early passage of *On Horsemanship*, he orders the rest of his discussion of the horse's ideal conformation, working up from this point (1.4).[39] Furthermore, where he describes the rider as standing astride the horse, with the legs set apart over the horse's back (7.5), if he stands in the same way as a hoplite, by extension he might also stand on the same *thing* as the hoplite, namely a firm foundation. This would make the horse itself the ὄχημα. A number of the accusations against the cavalry (for instance, cowardice) focused on the main difference between the hoplites who stood on the ground and the cavalrymen mounted on horseback: mounted on a horse, the cavalryman was supposedly removed above the dangers of battle, was protected by the body of the horse, and had the means for a swift escape.[40] Here, however, both despite being mounted on the horse *and* because of being mounted on the horse, one associates the cavalryman with the hoplite.

So, if the two representatives of the cavalry – the horse *and* the cavalryman – can both be aligned with the hoplite, then one might infer that the cavalry can embody similar attributes to the hoplites both on the battlefield (for instance, courage) and in the sociopolitical sphere. This can then lead to the further social integration of the cavalry into the democracy of classical Athens.

The extensive description of body parts is also useful in another way. It has already been established that a manual on the selection and training of a horse should contain extensive reference to the correct anatomy of the horse. However, the description of many individual body parts of the horse creates the overall picture of many different limbs working together. This is evocative of the 'body politic' image explored in Plato and Aristotle.[41] The horse working in the ideal way is described as 'with his body collected together':

> For in doing all these things with his body collected together, the horse will act in a safer way both for himself and for One who is mounted [that is, the rider], rather than leave out those behind [that is, the back legs] when he is either leaping across, or leaping up or leaping down.
>
> (8.5)[42]

Such a description is familiar to most modern riders and loosely involves getting the horse to carry himself in such a way that the muscular power of his hindquarters drives the frontquarters forwards. Such an example demonstrates that even if there is potential for metaphorical discussion in the work, *On Horsemanship* is still significant on a practical level of instruction. Xenophon's use of the word ἀθρόος (together) in this passage is interesting, since he and other authors use it to describe cavalry and troops formed up in close formation, which contributes again to the characterisation of the horse as hoplite.[43] Some authors use the word to differentiate the citizen body as a whole from the individual.[44] One now begins to understand that the horse represents not just the individual hoplite nor even the sociopolitical echelon for which the hoplite stands, but rather the city-state as a whole, which in turn enables Xenophon to use the horse in a metaphorical discussion of society. For example, the different parts of the horse's body working together in the most advantageous way is celebrated by Xenophon at Section 10 of *On Horsemanship* (especially 10.17). This image of different parts working together towards 'the whole' (the horse-city) is portrayed also in Xenophon's methodical descriptions of parts of the horse's anatomy (as we have seen earlier (139–140)), often using vocabulary more generally used for human body parts).[45] Although again it is not unexpected for a manual of horsemanship to contain descriptions of the horse's anatomy, Xenophon may have been able to use this to his advantage, considering his penchant in other works for similar imagery of many parts operating as a whole.[46] Such motifs, of the individual contributing to the collective, could again be viewed as having a place in a democratic framework.

σῶμα (body) and the ψυχή (mind, soul) are key elements of both Xenophon's advice on training the horse (3.7) in *On Horsemanship* and in Plato and Aristotle's theories of education.[47] Xenophon advises that the only gauge of the untrained young horse is the body, because the soul cannot be ascertained or developed before one trains the horse to be ridden (1.1), just as Aristotle opines that the education of a child should start with the body, and the soul will develop later.[48]

Furthermore, while it is not unexpected that part of Xenophon's advice on the maintenance of a horse and its body focuses on feeding the horse (4.1–3), he also discusses in the *Cavalry Commander* the responsibility of the commander to ensure the cavalry's horses are well fed, using the verb τρέφω (1.3, 1.13). Used also to describe the general training of the horse in *On Horsemanship* (11.13), τρέφω introduces a link between themes of nurture (for example, of children), the nutrition of food and the nourishment of education.[49] Furthermore, a link between food sharing and comradeship (and therefore again by extension the whole community) is a theme in much contemporaneous literature, including forensic speeches and philosophical dialogue.[50]

Therefore, there is a clear development of the training of the horse representing the education of the young.[51] Considering the use of the symbol of the horse to represent the wider community, comments of the training of the horses in *On Horsemanship* can be used allegorically to discuss the education of a younger generation and their integration into the political life of Athens. So, *On Horsemanship* as a whole becomes an elaborate metaphor for sociopolitical debate.

III

Although the hoplite would appear to be the best group with which to equate the horse in the attempted social integration of the cavalry discussed earlier, there are of course other significant groups to consider in the classical Athenian war machine and society. Classical Athens' naval identity was in some ways a social identity as much as a military one: as David Pritchard observes, 'the newly created large urban thetic class . . . was the prime mover in the creation and direction of the democracy'.[52] However, as Margaret Miller observes, 'in both public and private art hoplitic ideology continued to predominate well into the era of naval empire; martial valour was measured to a hoplitic norm'.[53] Nevertheless, another example of aligning the cavalry with a less aristocratic part of society would to some extent have the same advantages of aligning them with hoplites – it would be sociopolitical integration on a level of discourse and imagery. And certainly although there appears to be less direct comparison between the cavalry and naval forces as there is between the cavalry and the hoplites in sociopolitical interaction of the cavalry in classical Athens, works such as Aristophanes' *Knights*, with its rowing horse chorus, clearly demonstrate the potential receptiveness for such sentiment in the public conscious.[54] So, it is also worth exploring whether Xenophon can deploy an extended metaphor in *On Horsemanship* by relying on established literary and vocabularic links between ships and horses.

Sea and naval imagery has a rich pedigree before Xenophon. It was already established in Homeric poetry including images of warriors represented as waves.[55] Archilochus develops the image of the ship in a storm representing an army in combat, even possibly including a helmsman steering the ship as representative of a military army leader.[56] One can consider such Homeric and Archilochian imagery as a starting point for ship-of-state metaphors, especially considering the Homeric phrase ἕρμα πόληος (support of the city) used to describe a warrior as the support or pillar of a city or community.[57] Elsewhere ἕρματα (support) describes a pile of stones supporting a landed ship.[58] Authors of the seventh century develop further the metaphor: for example, Alcaeus' representations of a city or community in various conflicts as a ship in a storm and representations of leaders as men steering a ship.[59] By the mid-sixth century, Theognis develops the image further to represent sociopolitical revolution.[60] So, then, the tragic and comic dramas of fifth-century Athens – Aeschylus, Euripides, Sophocles, Aristophanes – come to express naval metaphors, especially the city in conflict (for instance, war, civic war, or plague) as ship in a storm, and leaders as helmsmen of ships.[61] Plato and Aristotle use similar metaphors to discuss leadership and citizens in society, as does Demosthenes.[62] It is fair to say then that by whatever date Xenophon produced *On Horsemanship*, nautical and ship-of-state metaphors would have been established in the public consciousness of Athens. Indeed, similar passages appear in Xenophon's *Oeconomicus* and *Memorabilia* – the naval analogy in the latter of which is then reinforced by being equated with a horsemanship analogy.[63] This has potentially significant implications for consideration of *On Horsemanship* as a military text: that the author utilises words and metaphors across his writings in this instance links the military treatise to his wider corpus, with the messages of each serving to reinforce the other. Therefore, *On Horsemanship* is not about horsemanship only; it is one piece of a wider reflection of the community with which Xenophon possibly seeks to engage across his entire oeuvre.

Vocabularic links between horses and ships also appear much earlier than classical Athens. The word ἵππος in the plural can mean a team of chariot-horses in Homer, and subsequently it comes to mean the chariot itself.[64] A ship as a chariot on the sea is seen in Homer and then also in Aristophanes' *Lysistrata*.[65] Horse bits (metal device in the horse's mouth) are referred to in Aeschylus as πηδάλια, used elsewhere to describe rudder oars.[66] Away from linguistic connections, the god Poseidon is linked to both the sea and horses, which is particularly interesting considering the connection between the cavalry and the worship of Poseidon Hippios.[67]

On Horsemanship also contains a number of examples of linguistic links between horses and the sea: χαλινός, used to describe horse bits but also anything that applies restraint, including an anchor or the rigging of a ship.[68] Many of the examples of language discussed thus far in this chapter regarding hoplite imagery also have naval connotations, and such examples of vocabulary overlapping different themes demonstrates the multi-themed potential for one vocabularic choice by Xenophon: for instance, the verb ἀναβαίνω translates also as 'to board/embark on a ship' in poetry and historical prose.[69] Predictably, the verb indicating opposite movement has connections: καταβαίνω, used by Xenophon and others to describe 'dismounting the horse', is also used for coming down from a ship.[70] Again, the word ὄχημα is used in tragedy and philosophical discourse, where the platform that bears something is a ship.[71] τρέφω can also refer to preparing ships.[72] Xenophon uses the word ἐχυρός to describe the stall (or safe place) for the horse (4.2). Alcaeus uses the image of a crew guiding a ship into a safe harbour (ἐχυρὸς λιμήν) as an allegorical exhortation to his soldier companions to face conflict, and the image recurs in Aeschylus and Euripides.[73] If the ship-of-state metaphor often has the added feature of the ship being steered into the safe ἐχυρός (harbour), then the horse led into the ἐχυρός environment is equated with the ship-of-state, rather like a 'horse-of-state'. So, as with the 'horse-hoplite' discussed earlier, the horse-of-state motif is able to introduce metaphorical exploration of larger civic matters.

Just as descriptions of parts of the horse's body allowed for the alignment of the horse with the hoplite (as discussed earlier, 138–140), so too are there linguistic links between descriptions of body parts and parts of a ship. πούς is not only used for the human foot and the foot or hoof of the horse, but elsewhere to describe parts of the sails/rigging and also the steering equipment of a ship.[74] γαστήρ can refer to the belly of a human and a horse, as well as the hollow of a shield (as seen earlier, 140); it can also mean the hollow of a receptacle and later by extension the hold of a ship.[75] The image of the different parts of the (horse's) body working together evokes the image of the crew working together on a ship, such as in Aristotle.[76] In Xenophon's *Oeconomicus*, Isomachus compares the ideal order or τάξις to the well-ordered chorus (8.3), an army (8.6), or a ship (8.11–16).[77] Similar sentiment is expressed in *On Horsemanship*: 'For of course a disobedient slave or a disobedient army is useless. Furthermore, a disobedient horse is not only useless, but he also frequently acts like a traitor' ~ ἄχρηστον μὲν γὰρ δήπου καὶ οἰκέτης καὶ στράτευμα ἀπειθές· ἵππος δὲ ἀπειθὴς οὐ μόνον ἄχρηστος, ἀλλὰ πολλάκις καὶ ὅσαπερ προδότης διαπράττεται (3.6). The key word is ἀπειθής (disobedient), which is elsewhere used of ships, such as by Thucydides, when describing ships not obeying the manoeuvring of the helmsmen (2.84.3). So, the symbol of the horse-ship can introduce and inform discussion of how different groups can cooperate within a community. Again, since the horse is synecdoche for the cavalry, it suggests the cavalry is able to be part of such a discussion, and therefore part of the society that is discussed. Furthermore, as in the image of the body parts working together in the first working example in this chapter (142), the motif of the individual elements contributing to the collective fits into the democratic framework of classical Athens.

In viewing the horse as a ship, the rider or cavalryman becomes the helmsman, who steers the horse-ship. Not only is such association simply implied here, but it also builds on earlier literature, such as when Bacchylides has the horse of Hiero obeying his rider who is described as a κυβερνήτης (helmsman).[78] Authors use the helmsman as an example of a figure with a level of individual power who rules for the benefit of others – Aristotle does this several times.[79] In the *Anabasis*, Xenophon explains the importance of discipline and communication on board a ship during a storm, and a similar sentiment appears again in Aristotle.[80] One equates the rider with the helmsman, then the skill of the rider to use the power of the horse to the benefit of himself and his city corresponds to the skill of the helmsman to use the ship to harness the power of the wind and the sea, and to guide the ship to the advantage of the helmsman, the crew, and the city for whom they sail. So, firstly the cavalryman is equated with elements of society that did not suffer from similar stigma of aristocratic connotations. Secondly, aristocratic individualistic power is to some extent justified by the fact that the skill of the rider makes the power of the horse useful, just as the skill of the helmsman makes the crew and ship useful.

In metaphors of the ship-of-state, the difference between the rider/horse and helmsman/ship relationship is that whereas the rider harnesses the power of the horse, the ship seems to act more like a middleman agent for the helmsman, who is ultimately using the power of the sea to his advantage. So the state, represented by the ship is an instrument for achieving the goal of ideal government. If one extends this interpretation to the rider/horse relationship in the same way, the rider uses the horse as a means for harnessing some other power for the benefit of the state. When the horse represents the *demos* working for the advantage of the state, the horse represents an instrument for the state. Now, equated to the ship-of-state, the horse-of-state, guided by the rider, represents the tool in achieving the goal of ideal government – whatever this tool is in discussion of theoretical government. If true, then Xenophon's treatise is a manual on how to be a cavalryman *and* an engaged city-citizen.

IV

The previous analysis notwithstanding, *On Horsemanship* is a manual of horsemanship for the military horse in the first instance. As mentioned earlier, the fact that equestrian communities still consult Xenophon is in and of itself is surely an endorsement of the continuing practical value of this military treatise. It is, arguably, the example of an ancient military treatise that has withstood the test of time. Be that as it may, the practical advice not only aims to advise the reader who seeks to refine his skill in the art of horsemanship, but also seeks to demonstrate to other possible readers outside of the cavalry the extent of commitment and skill required to select, train, and maintain a horse for military purpose – characterisation that works in the cavalry's favour and shows how even on this literal level Xenophon is able to use *On Horsemanship* as a device in the justification of the cavalry.

However, by using specific language to transmit this practical advice that allows him to suggest certain imagery and evocation of contemporary and long-established literary motifs, Xenophon introduces a wider range of levels of comment to this text. By equating the horse – and sometimes the rider – and by extension the cavalry with different groups of society (such as hoplites or thetes), Xenophon is able to explore and perhaps in a subtle way advocate for the socio-political integration of the cavalry. Furthermore, by extending these metaphors to investigate larger civic issues such as education and community infrastructure, the cavalry (represented by the horse) is seen at the centre of debates about the identity of the society into which they must integrate. Such imagery and literary interaction also permits Xenophon to explore a number of different genres – philosophical discourse, forensic speeches, drama and poetry, historiography. In doing so, he is able to expand the genre of military treatises itself to have relevance outside the narrow rubric.

These case studies also contribute to the discussion of the potential audience of *On Horsemanship*. With a multi-levelled potential function comes a much wider potential audience than might be initially anticipated. As well as catering for a specialist (or at least someone desiring to be specialist) audience from a specific echelon of society, the text becomes relevant to any producer or audience of the other genres and sociopolitical dialogue with which Xenophon interacts. Furthermore, such intertextual references as examined in this chapter suggest another characteristic of the audience of *On Horsemanship*: readers who have a certain level of education and literary awareness to appreciate such correspondences.[81]

Finally, it is worth returning to the earlier mention of horsemanship analogies in other works of Xenophon, such as the *Memorabilia*, in the discussions of what it is to be a good citizen and of leadership.[82] As Christopher Farrell observes,

> Both the allusion to the Socratic theory outlined in *Memorabilia* and the practical application of such theory in the *Cavalry Commander* and *On Horsemanship* reiterate that Xenophon understood his purpose in writing the treatises to make him a 'useful' Athenian citizen and thus sympathetic to democracy.[83]

So, perhaps this is another purpose of writing a treatise, military-based or otherwise: not only to make the reader more useful to society, but through the act of writing the treatise in the first place, for the writer to make himself more useful to society.[84]

Notes

1 Unless otherwise indicated, translations are my own and references are to *On Horsemanship*. The text comes from the volume of Xenophon's *Scripta Minora* in the Loeb Classical Library.
2 See Raulwing (2009) and McCabe (2007) 4 nn. See also the Introduction (4 n.37).
3 For example, Arrian's *Tactica*. See Dixon and Southern (1992) 121, 126–134 and Busetto (2015). I write 'subgenre' to suggest that writing about warfare and horses is both part of, and at the same time distinct from, the genre of military texts as a whole. The existence of a body of texts geared specifically towards cavalry warfare possibly indicates a perceived need for specialised texts that a broader military manual could not provide. That is, one might speculate that some – but not all – authors and readers of military texts need to have specialised interests. On the audience of *On Horsemanship*, see also later, 146–147.
4 For discussion, see McCabe (2007).
5 See Edwards, Enekel, and Graham (2011).
6 For overview, see Gray (2010) and Flower (2017).
7 For an excellent overview of the history of such scholarship and more recent revisions, see Konijnendijk (2017).
8 For an overview, see Gray (2010) 1–30 and Flower (2017) 1–14.
9 Spence (1993) 180.
10 ἐπειδὴ διὰ τὸ συμβῆναι ἡμῖν πολὺν χρόνον ἱππεύειν οἰόμεθα ἔμπειροι ἱππικῆς γεγενῆσθαι, βουλόμεθα καὶ τοῖς νεωτέροις τῶν φίλων δηλῶσαι ἧ ἂν νομίζομεν αὐτοὺς ὀρθότατα ἵπποις προσφέρεσθαι.
11 For example, *Memorabilia* 3.3, 4.1; *Oeconomicus*, 1.8, 3.11, 11.4f., 11.15f., 13.6f. See also later in this chapter, *passim*.
12 For example, Blaineau (2008); Fehr (2011); and Farrell (2012).

13 See Spence (1993) and Konijnendijk (2017).
14 For discussion of Xenophon's allegiances, see Farrell (2012).
15 Spence (1993) 164–230.
16 Spence (1993) 230. See also see Crowley (2012).
17 Spence (1993) 180. Unless otherwise indicated, the use of 'cavalry' in this chapter refers to the classical Athenian cavalry and Spence's 'cavalry class'.
18 For discussion, see Low (2002) and Spence (1993) 219.
19 μηρούς γε μὴν τοὺς ὑπὸ τῇ οὐρᾷ ἢν ἅμα πλατείᾳ τῇ γραμμῇ διωρισμένους ἔχῃ, οὕτω καὶ τὰ ὄπισθεν σκέλη διὰ πολλοῦ ὑποθήσει. τοῦτο δὲ ποιῶν ἅμα γοργοτέραν τε καὶ ἰσχυροτέραν ἕξει τὴν ὑπόβασίν τε καὶ ἱππασίαν καὶ ἅπαντα βελτίων ἔσται ἑαυτοῦ. τεκμήραιο δ' ἂν καὶ ἀπ' ἀνθρώπου· ὅταν γάρ τι ἀπὸ τῆς γῆς ἄρασθαι βούλωνται, **διαβαίνοντες** πάντες μᾶλλον ἢ **συμβεβηκότες** ἐπιχειροῦσιν αἴρεσθαι.
20 Homer, *Iliad* 12.458; Aristophanes, *Wasps* 588 and *Knights* 77.
21 Compare also the image of Archilochus Fr. 114 (West 1993). On Tyrtaeus and military literature, see Vela Tejada (2004) 132–133 and the Introduction (11 n.16).
22 ἀλλά τις εὖ **διαβὰς** μενέτω ποσὶν ἀμφοτέροισι
    στηριχθεὶς ἐπὶ γῆς, χεῖλος ὀδοῦσι δακών,
    μηρούς τε κνήμας τε κάτω καὶ στέρνα καὶ ὤμους
    ἀσπίδος εὐρείης γαστρὶ καλυψάμενος·
    δεξιτερῇ δ' ἐν χειρὶ τινασσέτω ὄβριμον ἔγχος,
    κινείτω δὲ λόφον δεινὸν ὑπὲρ κεφαλῆς·
    ἔρδων δ' ὄβριμα ἔργα διδασκέσθω πολεμίζειν,
    μηδ' ἐκτὸς βελέων ἑστάτω ἀσπίδ' ἔχων,
    ἀλλά τις ἐγγὺς ἰὼν αὐτοσχεδὸν ἔγχεϊ μακρῷ
    ἢ ξίφει οὐτάζων δήϊον ἄνδρ' ἑλέτω,
    καὶ πόδα πὰρ ποδὶ θεὶς καὶ ἐπ' ἀσπίδος ἀσπίδ' ἐρείσας,
    ἐν δὲ λόφον τε λόφῳ καὶ κυνέην κυνέῃ
    καὶ στέρνον στέρνῳ πεπλημένος ἀνδρὶ μαχέσθω,
    ἢ ξίφεος κώπην ἢ δόρυ μακρὸν ἑλών.
    ὑμεῖς δ', ὦ γυμνῆτες, ὑπ' ἀσπίδος ἄλλοθεν ἄλλος
    πτώσσοντες μεγάλοις βάλλετε χερμαδίοις
    δούρασί τε ξεστοῖσιν ἀκοντίζοντες ἐς αὐτούς,
    τοῖσι πανόπλοισιν πλησίον ἱστάμενοι.
    (This translation is that of Andrew Bayliss in *Brill's New Jacoby*).
23 Granted, Xenophon employs συμβαίνω elsewhere in On *Horsemanship* to convey other meanings: for example, see 1.15, 3.5, 9.1, 11.10.
24 Examples of εἰμί include: *Iliad* 11.340, 23.311, 23.480. Examples of ἵστημι include: *Iliad* 2.777, 5.196, 8.565, 10.520, 10.569, 14.308; Euripides, *Hippolytus* 1204; Herodotus 1.78, 8.98.
25 Again, see Crowley (2012).
26 μηροί γε μέντοι οἱ ὑπὸ ταῖς ὠμοπλάταις ἢν παχεῖς ὦσιν, ἰσχυρότεροί τε καὶ εὐπρεπέστεροι ὥσπερ ἀνδρὸς φαίνονται. καὶ μὴν στέρνα πλατύτερα ὄντα καὶ πρὸς κάλλος καὶ πρὸς ἰσχὺν καὶ πρὸς τὸ μὴ ἐπαλλὰξ ἀλλὰ διὰ πολλοῦ τὰ σκέλη φέρειν εὐφυέστερα. See also 1.4, 9.4, 9.7. Cf. 10.14.
27 ἀλλά τις εὖ **διαβὰς** μενέτω ποσὶν ἀμφοτέροισι
    στηριχθεὶς ἐπὶ γῆς, χεῖλος ὀδοῦσι δακών,
    μηρούς τε κνήμας τε κάτω καὶ στέρνα καὶ ὤμους
    ἀσπίδος εὐρείης γαστρὶ καλυψάμενος·
    . . .
    καὶ **πόδα πὰρ ποδὶ** θεὶς καὶ ἐπ' ἀσπίδος ἀσπίδ' ἐρείσας,
    ἐν δὲ λόφον τε λόφωι καὶ κυνέην κυνέηι
    καὶ **στέρνον στέρνωι** πεπλημένος ἀνδρὶ μαχέσθω.
    Translation: Bayliss (see earlier, 139 n.22).
28 This is also the view of Marchant (1925) 304–305.

29 ἀλλὰ μᾶλλον ὃς ἂν τὴν ὀσφῦν ὑγράν τε καὶ βραχεῖαν καὶ ἰσχυρὰν ἔχῃ, καὶ οὐ τὴν κατ᾽ οὐρὰν λέγομεν, ἀλλ᾽ ἣ πέφυκε μεταξὺ τῶν τε πλευρῶν καὶ τῶν ἰσχίων κατὰ τὸν κενεῶνα, οὗτος δυνήσεται πόρρω ὑποτιθέναι τὰ ὀπίσθια σκέλη ὑπὸ τὰ ἐμπρόσθια. ἢν οὖν τις ὑποτιθέντος αὐτοῦ ἀνακρούῃ τῷ χαλινῷ, ὀκλάζει μὲν τὰ ὀπίσθια ἐν τοῖς ἀστραγάλοις, αἴρει δὲ τὸ πρόσθεν σῶμα, ὥστε τοῖς ἐξ ἐναντίας φαίνεσθαι τὴν **γαστέρα** καὶ τὰ αἰδοῖα.

30 See Bugh (1948) 26, who also references Greenhalgh (1973) 146.

31 ἐπειδάν γε μὴν καθίζηται ἐάν τε ἐπὶ ψιλοῦ ἐάν τε ἐπὶ τοῦ ἐφιππίου, οὐ τὴν ὥσπερ ἐπὶ τοῦ **δίφρου** ἕδραν ἐπαινοῦμεν, ἀλλὰ τὴν ὥσπερ ὀρθὸς ἂν **διαβεβηκὼς** εἴη τοῖν σκελοῖν. τοῖν τε γὰρ μηροῖν οὕτως ἂν ἔχοιτο μᾶλλον τοῦ ἵππου, καὶ ὀρθὸς ὢν ἐρρωμενεστέρως ἂν δύναιτο καὶ ἀκοντίσαι καὶ πατάξαι ἀπὸ τοῦ ἵππου, εἰ δέοι.

32 For example, *On Horsemanship* 1.1, 3.4, 6.6, 6.16, 7.1, 7.2, 7.4, 8.5, 9.3. For other of Xenophon's works, see *Education of Cyrus* 4.1.7, 7.1.3, 3.3.27; *Cavalry Commander* 3.4, 1.4; Theopompus 2.

33 For example, 1.4, 1.8, 1.11, 3.3, 3.9, 3.12, 5.7, 6.6, 8.5, 8.7, 9.7, 11.5, 12.2, 12.8; see also Xenophon, *Hellenica* 5.3.1; Plato, *Critias* 119a–b.

34 For example, 1.1.

35 εἰ δέ τις ὑμῶν ἀθυμεῖ ὅτι ἡμῖν μὲν οὐκ εἰσὶν ἱππεῖς, τοῖς δὲ πολεμίοις πολλοὶ πάρεισιν, ἐνθυμήθητε ὅτι οἱ μύριοι ἱππεῖς οὐδὲν ἄλλο ἢ μύριοί εἰσιν ἄνθρωποι· ὑπὸ μὲν γὰρ ἵππου ἐν μάχῃ οὐδεὶς πώποτε οὔτε δηχθεὶς οὔτε λακτισθεὶς ἀπέθανεν, οἱ δὲ ἄνδρες εἰσὶν οἱ ποιοῦντες ὅ τι ἂν ἐν ταῖς μάχαις γίγνηται. οὐκοῦν τῶν γε ἱππέων πολὺ ἡμεῖς ἐπ᾽ ἀσφαλεστέρου **ὀχήματός** ἐσμεν· οἱ μὲν γὰρ ἐφ᾽ ἵππων κρέμανται φοβούμενοι οὐχ ἡμᾶς μόνον ἀλλὰ καὶ τὸ καταπεσεῖν· ἡμεῖς δ᾽ ἐπὶ γῆς βεβηκότες πολὺ μὲν ἰσχυρότερον παίσομεν, ἤν τις προσίῃ, πολὺ δὲ μᾶλλον ὅτου ἂν βουλώμεθα τευξόμεθα. ἑνὶ δὲ μόνῳ προέχουσιν οἱ ἱππεῖς ἡμᾶς . . . φεύγειν αὐτοῖς ἀσφαλέστερόν ἐστιν ἢ ἡμῖν.

36 Examples of metaphor: Pindar Fr. 124.1; Cf. Plato, *Phaedo* 85d. Chariots: Herodotus 5.21; Pi.Fr.106.6; Sophocles, *Electra* 740; Euripides, *Suppliants* 662. Ships: Aeschylus, *Prometheus Bound* 468; Sophocles, *Trachinae* 656; Euripides, *Iphigenia in Tauris* 410; Plato, *Hippias Maior* 295d. (Please note the connection here with the second section of this chapter).

37 Aristophanes, *Peace* 866.

38 See Dillery (2017) 213: 'If Xenophon has provided a didactic work that one could actually use, the core aspects of his larger worldview are also present: so, for instance, his devotion to the principle of order ("everything in its place," especially regarding the foundations of things or, in a horse, its footing'. Dillery also references *Memorabilia* 3.1.7. Consider *Oeconomicus* 8.3–16 and the second working example of this chapter.

39 It is interesting to note that of the little attention paid to the horse's conformation in *Cavalry Commander,* it is the feet on which Xenophon concentrates (1.4 and 1.16).

40 For details of these observations in literature, see Spence (1993) 220–221. As already mentioned, there are significant counterarguments to most of these accusations. For example, while the cavalryman might be raised above the general level of combat if fitting foot soldiers, the horse posed plenty of danger to his rider if he could not be controlled, as a large target and, if it died or fell, a potential crush risk to the rider. Many of these points are addressed in the *Anabasis* passage earlier.

41 For example, Plato, *Republic* 8.556e; Aristotle, *Politics* 1253a20.

42 ἀθρόῳ γὰρ τῷ σώματι ταῦτα πάντα ποιῶν καὶ ἑαυτῷ ὁ ἵππος καὶ τῷ ἀναβάτῃ ἀσφαλέστερον ποιήσει μᾶλλον ἢ ἂν ἐκλείπῃ τὰ ὄπισθεν ἢ διαπηδῶν ἢ ἀνορούων ἢ καθαλλόμενος.

43 Herodotus 6.112; Xenophon, *Anabasis,* 1.10.13. See also Xenophon, *Education of Cyrus* 3.3.22.

44 For example, Thucydides 2.60.2–3 and cf. 1.141.7.

45 On horse-human anatomical vocabulary in *On Horsemanship,* see Brodersen (2018) 12–13.

46 See again n.35.

47 See also Xenophon's comments elsewhere in his works: for example, *Memorabilia* 1.2.19 and 1.3.5. Cf. Aristotle, *Politics*, 1323a15–1342b30; Plato, *Republic* 376c–e.

48 Aristotle, *Politics* 1334b25–29.

49 Children: *Iliad* 8.283, 16.191; *Odyssey* 2.131, 19.354; Thucydides 2.46.1; Plato, *Statesman* 274a; Herodotus 2.2.2; Aeschylus, Libation Bearers 908, *Suppliants* 894. Food: Plato, *Republic* 568e; Xenophon, *Anabasis* 4.5.25; *Memorabilia* 4.3.10. Education: Hesiod Fr.19; Pindar, *Nemean* 3.53; Plato, *Republic* 534d.

50 For further discussion, see Lee (2007) 97–99.

51 Consider again the opening lines of the text regarding the older generation passing on information to the younger generation (1.1).

52 Pritchard (1998) 126, citing his earlier work (1994) 111–139.

53 Miller (2010), 332. For extensive discussion and references, see Pritchard (1994).

54 For example, see Anderson (2003).

55 For example, *Iliad* 2.144–146; 2.394–397; 3.60–62; 7.3–6; 13.390–391; 15.381–387; 15.410–412; 15.623–629; 16.483–484; 17.742–744. Cf. Euripides, *Orestes* 696–701 and *Hecuba* 606–608.

56 Fragments 105 and 106: West (1993). For further discussion, see Pritchard (1999) 169.

57 *Iliad* 16.548–551; *Odyssey* 23.121.

58 *Iliad* 1.486; 2.154. See Pritchard (1999), 168–169.

59 Fr. 208a, 73 and 306i Voigt (1971). Fr. 249 Voigt (1971). See Pritchard (1999) 169–170.

60 Theognis 667–682 and 855–856. See Pritchard (1999) 170–171.

61 For a detailed examination with extensive references, see Pritchard (1999) 163–223, especially 171–181.

62 Plato, *Republic* 488e–489d; Aristotle, *Politics* 1276b20–24; Demosthenes 19.135–136. See Corrêa (2016) 291–309.

63 *Oeconomicus* 8.3–16; *Memorabilia* 1 and 3.

64 For example, *Iliad* 16.370; 5.237; 8.41; 5.13; 5.19; 5.11; 5.163; 5.46.

65 *Odyssey* 4.708; Aristophanes, *Lysistrata* 19–63.

66 Aeschylus, *Seven Against Thebes* 206–207; Herodotus 2.96.3.

67 For example, Aristophanes, *Knights* 551–564; *Homeric Hymn 22 to Poseidon*, 5. See Spence (1993) 188–189.

68 For example, Pindar, *Pythian* 4.25.

69 For example, *Odyssey* 1.210 and 14.252; Pindar, *Pythian* 2.62; Thucydides 4.44.6.

70 For example, *Odyssey* 14.350.

71 For example, Aeschylus, *Prometheus Bound* 468; Euripides, *Iphigenia Tauris* 410; Sophocles, *Trachinae* 656; Plato, *Hippias Maior* 295d and *Phaedo* 113d.

72 For example, Xenophon, *Hellenica* 1.5.5, 4.8.9, 5.1.24; Thucydides 8.44.1.

73 Aeschylus, *Suppliants* 471–472; Euripides, *Andromache* 748–749 (see also 891–892) and *Heracleidae* 427–431; Alcaeus Frag. 6.

74 For example, *Odyssey* 5.260, 10.32; Pindar, *Nemean* 6.55; Euripides. *Orestes* 706–707; Aristophanes, *Knights* 436; Euripides, *Iphigenia Tauris* 1135; Sophocles, *Antigone* 715.

75 *Scholia in Thucydidem* 1.50.

76 *Politics* 1276b20–24.

77 Cf. *Anabasis* 5.8.13 and *Hellencia* 7.5.23.

78 Bacchylides 5.43–9.

79 Aristotle, *Politics* 1324b29–33 and 1278b40–1279a21.

80 Xenophon, *Anabasis* 5.8.20; Aristotle, *Politics* 1320b33.

81 This would seem to provide further support to the thesis set forth in Chapter 3: these texts could inform readers other than those seeking information specifically on horse-riding.

82 For example, *Memorabilia* 2.3.7, 2.6.7, 3.3.2–15, 4.2.25, 4.4.5.

83 Farrell (2012), 77.

84 I thank the editors for their comments on an earlier draft of this chapter.

Bibliography

Anderson, C. A. 2003. 'The Gossiping Theme in Aristophanes' *Knights*, 1300–1315'. *CJ* 99. 1–9.

Bayliss, A. 2017. 'Tyrtaios (580)', in *Brill's New Jacoby*. (http://dx.doi.org/10.1163/1873-5363_bnj_a580).

Blaineau, A. 2008. 'Le cheval, le cavalier et l'hippocentaure. Technique équestre, éthique et métaphore politique chez Xénophon'. *CEA* 45. 185–211.

Brodersen, K. 2018. *Xenophon: Ross und Reiter*. Berlin: De Gruyter.

Bugh, G. R. 1948. *The Horsemen of Athens*. Princeton: Princeton University Press.

Busetto, A. 2015. 'War as Training, War as Spectacle: The *hippika gymnasia* from Xenophon to Arrian', in G. Lee, H. Whittaker, and G. Wrightson, eds., *Ancient Warfare: Introducing Current Research*. Vol. 1. Newcastle-upon-Tyne: Cambridge Scholars. 147–171.

Corrêa, P. da C. 2016. '"The Ship of Fools" in Euenus 8b and Plato's *Republic* 488a–489a', in L. Swift and C. Carey, eds., *Iambus and Elegy: New Approaches*. Oxford: Oxford University Press. 291–309.

Crowley, J. 2012. *The Psychology of the Athenian Hoplite: The Culture of Combat in Classical Athens*. Cambridge: Cambridge University Press.

Dillery, J. 2017. 'Xenophon: The Small Works', in M. A. Flower, ed., *The Cambridge Companion to Xenophon*. Cambridge: Cambridge University Press. 195–220.

Dixon, K. R. and Southern, P. 1992. *The Roman Cavalry: From the First to the Third Century AD*. London: Batsford.

Edwards, P., Enekel, K. A. E., and Graham, E., eds. 2011. *The Horse as Cultural Icon: The Real and the Symbolic Horse in the Early Modern World*. Leiden: Brill.

Farrell, C. A. 2012. *Xenophon in Context: Advising Athens and Democracy*. PhD Dissertation. University of London (King's College).

Fehr, B. 2011. *Becoming Good Democrats and Wives: Civic Education and Female Socialization on the Parthenon Frieze, Hephaistos (Sonderband 2011)*. Munster: LIT Verlag.

Flower, M. A. 2017. 'Introduction', in *idem*, ed., *The Cambridge Companion to Xenophon*. Cambridge: Cambridge University Press. 1–12.

Gray, V. J. 2010. 'Introduction', in *eadem*, ed., *Xenophon*. Oxford: Oxford University Press. 1–30.

Greenhalgh, P. A. L. 1973. *Early Greek Warfare: Horsemen and Chariots in the Homeric and Archaic Ages*. Cambridge: Cambridge University Press.

Konijnendijk, R. 2017. *Classical Greek Tactics: A Cultural History*. Leiden: Brill.

Lee, J. W. I. 2007. *A Greek Army on the March: Soldiers and Survival in Xenophon's Anabasis*. Cambridge: Cambridge University Press.

Low, P. 2002. 'Cavalry Identity and Democratic Ideology in Early Fourth-Century Athens'. *PCPhS* 48. 102–122.

Marchant, E. C. 1925. *Xenophon: Scripta Minora*. Cambridge, MA: Harvard University Press.

McCabe, A. 2007. *A Byzantine Encyclopaedia of Horse Medicine: The Sources, Compilation, and Transmission of the Hippiatrica*. Oxford: Oxford University Press.

Miller, M. C. 2010. 'I am Eurymedon: Tensions and Ambiguities in Athenian War Imagery', in D. M. Pritchard, ed., *War, Democracy and Culture in Classical Athens*. Cambridge: Cambridge University Press. 304–338.

Pritchard, D. 1994. 'From Hoplite Republic to Thetic Democracy: The Social Context of the Reforms of Ephialtes'. *Ancient History: Resources for Teachers* 24. 111–139.

———— 1998. 'Thetes, Hoplites and the Athenian Imaginary', in T. W. Hillard, R. A. Kearsley, C. E. V. Nixon, and A. M. Nobbs, eds., *Ancient History in a Modern University:*

Proceedings of a Conference Held at Macquarie University, 8–13 July, 1993 to Mark Twenty-Five Years of the Teaching of Ancient History at Macquarie University and the Retirement from the Chair of Professor Edwin Judge. Grand Rapids, MG: Wm. B. Eerdmans. 121–127.

————— 1999. *The Fractured Imaginary: Popular Thinking on Citizen Soldiers and Warfare in Fifth-Century Athens*. PhD Dissertation. Macquarie University. Sydney.

Raulwing, P. 2009. *The Kikkuli Text: Hittite Training Instructions for Chariot Horses in the Second Half of the 2nd Millennium B.C. and Their Interdisciplinary Context*. Online publication (www.Irgaf.org/Peter_Raulwing_The _Kikkuli_Text_Masterfile_Dec_2009.pdf).

Seager, R. 2001. 'Xenophon and the Athenian Democratic Ideology'. *Classical Quarterly* 51. 385–397.

Spence, I. G. 1993. *The Cavalry of Classical Greece: A Social and Military History*. Oxford: Oxford University Press.

Tuplin, C. 2004. 'Xenophon and His World: An Introductory Review', in *idem*, ed., *Xenophon and His World: Papers from a Conference Held in Liverpool in July 1999*. Stuttgart: Franz Steiner. 13–31.

Voigt, E.-M. 1971. *Sappho et Alcaeus. Fragmenta*. Amsterdam: Polak and Van Gennep.

Vela Tejada, J. 2004. 'Warfare, History and Literature in the Archaic and Classical Periods: The Development of Greek Military Treatises'. *Historia* 53. 129–146.

West, M. L. 1993. *Greek Lyric Poetry: The Poems and Fragments of the Greek Iambic, Elegiac, and Melic Poets (Excluding Pindar and Bacchylides) Down to 450 BC*. Oxford: Oxford University Press.

8 Refighting Cunaxa

Xenophon's *Education of Cyrus* as a manual on military leadership

Jeffrey Rop

The *Education of Cyrus* (*Cyropaedia*) is a pseudo-historical biography of Cyrus II, founder of the Achaemenid Persian Empire, written by Xenophon of Athens in the fourth century BCE.[1] Following a brief discussion of rulership, Xenophon observes that Cyrus was an exceptional leader who brought under his control a great number of people and territories (1.1). Over the course of eight books, Xenophon investigates what made Cyrus successful, beginning with his education and moving through the rest of his life, from his service as a vassal to the Medes to his establishment of the Persian Empire. The work concludes with a section describing the decay of the Empire following Cyrus' death (8.8).[2]

Although Cyrus II was a historical figure and several events in the story are drawn from his life, many elements of the *Education of Cyrus* are fictional, and aspects of his personality are based on the late-fifth-century Persian prince Cyrus the Younger.[3] The result is a meditation on leadership that intersects a variety of genres, including biography, history, philosophy, romance, political theory, and even the military manual.[4] Like *How to Survive under Siege* of Xenophon's contemporary Aeneas Tacticus, moreover, the *Education of Cyrus* organises specialised military knowledge into principles and illustrates these principles with examples drawn from history and personal experience.[5] It also incorporates topics that would later become subgenres or cousins of the military manual, such as treatises on military technology and collections of stratagems (e.g., Athenaeus Mechanicus' *On Machines*, Apollodorus' *Siege Warfare*, and Polyaenus' *Stratagems*).[6]

In addition to writing for a broader civilian readership like Aeneas and other military manual authors, Xenophon has much to offer for experienced military commanders and modern specialists.[7] The *Education of Cyrus* was consulted by Roman generals such as Scipio Aemilianus (Cicero *Letters to Quintus* 1.1.23, *Tusculan Dispitations* 2.62) and Julius Caesar (Suetonius, *Life of Caesar* 87).[8] Moreover, scholars have observed that it includes lessons regarding heavy infantry and cavalry tactics applicable to understanding several momentous battles from Xenophon's lifetime, from Cunaxa in 401 to Leuctra in 371 and Mantinea in 362.[9]

This chapter examines the *Education of Cyrus* as both a manual outlining principles of military leadership and an implicit commentary on the generalship of Cyrus the Younger and Clearchus of Sparta, two commanders under whom

Xenophon served in 401. Xenophon identifies several principles that every military leader should follow in the opening book, and later shows Cyrus II applying them in his victories against the Assyrians and the Lydians. Reading his depiction of the *Education of Cyrus*' Cyrus against the portrayal of Cyrus the Younger and Clearchus in the *Anabasis*, it is clear that Xenophon not only intended his principles of leadership outlined in the *Education of Cyrus* to be universal, but that he also believed their implementation would have changed the outcome at the Battle of Cunaxa.[10]

I. Cunaxa and the principles of good military leadership

Xenophon drew on a wealth of experience when composing the military campaigns of the *Education of Cyrus*. He personally knew many of the most prominent generals of his time and had participated in several of the battles he wrote about in his *Anabasis* and *Hellenica*. Of all the engagements he recorded, none matched the magnitude of that at Cunaxa in 401. This was the earliest and almost certainly the largest of Xenophon's career. It pitted Cyrus the Younger's rebels, including roughly thirteen thousand Greeks under Clearchus, against a larger Persian royal army led by King Artaxerxes II himself (*Anabasis* 1.7.10–13).

Had Cyrus won, the Greeks who aided him would have gained substantial wealth and more access to a Persian monarch than had any others before or after (*Anabasis* 1.7.6–8). A close friend of Proxenus, one of Cyrus' Greek generals, Xenophon himself was in line for a substantial, life-altering share of these rewards (*Anabasis* 3.1.4). The rebellion's failure left him and his colleagues destitute and alone, forced to march home through hostile territory, difficult terrain, and harsh weather. The remnants of his famous Ten Thousand struggled to find reliable employment upon their return to the Greek world, and Xenophon himself was soon exiled from Athens.[11]

It should hardly be unexpected that Xenophon later imagined what could have been done to reverse the result at Cunaxa. At first glance, his account of Cyrus' rebellion and its aftermath in the *Anabasis* might seem to have been the appropriate venue for this. However, in order to justify his own participation in the revolt, Xenophon chose to portray Cyrus as a leader worthy of the Persian throne.[12] Thus, in the *Anabasis* he seems to imply that the fault for the defeat lay not with Cyrus, but with Clearchus' refusal to follow Cyrus' last-minute order to attack the King's position in the centre of the royal army (*Anabasis* 1.8.12–13).

Much of the debate over Cunaxa from antiquity to the present has revolved around the wisdom of this command, meaning that blame for the defeat either falls to Cyrus or Clearchus.[13] Yet, based on the lessons of military leadership that he provides in the *Education of Cyrus*, it appears Xenophon recognised that the reasons for the rebellion's failure ran much deeper than whether or not a simple tactical manoeuvre was executed. At the end of the first book, he records a conversation between Cyrus and his father Cambyses in which the former outlines principles of military leadership (1.6.10–44). These can be distilled into the following five categories, with quotations from the discussion included to illustrate each:[14]

1   *Planning.* A general must know planning and tactics:

And when you had made it clear to me that tactics was only a small part of generalship, I asked you if you could teach me any of those things. (1.6.14)[15]

2   *Relationships.* He must collaborate with fellow generals and earn the affection of soldiers through shared suffering:

And in his campaigns also, if they fall in the summer time, the general must show that he can endure the heat of the sun better than his soldiers can, and that he can endure cold better than they if it be in winter; if the way lead through difficulties, that he can endure hardships better. All this contributes to his being loved by his men. (1.6.25)[16]

3   *Maintenance.* A commander must always be concerned with the supply, health, and morale of his army:

And above all I beg you to remember this: never postpone procuring supplies until want compels you to it; but when you have the greatest abundance, then take measures against want. (1.6.10)[17]

"But at any rate, as regards the energetic general", said his father, "I can vouch for it that, unless some god do cross him, he will keep his soldiers abundantly supplied with provisions and at the same time in the best physical condition". (1.6.18)[18]

4   *Discipline.* Through competitions and wisdom, a general must train his soldiers to obey and he must obey his own superiors in turn:

"Yes", said Cyrus; "but at all events, as to practice in the various warlike exercises, it seems to me, father, that by announcing contests in each one and offering prizes you would best secure practice in them, so that you would have everything prepared for use, whenever you might need it". "Quite right, my son", said he; "for if you do that you may be sure that you will see your companies performing their proper parts like trained sets of dancers". (1.6.18)[19]

This, my son, is the road to compulsory obedience, indeed, but there is another road, a short cut, to what is much better – namely, to willing obedience. For people are only too glad to obey the man who they believe takes wiser thought for their interests than they themselves do. (1.6.21)[20]

5   *Intelligence.* A leader must use any means necessary to gain and exploit information about his enemy:

"But, father, what would be the best way to gain an advantage over the enemy?"

"By Zeus", said he, "this is no easy or simple question that you ask now, my son; but, let me tell you, the man who proposes to do that must be designing and cunning, wily and deceitful, a thief and a robber, overreaching the enemy at every point". (1.6.26–27)[21]

In the context of this conversation, the eulogy of Cyrus the Younger in the *Anabasis* is far less positive than it initially appears (1.9).[22] Asserting that he was the worthiest Persian leader since Cyrus II, Xenophon surveys Cyrus' early childhood education, in which he excelled in learning self-control, modesty, and obedience. He notes that Cyrus was always faithful to friends and to his oaths, and displayed a fondness for training horses and practising the bow and javelin.[23] By far, the bulk of his praise for Cyrus the Younger in the *Anabasis* is focused on practices designed to build a large and loyal following: he benefits friends and harms enemies, rewards the brave and shames the cowardly, promotes skilful administrators while demoting the greedy, gives greater gifts more eagerly than he receives, and so forth.

Despite this effusive praise, a number of traits from the *Education of Cyrus'* discussion of leadership are missing. *Education of Cyrus'* Cyrus is advised to win followers through superior wisdom and to share in their suffering (1.6.19–25). Cyrus the Younger is not praised for either of these qualities, and Xenophon even points out that the prince shares his own food only with close friends. He writes that Cyrus was obedient and practised horse riding, archery, and the javelin as a child, but offers no comment on his fighting skill as an adult or on his strategic or tactical aptitude. In total, his leadership fits the category of Relationships, and even then only partially. Cyrus the Younger is presented as an adept politician but not a talented general.

Clearchus also receives an encomium in the *Anabasis* (2.6.1–15). Unlike Cyrus the Younger, Clearchus is praised for procuring supplies, commanding obedience in battle, and facing danger alongside his soldiers. Yet he is criticised for being too fond of war and disobedient to superiors, a character flaw that led to his exile from Sparta and his refusal to follow Cyrus' command at Cunaxa. Moreover, he secures discipline in and obedience from his soldiers not through training, shared suffering, and wisdom, as Cambyses urges in the *Education of Cyrus*, but through fear and severe punishments. As a result, he could be a fine commander to follow in the midst of battle, but his soldiers had little affection for him once away from immediate danger and would seek out other generals.[24]

The dissonance between Xenophon's observations on the leadership and character of Cyrus the Younger and Clearchus in the *Anabasis* and his commentary on the principles of military generalship in the *Education of Cyrus* is not accidental. Further evidence of this can be found through an intertextual comparison of Xenophon's narrative of Cyrus the Younger's rebellion in the *Anabasis* with the two military campaigns recorded in the *Education of Cyrus*. In the battle against the Assyrians, *Education of Cyrus'* Cyrus parallels Clearchus as a general subordinate to Median King Cyaxares. In the Battle of Thymbrara of 546, he parallels Cyrus the Younger as the supreme command of the full Persian army. In each, Xenophon's idealised Cyrus II implements his principles of military leadership in ways that directly contrast with the behaviour of Clearchus and Cyrus the Younger on their march from Sardis to the battle at Cunaxa.

II. Clearchus and Cyrus against the Assyrians

The second book of the *Education of Cyrus* opens with Cyrus and thirty thousand Persians at the frontier between Assyria and Media in an army under the command of Cyaxares, King of the Medes. The Assyrian army is preparing an attack, Cyaxares explains, and significantly outnumbers the combined forces of the Medes and the Persians. Cyrus points out that the foot soldiers on both sides are equipped to fight at range with javelins, arrows, and slingshots, giving the more numerous Assyrians a significant advantage. He suggests that Cyaxares allow him to outfit and train the Persians as heavy infantry, which will allow them to engage at close quarters and drive the Assyrian light infantry off the battlefield (2.1.1–9).

Cyaxares agrees, and Cyrus procures battle-axes, swords, breastplates, and shields for his soldiers. He implements a training regimen designed to improve their physical fitness, teach them tactics, and instil confidence and an eagerness for battle. He tasks assistants with procuring supplies and fulfilling the quotidian needs of his men, allowing them to focus solely on military preparations. His soldiers embrace these methods and reveal affection for him at the dinners he shares with men of every rank (2.1.10–30). Cyrus then volunteers to lead his army into Armenia, where the king had withheld tribute and soldiers from Cyaxares (2.3.9–16). He quickly secures the allegiance of the Armenians, resolves their border dispute with the Chaldeans, and returns with soldiers from both peoples as well as his own now battle-hardened veterans (3.1.1–3.1).

Upon his arrival, Cyrus persuades Cyaxares to launch a preemptive invasion of Assyria. This allows him to supply his own army off enemy land rather than his own. It also prevents the development of unhealthy rivalries among his troops and emboldens them for the war to come (3.3.13–23). By the start of the invasion, Cyrus fulfils four (Planning, Relationships, Maintenance, and Discipline) of the five categories of military leadership: he has offered sound strategic advice regarding when and how to initiate the campaign, displayed excellent relationships with his subordinates and superiors, shown concern with how best to maintain the supply and morale of his army, and instilled discipline in his soldiers through constant training.

When the Assyrian army is ten days off, Cyrus reveals that he is also a master of category five, Intelligence. By eating dinner in the daylight and banning fires within camp after dark, he prevents the enemy from estimating the size of his force. Sometimes he sets fires in front of the camp, so any hostile approach can be seen, and other times he orders them lit in the rear of his camp. This trick deceives enemy scouts as to the actual location of his army, allowing his own advance guard to capture them for interrogation (3.3.24–25).

Upon their arrival near the Medes and Persians, the Assyrian army immediately constructs a fortified encampment surrounded by a trench with a narrow gate at the front. Xenophon notes that this provides protection for the army at the cost of intelligence, since Cyrus and Cyaxares now are able to judge the size and composition of the enemy force. By contrast, Cyrus makes sure to keep his Persian forces hidden, screening them with nearby villages and hills. When Cyaxares suggests marching against the Assyrian camp in full battle array to demonstrate the resolve of the Medes and Persians, Cyrus convinces him to hold back. He points out that doing so would reveal their plans and their numerical inferiority, and so boost the confidence of the enemy (3.3.26–33).

Thanks to Cyrus, on the eve of combat Cyaxares knows far more about the Assyrian army than they do about his own. Presumably this allows him to develop a successful tactical plan, but Xenophon shares nothing of how his army is arrayed. Instead, his narrative focuses narrowly on the actions of Cyrus and the Persians, leaving us with only two hints about the broader order of battle: the Persians are not intermingled with other allied ethnic forces, and, since their initial success spurs the Median army to attack, they advanced against the enemy first (3.3.65).

The reason that Xenophon does not include further tactical details is because his purpose is to demonstrate how the contributions of a loyal subordinate general can guide a less able superior commander to victory. The parallels between Clearchus and *Education of Cyrus*' Cyrus before and during the Assyrian war are striking. Both command a contingent of heavy infantry in an army invading enemy territory that was otherwise predominantly cavalry and light infantry. Both supplement this force of heavy infantry by securing soldiers from local rivals – for Cyrus, this was the Armenians and Chaldeans; for Clearchus, it was Thracians from the Chersonese (*Anabasis* 1.1.9, 2.9), and perhaps even horsemen from Paphlagonia.[25] Both are arrayed separately from other ethnic forces in the final battle order, and both were the first to advance at the start of the engagement. Finally, both serve under commanders less capable or experienced than themselves.

The similarities end there. While *Education of Cyrus*' Cyrus advises Cyaxares on the best ways to scout the enemy and maintain his own forces, in the *Anabasis* Clearchus is portrayed as silent on similar matters. This is not for lack of opportunity. On one occasion, Xenophon notes that Clearchus was considered the most honoured of the Greeks and was invited to Cyrus' tent as a counsellor (1.6.5). On another, when Cyrus presented himself to the Greek officers to discuss plans for the battle, he records Clearchus asking only whether the King would fight, but records nothing about any advice the general offered (1.7.1–9).

As a result, Cyrus the Younger and his army are not at all prepared when the King suddenly appeared. The battle is fought on a field chosen by his enemy, and Cyrus' army is barely in formation and forced to fight on empty stomachs (1.8.1–4, 10.19).[26] This last point is something Xenophon repeatedly emphasises in the *Education of Cyrus*: whereas the Greeks at Cunaxa do not eat before or after the battle because their camp had been plundered, *Education of Cyrus*' Cyrus takes care to make sure that his men are always well fed, and mentions their meals before and after the battle against the Assyrians (*Education of Cyrus*. 3.3.40, 4.1.7, 9).

Clearchus fails to implement the other principles of leadership identified in the *Education of Cyrus* as well. While *Education of Cyrus*' Cyrus encourages healthy competition between his soldiers and takes care to avoid the development of jealousy among the troops and officers (2.22–24), Clearchus mistreats colleagues and subordinates. He tries to force his troops into marching when they demand to know the campaign's true objective; after that does not work – and they almost stone him to death – he lies to them about it and his true loyalties, in the process winning over soldiers from his fellow Greek generals who were more honest about their loyalties (*Anabasis* 1.3, 1.4.7). Later, he goes out of his way to impose harsh discipline on the soldier of another Greek general, nearly provoking open fighting between their respective armies (1.5.11–17).

Clearchus is also disobedient to Cyrus the Younger himself. He does not advise Cyrus against the purportedly foolhardy order to attack the King's centre, or to even notify him that he would not carry it out, but simply disobeys his superior (1.8.13). The *Education of Cyrus'* Cyrus is confronted with a similar scenario prior to the battle against the Assyrians, when Cyaxares orders him to attack prematurely. Cyrus urges his commander to wait and remain patient rather than ignore him, but he dutifully begins the attack when Cyaxares sends again to insist upon it (*Education of Cyrus* 3.3.46–57).[27]

The final aspect of leadership that Clearchus lacks is the ability to instil sufficient discipline in his troops. This may come as a surprise to readers familiar with the *Anabasis*, since Xenophon inserts two scenes in the opening book that appear to showcase the superior discipline and training of the Greeks. The first is the troop review at Tyriaeum, where the troops under Clearchus and the other Greek generals march in formation for Cyrus and the queen of Cilicia, whom Xenophon reports marvels at their discipline and appearance (1.2.14–18). The second is during the Battle of Cunaxa itself, where the Greeks opened gaps in their lines to avoid the charges of scythed chariots and reminded each other to maintain their ranks in pursuit after the enemy broke (1.8.20).

Yet in each case there are complicating details that reveal a lack of discipline. During the troop review, the Greeks are ordered to advance in line. Once on the move, their pace quickens and, without being ordered, eventually turns into a headlong charge at the unarmed merchants and buyers in the army's marketplace, who flee in terror (12.17–18). At Cunaxa, the Greeks again charge without being commanded. The phalanx advances unevenly, forcing those who fall behind to run to catch up. Seeing their peers on the run, the rest of the soldiers in turn began to charge. Soon the entire Greek army was running before even coming in range of the arrows of the Persians opposite them, who, like the market-goers, flee before the Greeks reach them (1.8.18–19).

In the *Education of Cyrus*, Cyrus' soldiers practise formations and manoeuvres on several occasions while serving under Cyaxares, never once breaking into a sprint on their own (e.g., 2.3.17–21, 2.4.2–4, 3.3.11–13). Similarly, in the battle against the Assyrians, Cyrus' Persian heavy infantry outdo their Greek counterparts from the *Anabasis*. They advance into battle on Cyrus' order, quickening their pace – but not running – as they chant the paean, marching side-by-side with complete self-control, cheering each other on in a show of high morale (3.3.57–59).

As at Cunaxa, the Assyrian chariot charge fails, and the archers, slingers, and javelin-throwers spend their missiles before Cyrus and his Persian peers come into range. In the *Education of Cyrus*, however, Cyrus and his men do not begin to run before coming into range of enemy projectiles. Instead, the order to charge comes from Cyrus only as they are stepping over them, and the leaders – including Cyrus – are the first to run, setting an example for the rest of the phalanx. Much like the Persians at Cunaxa, the Assyrians here break before Cyrus and his men reach their lines (3.3.60–63).

In the *Anabasis*, Xenophon describes the Greeks as victorious at the moment their Persian adversaries fled (1.8.21). They give chase, and Xenophon writes that they shouted to one another to remain in formation, but tellingly never reports whether or not they actually did so (1.8.19). He also never states that the slow Greek hoplites were able to close the distance and inflict any casualties; indeed, his statement that the Greeks suffered only a single injury – from an arrow – is probably an implicit admission that they did not (1.8.20). When the battle ends, the Greeks are still in pursuit while the King and his forces plunder their camp (1.10.4). Several scholars have suggested that this is a sign that the Greeks were lured off the field of battle by a calculated retreat. Rather than being victorious in a losing effort, Clearchus may have been in fact outmanoeuvred at Cunaxa and contributed to Cyrus the Younger's defeat.[28]

Xenophon's description of the pursuit of the Assyrians in the *Education of Cyrus* further supports this thesis. There, *Education of Cyrus'* Cyrus and his Persian peers follow the enemy so far as it is safe, killing them at the walls of their camp (3.3.62–66). Before they force their entry into the camp itself and risk being surrounded, however, Cyrus calls off the attack and orders his men to reform. In doing so, Xenophon remarks that the Persians showed perfect discipline: word of the order is passed along the lines, and every man knows his place exactly (3.3.69–70). Their deeds have a direct impact on the outcome of the battle, since they inspire the rest of the Median army to charge alongside them and, in the process, to kill the Assyrian king (3.3.65, 4.1.8). In the *Anabasis*, by contrast, Cyrus the Younger is left to charge with only his elite cavalry, while his left wing never advances or even enters battle. As a result, the rebel prince is outnumbered, surrounded, and killed.

It is notable that in Xenophon's account of Cunaxa, Clearchus is nowhere to be found after he ignores Cyrus the Younger's order to attack the King's centre. The Greeks form their own lines (1.8.14), the watchword is passed along the ranks (1.8.16), the paean is struck and the soldiers advance without command (1.8.17), they break into a run without being ordered by anyone in particular (1.8.18), they shout to each other – as opposed to being ordered by their leaders – to maintain ranks (1.8.19), and they pursue the enemy without regard for their own position or what else is taking place on the battlefield (1.8.21). The image given is of an army acting on its own, without leadership.

The language of this episode in the *Anabasis* may appear innocuous when understood in a vacuum, but compared against the *Education of Cyrus* the difference is remarkable. Xenophon makes *Education of Cyrus'* Cyrus the subject of each sentence in the battle narrative, and in doing so leaves no doubt that he is in total control of his army throughout the battle and its aftermath. Cyrus marches at the head of his troops (3.3.57), he passes along the watchword (3.3.58), he begins the battle paean, he orders and leads the sprinting charge against the Assyrians (3.3.58–63), and he commands his soldiers to halt when they risk pursuing too far (3.3.69). He posts guards, sends scouts to learn the result of the engagement elsewhere, and seeks out his superior Cyaxares as soon as possible (4.1.1–6). Xenophon ends his account of the battle with *Education of Cyrus'* Cyrus observing to his soldiers that as heavy infantry they are not equipped to pursue the fleeing Assyrians because they lack their own cavalry (4.1.10–11).

The starkness of the contrast between Xenophon's description of Clearchus' leadership at Cunaxa and Cyrus' against the Assyrians is even more apparent when the two are compared side-by-side:

Clearchus and the Greeks at the Battle of Cunaxa (Anabasis *1.8, 1.10*)	Cyrus and the Persians during the defeat of the Assyrians (Education of Cyrus *3.3.46–4.1.6*)
Clearchus is noncommittal to Cyrus the Younger's order to attack the King's centre (1.8.12–13)	Cyaxares orders Cyrus to attack; Cyrus counsels him to delay (3.3.46–47)
The Greeks are still in the process of forming their line when the battle begins (1.8.14)	Cyaxares orders Cyrus to attack again; Cyrus obeys and arrays his army in perfect formation before battle (3.3.56–57)
The watchword is passed through the ranks; the Greeks begin the paean and advance (1.8.16-17)	Cyrus passes the watchword through the ranks, begins the paean, and leads the advance (3.3.58–59)
The Greek formation billows out, and the Greeks break into a run (1.8.18)	The Assyrian chariot charge fails; the Assyrian archers spend their missiles before the Persians enter range (3.3.60)
The Persian chariot charge fails, and the Persians expend their missiles and retreat before the Greeks enter range (1.8.19)	As the Persians step over the spent Assyrian missiles, Cyrus orders the Persians to charge and is the first to break into a run (3.3.61–62)
The Greeks give chase and suffer a single casualty in the battle from a wayward arrow (1.8.20)	The Assyrian line breaks before the Persians reach them; the Persians follow and slaughter many of them (3.3.63–64)
Seeing the Greeks give chase to the Persians, Cyrus the Younger leads his elite cavalry into battle and is killed (1.8.20–29)	Seeing the success of Cyrus and the Persians, the Medes charge and break the Assyrian cavalry (3.3.65), killing the Assyrian king (4.1.8)
Cyrus' left wing retreats to and is driven from its fortified camp without giving battle (1.10.1)	The Assyrians flee to their fortified camp (3.3.66–68)
The Greeks finally cease their pursuit of the Persians after learning the King is in their baggage train thirty stades away (1.10.4–5)	At Cyrus' order, the Persians reform their lines in order to avoid over-pursuing the Assyrians (3.3.69–70)
The Greeks return to their camp, finding it plundered (1.10.17–18)	Cyrus makes camp, posts guards and dispatches scouts, and rides to meet with Cyaxares (4.1.1–7)
The Greeks go without dinner and breakfast (1.10.19)	Cyrus eats dinner and breakfast with his men (4.1.7–9)

Based on the close similarities between Xenophon's battle descriptions in the *Education of Cyrus* and the *Anabasis*, it is obvious that he modelled Cyrus' defeat of the Assyrians upon his portrayal of the Greek experience at Cunaxa. As this table shows, the *Education of Cyrus*' Cyrus is demonstrably superior as a leader to Clearchus at every stage of the battle. In Xenophon's mind, the defeat at Cunaxa was not merely a result of Clearchus' refusal to march against Artaxerxes' position in the centre of the royal army, but due in large part to Clearchus' failure to follow the model of leadership identified and exemplified in the *Education of Cyrus*.[29]

III. The Battles of Thymbrara and Cunaxa

Cyrus and Cyaxares killed the Assyrian king and plundered the Assyrian camp in the first engagement of the war, but the bulk of the enemy army escaped from danger. Regrouping under the leadership of King Croesus of Lydia, opposition forces continued to vastly outnumber the Medes and Persians. Cyrus volunteered to continue the offensive and received full command of the army from Cyaxares, who warned that the campaign would only become more difficult:

> And now, when they had shut themselves up in their fortifications, they allowed us to manage things so as to fight as many at a time as we pleased. But if we go against them in an open plain and they learn to meet us in separate detachments, some in front of us (as even now), some on either flank, and some in our rear, see to it that we do not each one of us stand in need of many hands and many eyes.
>
> (*Education of Cyrus* 4.1.18)[30]

Xenophon devotes a significant portion of what follows in the *Education of Cyrus* to examining how Cyrus overcomes this challenge at the Battle of Thymbrara, demonstrating his mastery of all five qualities of leadership outlined in the opening book of the work. It is not a coincidence that the historical Cyrus the Younger faced the exact same predicament at Cunaxa, where he confronted a royal force reportedly at least double the size of his own (*Anabasis* 1.8.13). Just as he used *Education of Cyrus*' Cyrus' leadership during the campaign against the Assyrians to implicitly critique Clearchus, Xenophon narrates the campaign against Croesus to show how Cyrus the Younger could have defeated King Artaxerxes II.

J. K. Anderson previously detailed the tactical parallels between Cunaxa and the Thymbrara, the climax of Cyrus' campaign against Croesus in the *Education of Cyrus*.[31] Notably, the solution Xenophon implies would have turned the battle for Cyrus the Younger in the *Anabasis* – Clearchus leading the Greek hoplites against the King's position in the centre – is completely absent from the plan developed by *Education of Cyrus*' Cyrus. Instead of launching a desperate charge at the enemy centre after being outflanked, a manoeuvre that resulted in Cyrus the Younger's death at Cunaxa, in the *Education of Cyrus* Xenophon imagines his Cyrus II creating a plan to defeat the enemy wings first and foremost.

Aware that the Lydian king would try to use his superior numbers to surround Persian forces, *Education of Cyrus*' Cyrus previews his tactical countermeasure in a comment to a subordinate that Croesus "they may have an opportunity to find out whether the surrounders may not be surrounded" (6.3.20). By hiding infantry and cavalry behind the baggage train in the rear of his army, Cyrus is able to strike the vulnerable flanks of the Lydian army during their own flanking manoeuvre (7.1.26–28). Only after they are driven from the field of battle does he move against the Egyptian heavy infantry that were positioned in the centre of Croesus' army – much as they had been at Cunaxa for Artaxerxes (*Anabasis* 1.8.9) – and, by surrounding them on all sides, force their surrender (7.1.36–45).

Education of Cyrus' Cyrus accordingly meets the requirements of the first leadership category, Planning. Still, as clever as the imagined tactics at Thymbrara were, Xenophon suggests that their successful execution was only possible thanks to Cyrus' other leadership qualities. Immediately noticeable is category five, Intelligence: Cyrus dispatches spies into the enemy camp (6.1.31, 2.2–3, 2.9–11, 3.12–20), marches his army as fast as possible (6.3.2), and uses his cavalry to scout and capture enemy prisoners (6.3.2–11, 4.12). He knows in advance the composition and planned formation of the Lydian army, which allows him to prepare an ambush to counter Croesus' flanking manoeuvre. Meanwhile, Croesus never has the chance to prepare for Cyrus' own tactics, since the Persian army does not array itself in formation until the day of the battle and immediately engages upon coming into view (6.4.12, 7.1.4–8).

One reason the army was able to advance into enemy territory so quickly is that Cyrus is always concerned with its supplies – the key component of leadership category three, Maintenance. Prior to the campaign, he winters his forces in territory where provisions are abundant and personally leads raids into hostile regions nearby (6.1.23–24). He spends time acquiring other necessary items for the army, such as camels, horses, chariots, towers, and weapons (6.1.26–30, 50–55). He orders soldiers to discard superfluous items like wine and carpets, insisting that they bring only food, water, clothes, medicine, straps, and lumber for making weapons and repairing wagons. He also recruits merchants to supply a market and tasks suitable men to make roads for the army (6.2.25–40). He is careful to arrange his marching formation to protect the baggage train, and on the eve of battle reminds his men to double check their equipment before taking position (6.3.2–4, 21).

Cyrus is further successful because his leadership meets the requirements of leadership category four, Discipline. The core of his army had been drilled and trained prior to the first battle against the Assyrians. His soldiers gained further experience during an expedition into Assyrian territory immediately after this initial victory (5.3–4). During the following winter, he drilled his troops and led them on more raids (6.1.24). Cyrus also encouraged healthy competition and rivalry among his subordinates by leading them on hunts and instituting martial contests, spurring the soldiers to excel one another in skill with weapons, endurance, and bravery (6.2.4–6). The end result is that at Thymbrara his soldiers all know their places, follow his orders without hesitation, and implement his tactical plan to perfection.

Finally, Cyrus satisfies leadership category two, Relationships, by maintaining appropriate relationships with his superiors and subordinates. When Cyaxares becomes suspicious of his success and popularity, Cyrus publicly assures him of his continued loyalty (5.5.1–40). The ultimate reward for his success and fidelity is the daughter of Cyaxares in marriage (8.5.1–27). Similarly, he wins the devotion of his officers and generals, involving them in discussions of strategy and tactics (6.1.12–19, 6.2.13–24, 7.1.6–22). He looks out for their interests professionally and personally (6.1.31–52), and is known for keeping his promises (6.1.11, 6.2.4). His troops obey and love him because he participates in hunts and plundering expeditions (6.1.23–24), offers valuable prizes to those who display skill and bravery in martial competitions (6.2.4–6), and honestly addresses their anxieties about the campaign against Croesus (6.2.12–41). Indeed, the soldiers feel so confident in their preparations and in Cyrus' leadership that, even before setting out, 'the rank and file of the army generally cherished the feeling that the victory was already perfectly assured and that the enemy's side was as nothing' (6.2.8).

The contrast between *Education of Cyrus'* Cyrus' leadership and that of Cyrus the Younger in the *Anabasis* on each of these five leadership categories is damning to the latter. Perhaps the most important is his failure in Relationships. While Cyrus the Younger did manage to recruit a number of soldiers from the ranks of his friends and connections in the Greek world (*Anabasis* 1.1.6–11), his relationships were built on careerism and money, not affection won through wisdom or shared suffering. He rarely interacted with the Greeks personally, and did not lead them in any drills or hunts, institute games with prizes when the opportunity presented itself (1.2.10, 13–18), or go out of his way to ensure that they were well supplied and fed (1.5.6). He also lied to them about the purpose of the expedition before and during the march, leading to the defection of two of his longest-serving Greek generals (1.2.1, 1.3.1–21, 1.4.7).[32]

Cyrus the Younger's failure to build strong relationships with his Greek subordinates had a deleterious effect on other aspects of his leadership. The previous section already outlined their deficiency in Maintenance and Discipline at Cunaxa, responsibility for which he shared with Clearchus. Other shortcomings are Cyrus' alone, such as his policy of funding Greek armies all around the Aegean. While doing so allowed him to raise an army in secret,[33] it also meant that his soldiers took a long time to muster for the rebellion and had little experience with him prior to the campaign. This significantly slowed the pace of his march, as he was forced to wait seven days at Colossae for Menon to arrive, thirty days at Celaenae for Clearchus, and three at Issus for Cheirosophus (1.2.6–9, 1.4.2). Likewise, his progress was also stalled for twenty days at Tarsus and for five days at Thapsacus when the Greeks, angered at his lies concerning the true aim of the expedition, refused to advance without an increase in pay (1.3.1, 1.4.1).

Cyrus' efforts to establish strong relationships with other Persians – despite a concerted policy to do so (1.1.5) – were even less successful.[34] Unlike the *Education of Cyrus'* Cyrus, he did not make a genuine effort to reassure the King of his fidelity, and instead only pretended to do so in order to rebel (1.1.4–8). While his small inner circle demonstrated their loyalty by dying alongside him at Cunaxa,[35] Ariaeus, his second in command and the leader of his left wing, retreated almost immediately after Cyrus' charge (1.9.31–10.1). During the march to the battle, no major satraps or officials willingly deserted to Cyrus, and at least three of his own high-ranking subordinates attempted to betray him (1.2.20, 1.6.1–3).[36] Cyrus himself was nearly executed after his subordinate Tissaphernes accused him of planning a coup following the death of Darius II in 404 (1.1.3); later in 401, Tissaphernes provided Artaxerxes II advance warning of Cyrus' rebellion (1.2.1–5).

The slow pace of Cyrus' advance and his failure to recruit collaborators from the Persian provincial elite meant that Artaxerxes was his superior in Intelligence, since he had plenty of time and information to prepare his defences.[37] The King's cavalry burned supplies on his approach and undoubtedly spied on the rebels as well, a situation not improved by the attempted defection of the individual tasked with harrying the King's advance forces (1.6.1–4). Cyrus' decision to march in battle formation after entering Babylonia on the mistaken assumption that the main royal army was near also forfeited any advantages he may have held in Planning, since this preview of his planned battle array provided the King's generals an opportunity to formulate a tactical plan based on the rebel dispositions (1.7.1, 14). The royal army's failure to appear when expected then led Cyrus to compound his error on the mistaken assumption that the King would not fight, causing the rebel army to march carelessly and out of formation (1.7.19–20).

Consequently, Cyrus was completely unprepared for the sudden appearance of King Artaxerxes (1.8.1–2). Much like the force of *Education of Cyrus'* Cyrus at Thymbrara, the royal army marched into battle calmly, taking advantage of their enemy's lack of preparation (1.8.8–11, 14). The King's soldiers revealed their discipline by flawlessly executing flanking movements on the right and left (1.8.23, 1.10.7), where part of the wing under Tissaphernes also executed the feigned retreat that successfully drew Clearchus and the Greek hoplites out of position (1.8.19, 24). Just as victory on the wings led to victory in the centre at Thymbrara, at Cunaxa it forced Cyrus the Younger into making a fatal charge at Artaxerxes' position in the heart of his own battle line.

IV. Conclusion

The *Education of Cyrus* is a meditation on government and leadership that contains a military manual on leadership. Xenophon begins by identifying a series of principles that all commanders should follow. Beyond being an adept tactical planner, the ideal general must instil discipline in his army, collaborate effectively with superiors and subordinates, maintain the supplies, health, and morale of the soldiers, and be a master at deception and intelligence. Xenophon then offers (pseudo-)historical examples of these principles in action in the form of a narrative recounting of the military campaigns of Cyrus during his rise to power. It is a testament to the universal character of the military leadership lessons in the *Education of Cyrus* that it has served as a source of wisdom for ancient generals, a model for the ancient authors who developed the military manual into a distinct genre, and a tool for modern scholars to understand ancient battles from Cunaxa to Leuctra to Mantinea.

Even so, the ways in which Xenophon fictionalises the details of the campaigns of *Education of Cyrus'* Cyrus reveals that his account was also intended to be read as a subtle critique of the generalship of Cyrus the Younger and Clearchus at Cunaxa in particular. While he describes the latter as a competent commander and effusively praises the leadership of the former in the *Anabasis*, both fell far short of his idealised Cyrus II. The intertextual analysis of these leaders offered in this chapter suggests that Xenophon believed the failure of Cyrus the Younger's rebellion was caused not by disagreement over a single tactical order, but by the inability of Cyrus and Clearchus to collaborate as leaders from the start of their campaign. As a result, the rebel army – including the Greek hoplites often viewed as the best soldiers on the battlefield that day – arrived at Cunaxa low on supplies and morale, unprepared due to a lack of intelligence gathering, and without the discipline and sound tactical plan necessary to defeat the qualitatively and quantitatively superior army of King Artaxerxes II.

Notes

1 Unless otherwise stated, all dates are BCE.
2 On the authenticity of this passage, see Delebecque (1957) 405–409; Due (1989) 16–22; Tatum (1989) 217–239; Sage (1995); Nadon (2001) 139–146; Gray (2011) 246–263; and Tuplin (2013) 71–72.
3 Fictive elements, Sancisi-Weerdenburg (1985); Stadter (1991); Tuplin (2013) 70; and Gray (2016). Parallels with Cyrus the Younger, Anderson (1970) 165–191; Hirsch (1985) 75; Due (1989) 187–192; Sage (1991); Roscalla (2008); and Flower (2012) 51.
4 On Xenophon's interest in leadership, the scholarship is impressively vast: Wood (1964); Breebaart (1983); Dillery (1995) 164–194; Nadon (1996); Hutchinson (2000); Johnson (2005); Lendon (2006); Whidden (2007); Tuplin (2013); and Winter (2016). On the genre of the *Education of Cyrus*, see Delebecque (1957) 384–410; Due (1989); Tatum (1989); Gera (1993); Nadon (2001); Reichel (2007); Gray (2011); and Tuplin (2013) 67–68. On the treatise as a military manual, see Jähns (1889) 22; Pease (1934); Anderson (1974) 2–3; and Burliga (2008).
5 Burliga (2008) 98–99 notes that Aeneas cites forty-two historical anecdotes in total, thirty-four of which reference events contemporary to his lifetime, with the remaining eight drawn from past events. See also Brown (1981); David (1986); Hornblower (1995) 53; and Pretzler (2018). For Xenophon, Aeneas, and the emergence of the military manual genre, Vela Tejada (2004); Whitehead (1990) 35–38 and (2008); and Shipley (2018).
6 *Education of Cyrus* 6.1.29–30, 52–54 credits Cyrus II with inventing the scythed chariot – probably incorrectly; see Rop (2013) – and a chariot tower that could carry up to twenty men for use in open field battle. For stratagems, see the discussion of broader principles categorised under 'Intelligence' later. On the work as a source for the later military manual of Onasander, see Chapter 2. Notably, Xenophon was also the author of two other technical manuals related to military affairs, *The Cavalry Commander* and *On Horsemanship*; see Chapter 7 on the latter.
7 For further discussion of the intended audiences of military manuals, see Chapters 1, 3, 7, and 14.
8 As noted in Chapter 9 (117). See also the Introduction (7) and Chapter 1 (18–20).

9 Anderson (1970) 165–191; Tuplin (1994) 146–150; Azoulay (2004) 310–318; Christesen (2006); and Blaineau (2008).

10 Relatedly, Buxton (2016) observes that Xenophon's depiction of leadership in the *Hellenica* confirms his own performance as a leader of the Ten Thousand in books 5–7 of the *Anabasis*. Tamiolaki (2016) notes that Xenophon contrasts the leadership of several near contemporaries with those from previous generations, including Cyrus II, in the *Memorabilia*.

11 On his exile, see Delebecque (1957) 121–123; Anderson (1974) 147–149; and Dreher (2004).

12 For Xenophon's apologetic and literary strategies, see Flower (2012) 40–59. On his apology for Cyrus the Younger, see Tuplin (1994) 133–134; Bassett (1999); and Rop (2019b) 31–45.

13 Ctesias (F16.64), Diodorus (15.23.7), and Isocrates (5.90–2) blame Cyrus. Following Xenophon, Plutarch declares Clearchus to be at fault (*Artaxerxes* 8.3–5). For modern views, see Hewitt (1919) 237–238; Parke (1933) 23; Nussbaum (1967) 2; Rahe (1980) 94–95; Bigwood (1983) 342; Wylie (1992) 129–130; Whitby (2004) 227–228; Cawkwell (2005) 248; Waterfield (2006) 17–18; and Shannahan (2014).

14 Xenophon does not present these as a list *per se*, but perhaps his discussion here can be seen as a forerunner to summarising lists of later military manuals, as discussed in Chapter 11.

Hutchinson (2000) 180–239 examines the 'Ideal Commander' based on the full corpus of Xenophon's writings. The many similarities between the categories he defined and those identified here reveal that Xenophon's ideas on leadership remain remarkably consistent throughout his writings, as Gray (2011) has also argued. For a similar breakdown of leadership categories in the *Education of Cyrus*, though not as focused on military leadership, see Keim (2016).

15 ὡς δέ μοι καταφανὲς ἐποίησας ὅτι μικρόν τι μέρος εἴη στρατηγίας τὰ τακτικά, ἐπερομένου μου εἴ τι τούτων σύ με διδάξαι ἱκανὸς εἴης.

16 καὶ ἐπὶ τῶν πράξεων δέ, ἢν μὲν ἐν θέρει ὦσι, τὸν ἄρχοντα δεῖ τοῦ ἡλίου πλεονεκτοῦντα φανερὸν εἶναι: ἢν δὲ ἐν χειμῶνι, τοῦ ψύχους: ἢν δὲ διὰ μόχθων, τῶν πόνων: πάντα γὰρ ταῦτα εἰς τὸ φιλεῖσθαι ὑπὸ τῶν ἀρχομένων συλλαμβάνει.

17 τόδε δὲ πάντων μάλιστά μοι μέμνησο μηδέποτε ἀναμένειν τὸ πορίζεσθαι τὰ ἐπιτήδεια ἔστ' ἂν ἡ χρεία σε ἀναγκάσῃ: ἀλλ' ὅταν μάλιστα εὐπορῇς, τότε πρὸ τῆς ἀπορίας μηχανῶ.

18 τὸν δέ γε ἐργάτην στρατηγὸν ἐγώ, ἔφη, ἀναδέχομαι, ἢν μή τις θεὸς βλάπτῃ, ἅμα καὶ τὰ ἐπιτήδεια μάλιστα ἔχοντας τοὺς στρατιώτας ἀποδείξειν καὶ τὰ σώματα ἄριστα ἔχοντας παρασκευάσειν.

19 ἀλλὰ μέντοι, ἔφη, τό γε μελετᾶσθαι ἕκαστα τῶν πολεμικῶν ἔργων, ἀγῶνας ἄν τίς μοι δοκεῖ, ἔφη, ὦ πάτερ, προειπὼν ἑκάστοις καὶ ἆθλα προτιθεὶς μάλιστ' ἂν ποιεῖν εὖ ἀσκεῖσθαι ἕκαστα, ὥστε ὁπότε δέοιτο ἔχειν ἂν παρεσκευασμένοις χρῆσθαι. κάλλιστα λέγεις, ἔφη, ὦ παῖ: τοῦτο γὰρ ποιήσας, σάφ' ἴσθι, ὥσπερ χοροὺς τὰς τάξεις ἀεὶ τὰ προσήκοντα μελετώσας θεάσῃ.

20 καὶ ἐπὶ μέν γε τὸ ἀνάγκῃ ἕπεσθαι αὕτη, ὦ παῖ, ἡ ὁδός ἐστιν: ἐπὶ δὲ τὸ κρεῖττον τούτου πολύ, τὸ ἑκόντας πείθεσθαι, ἄλλη ἐστὶ συντομωτέρα. ὃν γὰρ ἂν ἡγήσωνται περὶ τοῦ συμφέροντος ἑαυτοῖς φρονιμώτερον ἑαυτῶν εἶναι, τούτῳ οἱ ἄνθρωποι ὑπερηδέως πείθονται.

21 πλέον δ' ἔχειν, ὦ πάτερ, πολεμίων πῶς ἄν τις δύναιτο μάλιστα; οὐ μὰ Δί', ἔφη, οὐκέτι τοῦτο φαῦλον, ὦ παῖ, οὐδ' ἁπλοῦν ἔργον ἐρωτᾷς: ἀλλ' εὖ ἴσθι ὅτι δεῖ τὸν μέλλοντα τοῦτο ποιήσειν καὶ ἐπίβουλον εἶναι καὶ κρυψίνουν καὶ δολερὸν καὶ ἀπατεῶνα καὶ κλέπτην καὶ ἅρπαγα καὶ ἐν παντὶ πλεονέκτην τῶν πολεμίων.

22 Braun (2004) argues that Xenophon intended the obituaries of Cyrus the Younger and Clearchus to be subtly critical. See also Flower (2012) 188–194 and (2016) 102–105. Gray (2011) 71–79 disagrees, while Reichel (2007) suggests that *Education of Cyrus'*

Cyrus is an idealisation of the best traits of Cyrus the Younger and other figures from Xenophon's life.

23 Tuplin (1994) 150–161 notes the similarities in the education of his Cyrean protagonists in the *Anabasis* and *Education of Cyrus*.

24 For a character analysis of Clearchus, see Tritle (2004). Roisman (1985) persuasively argues that Xenophon's broader character evaluation of the general in his encomium does not accurately reflect his actions as a general in many respects throughout the narrative of the *Anabasis*. Bassett (2001) notes that Xenophon favorably misrepresents the reasons for Clearchus' exile.

25 They are stationed next to Clearchus in the battle (1.8.5). Hyland (2008) suggests that Pharnabazus was complicit in the rebellion and may have provided the Paphlagonians.

26 For a recent analysis of Cyrus' battle plans and the King's countermeasures, see Rop (2019b) 45–63.

27 Buxton (2016) 176–189 notes that Xenophon emphasises the importance of willing obedience by subordinate officers in the *Anabasis* and the *Hellenica*. See also Keim (2016) 137–151.

28 Wylie (1992) 129–130; Ehrhardt (1994) 1–2; and Waterfield (2006) 17–18. Briant (2002) 629 acknowledges that the Greeks over-pursued but does not attribute this to a feigned retreat by the Persians. Shannahan (2014) 68–76 challenges the idea of an intentional retreat, but see Rop (2019b) 52–63 for a response.

29 Cawkwell (2005) 48 notes, 'there is no knowing when he precisely wrote the *Education of Cyrus*', but its dependence upon the *Anabasis* here supports the timeline of Delebecque (1957).

30 καὶ νῦν μὲν κατακλείσαντες ἑαυτοὺς εἰς ἔρυμα παρέσχον ἡμῖν ταμιεύεσθαι ὥσθ' ὁπόσοις ἐβουλόμεθα αὐτῶν μάχεσθαι· εἰ δ' ἐν εὐρυχωρίᾳ πρόσιμεν αὐτοῖς καὶ μαθήσονται χωρὶς γενόμενοι οἱ μὲν κατὰ πρόσωπον ἡμῖν ὥσπερ καὶ νῦν ἐναντιοῦσθαι, οἱ δ' ἐκ πλαγίου, οἱ δὲ καὶ ὄπισθεν, ὅρα μὴ πολλῶν ἑκάστῳ ἡμῶν χειρῶν δεήσει καὶ ὀφθαλμῶν.

31 Anderson (1970) 165–191.

32 On Cyrus' recruitment of Greeks and the reasons for their loyalty to him, see Rop (2019b) 70–87.

33 With debatable results, as Tissaphernes apparently recognised Cyrus' true purpose upon its muster (1.2.4). See Rop (2019a) 81–82.

34 For further discussion, see Rop (2019b) 75–77.

35 They also enthusiastically remove wagons stuck in mud on one occasion, but he himself does not join in (1.5.7–8).

36 Syennesis of Cilicia did meet Cyrus and exchange gifts, but under duress (1.2.25–27).

37 According to Wylie (1992) 120, the entire march inland took 181 days, during which the army was on the move only eighty-five. On the timing and chronology of Cyrus' rebellion, see Rop (2019a).

Bibliography

Anderson, J. K. 1970. *Military Theory and Practice in the Age of Xenophon*. Berkeley: University of California Press.

——— 1974. *Xenophon*. New York: Longwood Press Ltd.

Azoulay, V. 2004. *Xénophon et les Grâces du Pouvoir: de la Charis au Charisme*. Paris: Publications de la Sorbonne.

Bassett, S. R. 1999. 'The Death of Cyrus the Younger'. *CQ* 49. 473–483.

——— 2001. 'The Enigma of Clearchus the Spartan'. *AHB* 15. 1–13.

Bigwood, J. M. 1983. 'The Ancient Accounts of the Battle of Cunaxa'. *AJPh* 104. 340–357.

Blaineau, A. 2008. 'Le cheval, le cavalier et l'hippocentaure. Technique équestre, éthique et métaphore politique chez Xénophon'. *CEA* 45. 185–211.

Braun, T. 2004. 'Xenophon's Dangerous Liaisons', in R. L. Fox, ed., *The Long March: Xenophon and the Ten Thousand.* New Haven: Yale University Press. 97–130.

Breebaart, A. B. 1983. 'From Victory to Peace: Some Aspects of Cyrus' State in Xenophon's "Cyrupaedia"'. *Mnemosyne* 36. 117–134.

Briant, P. 2002. *From Cyrus to Alexander: A History of the Persian Empire.* Engl. trans. Winona Lake: Eisenbrauns.

Brown, T. S. 1981. 'Aeneas Tacticus, Herodotus and the Ionian Revolt'. *Historia* 30. 385–393.

Burliga, B. 2008. 'Aeneas Tacticus between History and Sophistry: The Emergence of the Military Handbook', in J. Pigon, ed., *The Children of Herodotus: Greek and Roman Historiography and Related Genres.* Newcastle-upon-Tyne: Cambridge Scholars Publishing. 92–101.

Buxton, R. F. 2004. 'When, How, and Why Did Xenophon Write the Anabasis?', in R. L. Fox, ed., *The Long March: Xenophon and the Ten Thousand.* New Haven. 47–67.

———— 2016. 'Novel Leaders for Novel Armies: Xenophon's Focus on Willing Obedience in Context', in R. F. Buxton, ed., *Aspects of Leadership in Xenophon.* Histos Supplement. 163–198.

Cawkwell, G. 2005. *The Greek Wars: The Failure of Persia.* Oxford: Oxford University Press.

Christesen, P. 2006. 'Xenophon's "Education of Cyrus" and Military Reform in Sparta'. *JHS* 12. 47–65.

David, E. 1986. 'Aeneas Tacticus, 11.7–10 and the Argive Revolution of 370 B.C.'. *AJPh* 107. 343–349.

Delebecque, E. 1957. *Essai sur la vie de Xénophon.* Paris: Klincksieck.

Dillery, J. 1995. *Xenophon and the History of His Times.* London: Routledge.

Dreher, M. 2004. 'Der Prozess gegen Xenophon', in C. J. Tuplin and V. Azoulay, eds., *Xenophon and His World.* Stuttgart: Franz Steiner. 55–70.

Due, B. 1989. *The Cyropaedia: Xenophon's Aims and Methods.* Aarhus: Aarhus University Press.

Ehrhardt, C. T. H. R. 1994. 'Two Notes on Xenophon, Anabasis'. *AHB* 8. 1–4.

Flower, M. A. 2012. *Xenophon's Anabasis, or the Expedition of Cyrus.* Oxford: Oxford University Press.

———— 2016. 'Piety in Xenophon's Theory of Leadership', in R. F. Buxton, ed., *Aspects of Leadership in Xenophon.* Histos Supplement. 85–120.

Gera, D. L. 1993. *Xenophon's Education of Cyrus: Style, Genre and Literary Technique.* Oxford: Oxford University Press.

Gray, V. J. 2011. *Xenophon's Mirror of Princes: Reading the Reflections.* Oxford: Oxford University Press.

———— 2016. 'Herodotus (and Ctesias) Re-Enacted: Leadership in Xenophon's Education of Cyrus', in J. Priestley and V. Zali, eds., *Brill's Companion to the Reception of Herodotus in Antiquity and Beyond.* Leiden: Brill. 301–321.

Hewitt, J. W. 1919. 'The Disobedience of Clearchus at Cunaxa'. *CJ* 14. 237–249.

Hirsch, S. W. 1985. *The Friendship of the Barbarians: Xenophon and the Persian Empire.* Hanover, NH: University Press of New England.

Hornblower, S. 1995. 'The Fourth-Century and Hellenistic Reception of Thucydides'. *JHS* 115. 47–68.

Hutchinson, G. 2000. *Xenophon and the Art of Command.* London: Stackpole Books.

Hyland, J. 2008. 'Pharnabazos, Cyrus' Rebellion, and the Spartan War of 399'. *ARTA*. 1–27.

Jähns, M. 1889. *Geschichte der Kriegswissenschaften: vornehmlich in Deutschland*. Munich: R. Oldenbourg.

Johnson, D. M. 2005. 'Persians as Centaurs in Xenophon's "Education of Cyrus"'. *TAPA* 135. 177–207.

Keim, B. 2016. 'Honour and the Art of Xenophontic Leadership', in R. F. Buxton, ed., *Aspects of Leadership in Xenophon*, Histos Supplement. 121–162.

Lendon, J. E. 2006. 'Xenophon and the Alternative to Realist Foreign Policy: "Education of Cyrus" 3.1.14–31'. *JHS* 126. 82–98.

Nadon, C. 1996. 'From Republic to Empire: Political Revolution and the Common Good in Xenophon's Education of Cyrus'. *American Political Science Review* 90. 361–374.

——— 2001. *Xenophon's Prince: Republic and Empire in the Education of Cyrus*. Berkeley: University of California Press.

Nussbaum, G. B. 1967. *The Ten Thousand a Study in Social Organization and Action in Xenophon's Anabasis*. Leiden: Brill.

Parke, H. W.1933. *Greek Mercenary Soldiers: From the Earliest Times to the Battle of Ipsus*. Oxford: Oxford University Press.

Pease, S. J. 1934. 'Xenophon's Education of Cyrus, "The Compleat General"'. *CJ* 29. 436–440.

Pretzler, M. 2018. 'Aineias and History: The Purpose and Context of Historical Narrative in the *Poliorketika*', in M. Pretzler and N. Barley, eds., *Brill's Companion to Aineias Tacticus*. Leiden: Brill. 68–95.

Rahe, P. A. 1980. 'The Military Situation in Western Asia on the Eve of Cunaxa'. *AJPh* 101. 79–96.

Reichel, M. 2007. 'Xenophon als Biograph', in M. Erler and S. Schorn, eds., *Die griechische Biographie in hellenistischer Zeit*. Berlin: De Gruyter. 25–43.

Roisman, J. 1985. 'Klearchos in Xenophon's *Anabasis*'. *SCI* 8–9. 30–52.

Rop, J. 2013. 'Reconsidering the Origin of the Scythed Chariot'. *Historia* 62. 167–181.

——— 2019a. 'The Outbreak of the Rebellion of Cyrus the Younger'. *GRBS* 59. 57–85.

——— 2019b. *Greek Military Service in the Ancient Near East, 401–330 BCE*. Cambridge: Cambridge University Press.

Roscalla, F. 2008. 'Vicissitudes de la souveraineté: Xénophon à la lumière de l'intertexte'. *Europe: revue littéraire mensuelle* 86. 212–235.

Sage, P. W. 1991. 'Tradition, Genre, and Character Portrayal: Education of Cyrus 8.7 and Anabasis 1.9'. *GRBS* 32. 61–79.

——— 1995. 'Dying in Style: Xenophon's Ideal Leader and the End of the "Education of Cyrus"'. *CJ* 90. 161–174.

Sancisi-Weerdenburg, H. 1985. 'The Death of Cyrus: Xenophon's Education of Cyrus as a Source for Iranian History'. *Acta Iranica* 25. 459–471.

Shannahan, J. 2014. 'Two Notes on the Battle of Cunaxa'. *AHB* 28. 61–81.

Shipley, D. G. J. 2018. 'Aineias Tacticus in His Intellectual Context', in M. Pretzler and N. Barley, eds., *Brill's Companion to Aieneias Tacticus*. Leiden: Brill. 49–67.

Stadter, P. A. 1991. 'Fictional Narrative in the Cyropaideia'. *AJPh* 112. 461–491.

Tamiolaki, M. 2016. 'Athenian Leaders in Xenophon's Memorabilia', in R. F. Buxton, ed., *Aspects of Leadership in Xenophon*. Histos Supplement. Durham, UK. 1–50.

Tatum, J. 1989. *Xenophon's Imperial Fiction: On the Education of Cyrus*. Princeton: Princeton University Press.

Tritle, L. 2004. 'Xenophon's Portrait of Clearchus: A Study in Post-Traumatic Stress Disorder', in C. J. Tuplin and V. Azoulay, eds., *Xenophon and His World*. Stuttgart: Franz Steiner. 325–340.

Tuplin, C. J. 1994. 'Xenophon, Sparta, and the Education of Cyrus', in S. Hodkinson and A. Powell, eds., *The Shadow of Sparta*. London: Routledge. 127–182.

———— 2013. 'Xenophon's Education of Cyrus: Fictive History, Political Analysis, and Thinking with Iranian Kings', in L. Mitchell and C. Melville, eds., *Every Inch a King: Comparative Studies on Kings and Kingship in the Ancient and Medieval Worlds*. Leiden: Brill. 67–90.

Vela Tejada, J. 2004. 'Warfare, History and Literature in the Archaic and Classical Periods: The Development of Greek Military Treatises'. *Historia* 53. 129–146.

Waterfield, R. 2006. *Xenophon's Retreat: Greece, Persia and the End of the Golden Age*. Cambridge, MA: Belknap Press.

Whidden, C. 2007. 'The Account of Persia and Cyrus's Persian Education in Xenophon's "Education of Cyrus"'. *The Review of Politics* 69. 539–567.

Whitby, M. 2004. 'Xenophon's Ten Thousand as a Fighting Force', in R. L. Fox, ed., *The Long March: Xenophon and the Ten Thousand*. New Haven: Yale University Press. 215–242.

Whitehead, D. 1990. *Aineas the Tactician: How to Survive under Siege*. Oxford: Oxford University Press.

———— 2008. 'Fact and Fantasy in Greek Military Writers'. *AAHung* 48. 139–155.

Winter, J. A. 2016. 'The Rhetoric of Leadership in Xenophon's *Anabasis*'. Royal Holloway. Ph.D thesis. University of London.

Wood, N. 1964. 'Xenophon's Theory of Leadership'. *C&M* 61. 33–66.

Wylie, G. 1992. 'Cunaxa and Xenophon'. *L'Antiquité Classique* 61. 110–134.

Xenophon 1914. *Xenophon: Education of Cyrus Volume V: Books 1–4*. Translated by W. Miller. Cambridge, MA: Harvard University Press.

9   The lost *Tactica* of Lucius Papirius Paetus

Murray Dahm

In February of 50, midway through his year-long tenure as proconsular governor of Cilicia, Cicero wrote to his friend, Lucius Papirius Paetus, from Laodicea:

> Your letter has made me a first rate general. I had not the slightest idea that you were such an expert in military matters. I see that you have read the books of Pyrrhus and Cineas. I think that I will act in accordance with your precepts and thus even further to have some little boats on the seacoast. They say no better arm can be found against Parthian horsemen. But why joke? You do not know with what an imperator you are dealing. The *Education of Cyrus*, which I wore away with reading, I have totally incorporated in this command.[1]

This letter has interesting ramifications on several levels: the approach and preparation of imperial magistrates before taking up military appointments; the practical application of works from the genre of didactic military handbooks; and the ancient approach to Xenophon's enigmatic *Education of Cyrus* (*Cyropaedia*), described, even today, as a 'Historical Romance'.[2] The letter also raises the possibility of a peculiar environment of literary production in the late 50s when, in the aftermath of the disastrous defeat at the battle of Carrhae, in summer 53, didactic military handbooks which claimed to teach the art of generalship may have been in great demand. Several other military handbooks were produced specifically for particular campaigns or periods of crisis, especially when there was a campaign against the Parthians, and when the example of Alexander the Great could be evoked, there always seems to have been a flurry of literary activity taking advantage of the opportunity such as during the campaigns of Trajan and Lucius Verus. In times of crisis, works evoking generals of the past were produced (such as Plutarch's *Sayings of Kings and Commanders* for Trajan or the plethora of Alexander-themed material produced for Marcus Aurelius and Lucius Verus in advance of their Parthian War, which possibly included Arrian's *Anabasis*).[3] Other examples of didactic military handbooks produced at these times include Polyaenus of Macedon's *Stratagems* (for Marcus Aurelius and Lucius Verus' Parthian War), and Aelian's *Tactics* (for Trajan's forthcoming Parthian campaign). Other surviving didactic military handbooks fall into similar patterns: Vegetius'

On Military Affairs (*De Re Militari*), which responds to a perceived military crisis at the close of the fourth century CE; the anonymous *On Military Matters* (*De Rebus Bellicis*) also fits into a similar (if not the same) fourth-century-CE context. Likewise, Onasander's *The General* (*Strategikos*) (for Quintus Veranius' governorship of Britain in 57 CE) was produced at a time when another Parthian War was brewing and Venutius of the Brigantes was causing trouble in Britain.[4] In this regard, Paetus' material and the circumstances in which it was produced and sent to Cicero add to our understanding of how such works came into being time and again. In combination with the context in which Paetus' material was produced, the letter suggests that we may be able to identify a lost Roman military treatise and, what is more, surmise with some confidence what that treatise contained.[5]

This paper seeks to explore the possibilities of what it was that Paetus sent to Cicero. This consideration allows us to explore the context and purpose of authors supplying similar didactic military material for commanders, or addressing such works to emperors, at a time considered most relevant. This further allows us to explore and reinforce the intended practical nature of such works within the genre of military handbooks. Cicero's letter makes it clear that Paetus had sent him some material rather than that it simply displayed his own knowledge of military handbooks. A starting point for this investigation lies in Cicero's mention of Cineas and Pyrrhus.

I

Our knowledge of Paetus is, at best, patchy. The only record of him comes from *Letters to Friends* 9.15–26 and a few other letters (*To Atticus* 1.20.7; 2.1.12; 4.9.2; 14.16). Commenting on *Letters to Atticus* 1.20.7, Shackelton Bailey sums up our knowledge: a wealthy contemporary of Cicero's, normally resident in Naples, like Atticus 'a man of affairs' and an Epicurean.[6] *Letters to Family* 9.25 is the earliest letter addressed to Paetus to survive and the only one to date to before July 46 (9.16/190). It is clear from *To Atticus* 1.20 and 2.1, however, that the two men knew each other well enough by June 60 that Paetus could send Cicero all the books that Paetus' brother, Servius Claudius, had left to him. Cicero had visited Paetus at home in Naples in April 55 (*To Atticus* 4.9.2). It has been argued that Paetus was one of only a few contemporaries who really 'knew' Cicero.[7]

Our evidence of the works of both Cineas and Pyrrhus comes from Aelian's *Tactics* (Preface 1.2), written under the emperor Trajan.[8] Aelian discusses those who have written on Greek military theory and states that Cineas the Thessalian made an epitome of Aeneas Tacticus' 'considerable number of military books' and that Pyrrhus of Epirus composed a *Tactics*, a manual on the art of marshalling the Macedonian phalanx. Of Aeneas' works, only one survives, *How to Survive Under Siege*, but we know from internal and external evidence that he wrote several other works including one on the laying of sieges.[9] Aelian states (*Tactics* 3.4) that 'the definition of tactics laid down by Aeneas is that it is the science of military movements'. No such definition occurs in the surviving treatise, however, and it is probable that Aeneas also wrote a *Tactics*.

We know that Pyrrhus wrote a *Tactica*, although he also wrote a work on siegecraft referred to by Athenaeus Mechanicus (*On Machines* 6.1 and 31.6–10). Frontinus also claims Pyrrhus' *Precepts* as a source (2.6.10). Although the inclusion of precepts in a *Tactics* is possible, none of the examples which survive include them. It is most likely, therefore, that Pyrrhus' *Tactics* closely resembled the form in which we have it today, namely in a form close to the surviving *Tactics* of Asclepiodotus, Aelian, and Arrian. We have no trace that any *Tactics* ever included anecdotes illustrating the treatise's practicality.[10] Cicero's reply to Paetus that he will 'act in accordance with your precepts' might suggest, however, that whatever Paetus sent, it contained 'precepts' or anecdotes of other generals' actions such as those of Pyrrhus to which Frontinus refers.

Aelian's preface goes on to list several other authors of *Tactics* whose works have not survived: Pyrrhus' son Alexander, Clearchus, Pausanias, Evangelus, Polybius of Megalopolis, Eupolemus, Iphicrates, the Stoic Posidonius, 'and many others'. Although Asclepiodotus' work does not include any list of previous writers, Arrian's treatise begins midway through a list which looks remarkably similar to Aelian's. Arrian (*Tactics* 1) lists works by Pyrrhus of Epirus, Alexander, Pyrrhus' son, Clearchus, Pausanias, Evangelos, Polybius, Eupolemus, Iphicrates, and Posidonius. Despite the direct line of descent claimed by Aelian and Arrian by including such lists, and one Asclepiodotus is also evidently part of, Aelian does not mention Asclepiodotus, and Arrian does not mention Aelian or Asclepiodotus. The three authors of the *Tactica* we have are also rarely named in any subsequent lists of military writings; Aelian and Arrian only are named once, by Johannes Lydus (*De Magistratibus* I.47). Such writings therefore did not name all their predecessors, and there must also have been works of which they were unaware. The name of Asclepiodotus is, in fact, not connected with a *Tactica* at all until the tenth or eleventh century when the Florentine MS, *Laurentianus LV 4*, mentions him as the author of the *Techne Tactica*. And it is this *Tactics* and its relationship to that of Posidonius of Apamea, and further, the relationship of the author Asclepiodotus to Posidonius, that have further intriguing possibilities for Paetus' letter. Before we can come to consider those, however, we must conclude our considerations on what it was Paetus actually sent to Cicero.

The similarities between Aelian's *Tactics*, that of Asclepiodotus, and the passage of Polybius at 18.29.1–18.30.4 reveal that claims of a direct line of descent for these treatises are valid. It has been argued that either Polybius or Posidonius' *Tactics* was the original model for our three surviving examples, although the story is more complex.[11] It is possible that Polybius himself drew on Pyrrhus' work. Certainly, Polybius was respected as a tactical writer; the fact that he had experience of command and that his histories are respected give him more credibility than the philosophers Aelian and Asclepiodotus. And yet the material that all three produced, along with Arrian, is much the same. What is more, the lists of previous writers include both men with military experience and philosophers (like Posidonius) without drawing any distinction between the quality or practicality of their output. We should not begrudge the survival of the *Tactics* we have even though they were written by men without military experience; antiquity on the whole seems to have considered that such works belonged to the same, practical tradition.

II

Cicero does not seem to be saying that what he has received is an epitome or abbreviation of Cineas or Pyrrhus but rather that what he has received reminds him of them. It is likely Cicero was showing his erudition for similar works that he knew to exist (and thus paying a learned compliment to his friend) and (possibly) correctly surmising the source of Paetus' material. It would seem that there are several possibilities for what Paetus sent – precepts, material relating to sieges, a *Tactics*, or a combination of such material. It is highly unlikely that Paetus would have only sent Cicero material on sieges – this would certainly not have elicited the comment that Paetus was an 'expert in military matters.' Cicero did, however, conduct a siege during his tenure against the 'Free Cilician' stronghold of Pindenissum, and so the possibility Paetus' letter may have included poliorcetic material cannot be discounted.[12] Cicero's reply to Paetus, that he will 'act in accordance with your precepts', might suggest that, whatever Paetus sent, it did contain 'precepts'. We cannot know if they were included in a single treatise or if Paetus wrote a collection of works he thought relevant. It seems most likely, however, that what Paetus wrote included a *Tactics* since Cicero was expected to face the Parthians/Persians, and one probably closely resembling those which have come down to us. The fact that Paetus' name is not in surviving lists of Greek or Latin writers found in Aelian and Arrian and elsewhere should not be considered overly problematic since none of Asclepiodotus, Aelian, or Arrian feature prominently either. Indeed, we do not know if Paetus' work was ever published, and it certainly appears to have been intended as private correspondence.

Cicero's mention of defence against Parthian cavalry adds further reason to conclude that Paetus' letter most likely included a *Tactica* for it was recognised that the main strength of the Macedonian phalanx was to be able to withstand 'Parthian' cavalry – and that is what Cicero's force had to be prepared to do. The equation of the Parthians with the Persians Alexander had faced was standard practice. Therefore, there was no better exemplar of success against the Parthians than Alexander, and no better exemplification of Alexander's successful tactics than his phalanx. What is more, the strength and success of similar tactics against enemies similar to the Parthians was recognised.[13] It seems possible to suggest therefore that when faced with Parthian foes (who were predominately associated with cavalry, either mounted archers or heavy cataphracts), the value of the phalanx was recognised well into the Roman Empire and especially when 'Roman' tactics were ineffective. This ineffectiveness was most forcefully shown by the annihilation of Crassus' forces at Carrhae in 53.[14] Certainly the *Tactics* of Aelian and Arrian seem to have been composed with specific Parthian campaigns in mind: Aelian's for Trajan's imminent Parthian campaign in 113 CE, and Arrian's work, composed in 136/7 CE, written with a campaign similar to that related in his *Ektaxis kata Alanon* (composed in 134 CE) in mind.

In all likelihood, therefore, what Paetus sent to Cicero most probably included a *Tactics* closely resembling those that have come down to us. Paetus may also have included poliorcetic material and precepts, although, since our surviving examples of *Tactics* do not include such material, it is difficult to argue for such a combination into a single work. It is more likely therefore that Paetus sent a *Tactics* (as it seems to have been a stand-alone genre) and possibly additional material which contained precepts and possibly material on sieges. Cicero's reference to Cineas, who epitomised Aeneas' works, and Pyrrhus, who may have written more than one treatise himself, may itself refer to Paetus sending more than one treatise. It is also possible that Paetus simply sent copies of treatises in his (clearly extensive) library, although the tenor of the letter seems to suggest that Paetus edited or excerpted material specifically for Cicero rather than just sent copies.

III

The province of Cilicia had a garrison of two legions plus auxiliaries, although they seem to have been allowed to lapse into low standards of discipline by Appius Claudius Pulcher (see *To Family* 8.5.1) before Cicero's arrival in July 51. Cicero hoped (*To Atticus* 5.9) that the Parthians would remain quiet during his tenure but report reached him that they had invaded Syria. Cicero's mention of the *Education of Cyrus* suggests that he must have had to re-discipline his eastern legions in some way since disciplining troops was a focus of that work.[15] Cicero was not the first or the last to have to re-discipline eastern legions, although the laxity and ill discipline of eastern legions was a standard literary trope.[16] As Cicero approached the Amanus mountains with his force, he received news that the Parthians had turned back and had been defeated by Gaius Cassius Longinus in Syria. Parthian invasion fears revived in 50 but proved unfounded. Instead, in October 51, Cicero attacked the 'Free Cilicians', a large gang of ruffians and the perfect allies for the Parthians, agitating dissent within the Roman province, who inhabited the mountains. Cicero destroyed various settlements, was hailed 'imperator' and then besieged the Free Cilician stronghold at Pindenissum. The siege lasted eight weeks and was probably enough to earn Cicero a Triumph – it did earn him the precursor of a *supplicatio*. After the 19th of December, Cicero handed over his army to his brother Quintus and returned to Laodicea from where he wrote to Paetus in February. It would seem Cicero was thanking Paetus flushed with the eventual success of his campaigning season (and perhaps giving him some small credit).

Regardless of the merit of Cicero's generalship, his army was still active and had to be prepared to face the Parthians. *Tactics*, designed to explain how to draw up a Macedonian phalanx, the only infantry tactic with a proven track record against 'Parthians', were (understandably) produced especially at times when Parthian campaigns were *de rigueur*. So many Parthian campaigns were proposed and undertaken during the period of the 50s, 40s, and 30s that they would have provided ample opportunity for *litterati* of all qualities to write *Tactics*. The association of the Parthians and Alexander was unavoidable and indeed courted by some with varying degrees of success. *Tactics* continued to be written, and they (and other works) continued to invoke the memory of Alexander well into the imperial period.[17]

In this context, it may therefore be possible to suggest that Posidonius' and/or Asclepiodotus' *Tactics* were also written specifically for a real or proposed Parthian campaign, possibly Crassus', Caesar's, or Mark Antony's. Extant sources are virtually silent on the readership or anyone actually reading *Tactics* and other military handbooks. Our knowledge of the reading of these works is slim and not always complimentary.[18] Polybius (11.8.1–3) describes how one might learn the art of generalship by 'following the systematic instruction of experienced men' (11.8.1–3). Cicero records that Scipio Aemilianus always had the *Education of Cyrus* in his hand – something Cicero evidently emulated in Cilicia (*To Quintus* 1.1.23, *Tusculan Disputations* 2.62). We also know Caesar had read the *Education of Cyrus* (Suetonius, *Life of Caesar* 87).[19] Marius attacks the noble generals (whom he calls 'preposterous creatures' *praeposteri homines*) for attempting to learn how to be a general from history and the military treatises of the Greeks (Sallust, *Jugurthine War* 85.12). Diodorus (38/39.9) tells how Pompey used to spend his sleepless hours in the study of works on strategy. Plutarch records how Alexander seriously considered the *Iliad* a handbook on the art of war (*Alexander* 8.2 and 26.1).[20] Plutarch also records that Philopoemen, the great commander of the Achaean League in the third and early second century, was 'devoted to the *Tactica* of Evangelos' (*Philopoemen* 4.4–6). Perhaps the best example of a practical application of a *Tactica* can be seen in Plutarch's next statement, that Philopoemen explored

> all the vicissitudes and shapes of the phalanx when it is elongated and contracted again in the vicinity of ravines or ditches or narrow defiles, these he would investigate by himself as he wandered about, and discuss them with his companions.

Plutarch notes that Philopoemen 'brought more zeal than was necessary to the study of military science.' We also know that Frontinus was well respected as a military writer and that even Polyaenus' *Stratagems* was read by the emperors Marcus Aurelius and Lucius Verus (*Preface* 5).[21] The works sent to generals were not just about military strategy; the *Itinerarium Alexandri* (6) records that Terentius Varro sent Pompey an *Ephemeridos* before the latter's Spanish campaign, enabling him to determine latitude and navigate by the stars. We also know of a whole litany of works addressed to Marcus Aurelius or Lucius Verus (or both) before (and during) their Parthian campaign in 162 (in Lucian's *How to Write History* and elsewhere).[22]

We have seen claims of a direct line of descent for *Tactica* from Pyrrhus' treatise onwards. The exact relationship of each to its fellows is, in fact, impossible to decipher, especially since so few survive. Similarities of material, vocabulary, and even individual phrases should be expected in treatises with the exact same (narrow) topic. The most commonly held opinion is that Asclepiodotus' *Tactica* was an abbreviation of Posidonius' work.[23] We know that Asclepiodotus was the pupil (*auditor*) of Posidonius from the references to him as such in Seneca the Younger's *Natural Questions* (2.26.6 and 6.17.3).[24] These references also give us the title of a work, *Causes of Natural Phenomena*, and a work on meteorology. The word *kephalaia* 'outline', in the title of the Asclepiodotus' monograph, also suggests it was an epitome or an abridgement, and some name Asclepiodotus' treatise *Tactica Kephalaia*. Another possibility is that the treatise reflects a series of lectures (or notes of Posidonius' own lectures) which Asclepiodotus subsequently wrote up. As a summary of Posidonius' work, or a writing-up of his master's notes, it is perhaps understandable that Asclepiodotus' name would not be recorded, and his work may have been disseminated as Posidonius' (although we should note that Cineas' own summaries are recorded separately from the originals they summarised). At the same time, however, it is possible that Asclepiodotus' *Meterology* and Posidonius' *Meterology*, which were separate treatises, differed from each other. Thus, it is possible that there were two separate *Tactica*, one written by Posidonius and one by Asclepiodotus, and that they may have differed slightly. Similar interests in teacher and student should be no surprise, but that does not mean that their approaches will be the same; alas, we cannot know when one of the works does not survive. As such, it is possible that Posidonius' treatise and earlier works did include other material (such as anecdotes, precepts, or siege material), although Polybius' 18.29.1–18.30.4 (from earlier still) does not.

Posidonius, Asclepiodotus, and Paetus, as far as we know, were not military men. Cicero by contrast (and a facet of his career which is often overlooked) was, as is revealed by his proconsulship.[25] Regardless of questions of Cicero's merit as a commander, there can be no denying he had commanded men in the field. As such, he and many of his contemporaries were the likely targets of literature, both solicited and otherwise, covering all manner of subjects which would help an official in their role, and often aimed at furthering the career of the men who wrote them. Posidonius, Asclepiodotus, and Paetus all fit with the traditions of philosophers having a role to play in the teaching of military 'theory' (and other administrative tasks) which had traditions reaching back to at least the fifth century.

What is more, at least we know that Cicero had actually read Paetus' material. Although the tone is jocular (and Cicero is self-deprecating towards his own 'generalship'), Cicero must have found Paetus' letter in some way useful – Cicero's compliment cannot be explained away as simple politeness to, or jocularity with, a friend. Therefore, Cicero definitely read Paetus' work and found it in some way useful, and it is more than likely Paetus' material resembled a *Tactics* in the same form as those that have come down to us.

The possibility that *Tactics* and other military treatises inspired some kind of practical outcome in the context of an actual campaign has not given many historians pause for thought. There are, however, examples of such inspiration; in December 1594, William Louis of Nassau was inspired to develop a new countermarch tactic for firearms whilst reading Aelian's *Tactics*. He wrote immediately to his brother Maurice with his (highly successful) idea.[26] There may be countless unnamed examples of similar inspiration, but it is nigh on impossible to argue against the contradiction of silence. We must also remember not to cling too tightly to the few explicit examples we do have and unwarrantedly extrapolate those few examples into a plethora of practicality.

Paetus' material was probably produced for Cicero at some time in early 51 in order for Cicero to reply in February of 50. It seems most likely that Paetus would have sent his material to Cicero before he departed for Cilicia in June 51. This is the same year it is normally assumed that Asclepiodotus wrote the *Tactica* attributed to him.[27] It might be considered a coincidence that Paetus' treatise may have been written in the same year as Asclepiodotus'. In fact, the only reason for ascribing that date to Asclepiodotus' work is that it is one possible date for Posidonius' death (dates for Posidonius are given as 135–51 or 130–46). There is no real reason to assume that Asclepiodotus wrote the *Tactics* that year, although the idea that more than one such work was being written in precisely that period, when the Parthian threat in the aftermath of Carrhae was at its height, does open up remarkable possibilities. If Posidonius' own *Tactics* can be dated to the end of his life, then it is possible that three treatises (by Posidonius, Asclepiodotus, and Paetus) on exactly the same subject were produced at a time when they would more than likely have been in great demand.

It is clear from *To Atticus* 1.20.7 that Paetus spoke Greek – the books of his brother which he gave to Cicero were written in both Greek and Latin, and Cicero made a special point of asking Atticus to ensure 'not a leaf is lost' and that he had 'urgent necessity of the Greek works', implying that Paetus' brother's collection was a good one. Later, Cicero wrote that he wanted the Latin works as well as the Greek (*To Atticus* 2.1). We might therefore assume that Paetus too had a superior library, especially when he offered all of his brother's book to Cicero. Such a quality library would have probably meant Paetus' sources for the material he sent to Cicero in 51 were close to hand. He may have had original *Tactica* and other treatises (such as those of Pyrrhus and Cineas) from which to cull what he considered relevant material. We know that libraries like Paetus', his brother's, and Cicero's as well as many later individuals were well stocked and the owners were well placed to excerpt appropriate material from them for all sorts of situations. Sending such material to various correspondents may have been more than just being helpful or reinforcing various intellectual, political, or social bonds but may also have been an opportunity to show off erudition or even the extent of a collection (in much the same way as Athenaeus' *Learned Banqueters* or Aulus Gellius' *Attic Nights* do, or, in a different way, as do Galen's *On My Own Books* and *On the Order of My Own Books*). Knowing that such an individual's library was well stocked and might be the go-to place for a particular volume could well yield future fruit in all sorts of situations. If this was the case here, then Cicero's erudition at surmising the source of Paetus' material was indeed showing off. Perhaps he knew the contents of Paetus' library, or perhaps the two libraries shared such material and it was even more of an in-joke. Although we cannot know whether Paetus wrote his letter in Greek or Latin, it is likely that, as a *Tactica*, it would have been written in Greek. If it had been written in Latin (and there is no ancient precedent for a *Tactica* in Latin), Cicero would not have been reminded of Greek antecedents but would most likely have listed Latin ones or remarked on Paetus' originality. That we know of, by the 50s there was a *De Re Militari* by Cato the Elder written in the second century,[28] and possibly a *De Re Militari* in at least six books by Cincius.[29] A Cincius is mentioned by Cicero several times, and although we have as little information on Cincius as we do on Paetus, if these are the same man then it is clear that he, Paetus, and Cicero all circulated in the same circles and all possibly engaged in composing military treatises in roughly the same period as Posidonius and Asclepiodotus.[30] Cicero himself may have written a *De Re Militari*, a supposition based on the problematic *Rhetorica ad Herennium* 2.3.[31] A surviving treatise was long attributed to him, but is now considered to be the work of Modestus.[32] On the other hand, if Cicero did write a *De Re Militari*, the most probable date for it would have been after his return from Cilicia. It is possible that he was inspired to write such a treatise by the literary environment of didactic military literature that produced works by Paetus, Asclepiodotus, Posidonius, and possibly Cincius.

Cicero also knew Posidonius. The two met on Rhodes when Cicero travelled there in the 70s, and we know they corresponded. Cicero asked Posidonius to write treatises for him (*To Atticus* 2.1.1). It is in this same letter to Atticus that Paetus is mentioned.[33] It is possible that Paetus, clearly a fellow Philhellene, sent his letter to Cicero knowing of his connection to the philosopher and possibly knowing Posidonius', or both men's, interest in military science. Perhaps Paetus wanted to show that such treatises were not the sole realm of Greek philosophers or the Stoics. He may even have attended Posidonius' lectures and written up what he had heard (in which case his material may have looked very similar to Asclepiodotus' (if his master's lectures were the origins of Asclepiodotus' treatise)).[34] Cicero and Paetus are silent on all these possibilities, and so the exact implications of all these connections are tantalising but their precise nature cannot be known.

Nonetheless, in Paetus' letter to Cicero we can identify with some confidence (at least) a lost *Tactica*, probably in the form with which we are familiar, written by a Roman author at a period when tactics against the Parthian threat were a hot topic and being written about by Greek contemporaries, and when it would seem other works of military science, both old and new, were in demand.

Notes

All dates are BCE unless otherwise noted. I am grateful to Peter Brennan for his advice and suggestions.

1 *Letters to Family 9.25/114: Summum me ducem litterae tuae reddiderunt. Plane nesciebam te tam peritum esse rei militaris; Pyrrhi te libros et Cineae video lectitasse. Itaque obtemperare cogito praeceptis tuis; hoc amplius, navicularum habere aliquid in ora maritima. Contra equitem Parthum negant ullam armaturam meliorem inveniri posse. Sed quid ludimus? Nescis quo cum imperatore tibi negotium sit. Παιδείαν Κύρου, quam contrieram legendo, totam in hoc imperio explicavi. Sed iocabimur alias coram, ut spero, brevi tempore.* The translation (adapted) is that of W. Glynn Williams.
2 Walter Miller's summary in the introduction to his Loeb translation ('we may best call it an historical romance', viii) Xenophon Volume V. It was termed an 'historical novel' by Derek Mosley in *OCD²*. See now, however, Tamiolaki (2016) and Rop's chapter in this volume.
3 Dahm (2002) 31–32 and 36–37.
4 Tacitus, *Annals* 12.40 on Venutius, and 12.44–51 and 13.7–9 on the situation in the East.

5 I disagree with Leach (1999) 169: 'At the beginning of this letter, Cicero, still serving out his proconsular governorship, teases Paetus about the knowledge of the literature of military science he had displayed in a letter congratulating Cicero upon his little victory.' She writes of the reciprocal jocularity as being the consistent mode of their correspondence. Elsewhere (141) she notes Cicero's letters to Paetus being 'jocular and familiar in their characterization of the recipient' and that they 'sparkle with wit.' She does note (142) that there are serious passages in these letters, and I would simply argue that 9.25/114 can be included in such a context rather than being all frivolous jocularity despite the small jokes about having boats on the coast by which to flee the Parthian threat and wearing out the *Education of Cyrus*.

6 Shackelton Bailey (1965) 342.

7 Leach (1999) 142.

8 See Devine (1989) and Matthew (2012).

9 See Whitehead (2001) 4–17.

10 The *Tactics* of Leo VI from the tenth century is, by contrast, a work which excerpted sections from Aelian, Onasander's *The General*, and Maurice's *Strategikon* written in the late sixth century. See Dennis (2010) and Chapter 14 in this volume (specifically 245–251).

11 See Devine (1995).

12 See later, 176.

13 See Wheeler (1979), who explores the continuity of phalanx tactics employed by Roman legions against Parthian cavalry, indeed against any mounted horde.

14 See Plutarch, *Crassus* 20–25; Cassius Dio 40.21–30.

15 See, for instance, 1.6.25. On discipline, see Chapter 8 (specifically 155, 159–163).

16 See Wheeler (1996).

17 See Lendon (2005) 277–279, 288, 313.

18 On readers of manuals, see Chapter 3.

19 See also the Introduction, 7.

20 See also the Introduction, 2, 4.

21 Polyaenus thanks the emperors for their attention immediately after writing his book on Macedonian stratagems.

22 See Dahm (2002) 34–37.

23 See Pauly-Wissowa *RE* ii Asklepiodotus (10), 1637–1641, especially 1638; Edelstein and Kidd (1972–88).

24 Other references can be found at 2.30.1, 5.15.1, and 6.22.2.

25 Such as the view of Everitt (2003) 195: 'he did not take himself seriously as general'; or Peterssen (1963) 470.

26 The Hague Koninklijke Huisarchief, MS. A22-IXE-79. See Parker (1996) 19–20 and plate 5; van Nimwegen (2010) 289 and (2014) 121–151.

27 See Pauly-Wissowa *RE* ii Asklepiodotus (10), pp. 1637–1641 and Edelstein and Kidd (1972–1988) 32–33.

28 See Vegetius 2.3; and Frontinus 4.1.16; 1.1.1; 1.2.5; 2.4.4; 2.7.14; 3.1.2; 3.10.1; 4.1.33; 4.3.1; 4.7.12; 4.7.31; 4.7.35. See Jordan (1967) 80–82 for the collected fragments.

29 Aulus Gellius, *Attic Nights* 16.4.1–6.

30 *To Atticus* 1.1, 1.7, 1.8, 1.16, 1.20 (Paetus is also mentioned in this letter at 1.20.7), 4.4, and 6.2.

31 Especially 'I shall discuss at a more appropriate time, if ever I attempt to write on the art of war or on state administration' ~ *idoneo tempore loquemur si quando de re militari aut de administratione rei publicae scibere velimus*. Since Cicero did indeed write on state administration, it might be assumed that he also wrote a *De Re Militari*.

32 See Reeve (2003) and Liong (2011).

33 See also *On Duties* 1.159, and 3.8, and 10.

34 It would be pushing speculation too far to contend that, perhaps, what we know as by Asclepiodotus is, in fact, by Paetus.

Bibliography

Dahm, M. K. 2002. 'Polyaenus of Macedon and Alexander the Great'. *Archaiognosia* 11. 29–46.

Dennis, G. T. 2010. *The Taktika of Leo VI.* Washington, DC: Dumbarton Oaks.

Devine, A. 1989. 'Aelian's Manual of Hellenistic Military Tactics. A New Translation from the Greek with an Introduction'. *The Ancient World* 19. 31–64.

——— 1995. 'Polybius' Lost *Tactica*: The Ultimate Source for the Tactical Manuals of Asclepiodotus, Aelian, and Arrian?'. *AHB* 9. 40–44.

Edelstein, L. and Kidd, I. 1972–1988. *Posidonius.* 2 vols. Cambridge: Cambridge University Press.

Everitt, A. 2003. *Cicero: The Life and Times of Rome's Greatest Politician.* London: Random House.

Jordan, H., ed. 1967. *M. Catonis Praeter Librum De Re Rustica Quae Exstant.* Stuttgart: Teubner.

Leach, E. 1999. '"Bi-Marcus": Correspondence with M. Terentius Varro and L. Papirius Paetus in 46 B.C.E.'. *TAPA* 129. 139–179.

Lendon, J. E. 2005. *Soldiers and Ghosts.* New Haven and London: Yale University Press.

Liong, K. A. 2011. *Cicero de re militari: A Civilian Perspective on Military Matters in the Late Republic.* PhD thesis. Edinburgh.

Matthew, C. 2012. *The Tactics of Aelian.* Barnsley: Pen and Sword.

Parker, G. 1996. *The Military Revolution.* Cambridge: Cambridge University Press.

Peterssen, T. 1963. *Cicero: A Biography.* New York: Biblo & Tannen Publishers.

Reeve, M. D. 2003. 'Modestus, *scriptor rei militaris*', in P. Lardet, ed., *La tradition vive: Mélanges d'histoire des textes en l'honneur de Louis Holtz.* Turnhout: Brepols. 417–432.

Shackelton Bailey, D. R. 1965. *Cicero's Letters to Atticus.* Vol. 1. Cambridge: Cambridge University Press.

Tamiolaki, M. 2016. 'Xenophon's Cyropaedia: Tentative Answers to an Enigma', in M. A. Flower, ed., *The Cambridge Companion to Xenophon.* Cambridge: Cambridge University Press. 174–194.

van Nimwegen, O. 2010. *The Dutch Army and the Military Revolutions 1588–1688.* Translated by A. May. Woodbridge, Suffolk: Boydell and Brewer.

——— 2014. 'The Tactical Military Revolution and Dutch Army Operations during the Era of the Twelve Years Truce (1592–1618)', in R. Lesaffer, ed., *The Twelve Years Truce (1609).* Leiden: Brill. 121–151.

Wheeler, E. 1979. 'The Legion as Phalanx'. *Chiron* 9. 303–318.

——— 1996. 'The Laxity of Syrian Legions', in D. L. Kennedy, ed., *The Roman Army in the East* [*Journal of Roman Archaeology* Supplement 9]. Ann Arbor: University of Michigan Press. 229–276.

Whitehead, D. 2001. *Aineias the Tactician: How to Survive under Siege.* Second edition. London: Bristol Classical.

10 Defeat as stratagem

Frontinus on Cannae

James T. Chlup

> If it is true, which no one doubts, that the Roman people exceed all races in virtue,
> it must not be denied that Hannibal surpassed other commanders in practical judg-
> ment to the same degree as the Roman people exceed all nations in sagacity. For
> whenever one engaged him in Italy, he always came away victorious.
>
> Nepos, *Hannibal* 1.1[1]

The Battle of Cannae, fought on 2 August 216 BCE, was, arguably, the greatest
defeat experienced by ancient Rome hitherto and remains one of the most famous
defeats in world history: 'the Roman army perhaps suffered higher casualties in a
single day's fighting than any other western army before or since'.[2] The Romans
indeed took pride in their perceived near-perfect record of winning in battle after
battle after battle and war after war after war, but this did not mean that they
avoided acknowledging their catastrophic defeats.[3] In his narrative of Cannae and
its immediate aftermath, for instance, Livy records the Romans' anxiety at the
catastrophe (*clades*), which expeditiously – almost instantly, in fact – yields to a
determination to continue the war with renewed vigour (22.38–60).[4]

Sextus Iulius Frontinus (c. 40–105 CE) provides no fewer than eight *exempla*
that feature Cannae in his *Stratagems* (*Strategemata*). Of the other *clades* that
Rome suffered up to Frontinus' time – Allia at hands of the Gauls; Caudium at the
hands of the Samnites; Carrhae at the hands of the Parthians; and the Teutoburgwald at the hands of the ancient Germans – Cannae receives the most discussion
to a significant degree.[5] Some of the *exempla* explore Hannibal's tactical decisions: they seem, at least initially, more stratagems of Hannibalic military acumen
rather than Roman failure. Other *exempla*, however, explore the event from the
Roman perspective, and do so in a way that reflects positively on the Romans.
Thus, while at the zero-sum level of the examples that feature Hannibal effecting
victory establish his *bona fides* as a general worthy of study, if not quite emulation, closer examination demonstrates that the author subtly negates Hannibal's
victory and at the same time mitigates the Romans' defeat, which indirectly conveys confidence in the Romans as a – if not *the* – superior military power. In so
doing, Frontinus couples the tactical with the ideological, providing his readers
with the option of understanding Hannibal as a skilled general on the one hand,
and (re)affirming the message of Roman exceptionalism on the other hand.[6]

How does Frontinus achieve this? While his treatise has a clearly articulated macrostructure, logically divided into books, each of which pertains to a specific phase of battle, and a clearly laid out microstructure, with each book subdivided into sections on particular kinds of stratagems, there is also a 'sub-microstructure' within each of these sections, as each stratagem exists in an implied relationship with others in its section: this is what Frontinus uses to angle the text to the Romans' favour – and in this case to Hannibal's detriment. That is, Hannibal's tactical advantage yields, almost instantaneously, to a textual disadvantage. The Cannae *exempla* turn out to be deftly positioned within their respective chapters, diluting stealthily Hannibal's victory and ameliorating Roman failure therefrom.

I. When a victory is (not quite) a victory

The *exempla* that feature Cannae appear in books two and four of the *Stratagems*. The first three examples appear in book two, the focus of which is tactics in (2.1–8), and immediately following (2.9–13), battle.[7] The first stratagem is particularly important in that it sets the tone for the author's approach vis-à-vis Cannae in the treatise henceforth. This raises what one ought to consider is a very important point about a military treatise in the form of a catalogue of stratagems: does the reader proceed from start to finish or does he seek out the section(s) that he believes are immediately pertinent to his real or imagined situation? If the former, then the first Roman defeat that the reader encounters is the only stratagem that features Caudium (1.5.16); if the latter, then the reader may encounter Cannae first, whether searching for stratagems on this particular battle or out of interest in Hannibal.

Frontinus' focus in the first instance is on Hannibal as the sagacious strategist: the general providently evaluates the terrain of Cannae, resolving to use the wind sent forth by the river to his advantage (2.2.7).[8] As Vegetius observes,

> the good general should know that a large part of victory depends on the actual place in which the battle is fought. Be at pains therefore when you are going to engage in combat, to get help first from the location.
>
> (3.13.1)

Hannibal is Vegetius' – and presumably Frontinus' – good leader (*bonus dux*), and he may have this specific tactic of Hannibal's in mind in this instance.[9]

The verb choice of *comperisset* ('he ascertained') to convey Hannibal's evaluation of the terrain suggests more than a precursory study: Hannibal appears as an erudite strategist, studying, then arranging (*direxit*) his battle-line to achieve *memorabilem . . . victoriam*.[10] Hannibal thinks, then acts, and achieves victory. As Ida Östenberg observes, narratives of Roman defeat normally include descriptions of difficulties posed by the terrain and weather.[11] This has two effects: it serves to explain in part, and therefore to absolve slightly, the Roman defeat on the one hand, and it serves to magnify the eventual Roman victory in the war on the other – Frontinus would appear to be in alignment with this approach. While the treatise's macrostructure necessitates the placement of this stratagem here, the subtext of the *exemplum* is that the terrain is the major, if not quite the only, factor in Hannibal's victory. Without further analysis of the battle as a whole, as a historian could provide, one cannot understand the battle's complex nature, where the reality of Hannibalic victory and Roman defeat appears unambiguous. In fact, Frontinus cautions the reader that his arrangement of the *exempla* are deliberate with regards to the division of material into books and topics (1 Preface 2): in this instance one may argue that there is a more intricate arrangement than the author admits: 'rather than far-flung anecdotes or an overwhelming mass of material (the imagery here makes quite an impression), he promises a collection of examples that has been organised, fittingly, with military precision'.[12]

While this statement refers to the overarching macro- and microstructure structure of the treatise, the sub-microstructure demonstrates that Frontinus reflected carefully about the specific placement of each stratagem in relation to those that precede and follow it. The impact of this stratagem changes – that is, Frontinus pre-negates it at least partially – when one considers the *exemplum* that antecedes it: Hannibal using the terrain at Numistro against Claudius Marcellus (2.2.6).[13] On the one hand, Cannae does not appear as an isolated incident, but rather as part of the general's normal strategy, one that consistently delivers a reliable (positive) outcome. However, Numistro occurs *after* Cannae (210 BCE), which suggests that Frontinus seeks to present Hannibal's victories in an ascending order of greatness: Cannae is the superior instance of this tactic.[14] He begins the Cannae section 'the same thing (occurred) at Cannae' (*idem apud Cannas*), which tethers the two examples.

Taken in their textual order, however, Cannae loses its uniqueness and therefore (some of) its impact. Had Frontinus placed the examples in the reverse order, it would convey that the Romans would not learn from Hannibal's strategy, and thereby presume that Hannibal successfully deployed this precise strategy again and again and again in the six-year interval between these events; this would not diminish, but in fact fortify, Hannibal's position within the treatise, inviting a counterfactual reflection that perhaps at this precise moment in the text, Hannibal is heading towards victory in the war. However, the identification of Claudius Marcellus, one of the architects of the successful Roman counteroffensive against the Carthaginians, implies Roman success in the war in spite of Hannibal's success at Cannae. Frontinus' description of Marcellus as 'a most famous general' (*clarissimum ducem*) is surely a deliberate qualifier: it eclipses, at least partially, Hannibal's *memorabilis victoria*, since the superlative used to describe Marcellus denies Hannibal the opportunity to rise higher in distinction. In fact, this salutation of Marcellus proves prophetic: in the next section, Marcellus defeats Hannibal *several* times (*frequentibus proeliis*; 2.3.9).[15] In this instance, the use of the passive voice (*superaretur*) underscores Hannibal's shift from the active to the reactive general, in this instance responding – unsuccessfully – to the action of a Roman general.

The second reference to Cannae appears in the next chapter on formations in battle (*de acie ordinanda*): Hannibal contrives to encircle the Roman army in a pincer movement (2.3.7) – *that* tactic for which he receives recognition as a great general; surely this stratagem is a *sine qua non* in a treatise that consists of stratagems.[16] The specific credit here goes to Hannibal's soldiers, whom Frontinus compliments: 'for hardly anything but a trained army, responsive to every instruction, can carry out this sort of tactic'. The verb 'carry out' (*exsequi*) surely conveys more than an ability to follow simple instruction only; it arguably conveys ability of thought on *how* to implement the tactic.

Hannibal's victory occurs against '[an] impetuously attacking enemy' (*avide insequentem hostem*), which, clearly not complimentary (to either side), nevertheless reduces slightly the impression of Hannibalic tactical acumen. Describing the Romans obliquely as *hostes* is surely not insignificant. Frontinus provides the Romans with the partial cover of anonymity: thus Hannibal's victory does not appear as over the Romans explicitly. Moreover, Hannibal cannot claim originality in deploying this manoeuvre: in the previous section, Artaxerxes successfully performs a variation of this stratagem at Cunaxa (2.3.6).

Frontinus accentuates Hannibal's textual disadvantage when he describes Livius Salinator and Claudius Nero employing the same tactic against Hasdrubal (2.3.8), which he immediately reinforces by recording Hannibal suffering defeat several times (*frequentibus proeliis*) against Claudius Marcellus (2.3.9): thus Frontinus shifts almost imperceptibly from Hannibal's victory to Hasdrubal's defeat. The lack of a reference to when and where Hannibal seeks refuge appears to resituate the Carthaginian in a time and place of perpetual defeat, undermining the safety he secures amongst mountains and marshes. In other words, Hannibal shifts from using terrain as a means to effect victory to using it to avoid further defeats. Moreover, the shift from Hannibal as the subject of an active to a passive verb (*protrusit . . . superaretur*) mimics the alteration of his situation, from a general executing a manoeuvre to achieve victory to having a stratagem deployed effectively against him.

Most significant is the fact that the Roman victory over Hasdrubal marked the beginning of the end of the war. From Hannibal's perspective, it confirmed that he would lose the war.[17] Frontinus constructs a highly telescoped 'narrative' in which Hannibalic victory yields almost instantaneously and seamlessly to Roman advantage. This balancing of stratagems – discussing Hannibal and the Carthaginians first, then the Romans, implying that the latter cancels out the former – recurs in the same book where Frontinus describes Hannibal's arrangement of his forces at Zama followed by that of Scipio (2.3.16). The final Roman victory over Hannibal situated so textually close to Hannibal's greatest victory surely indicates that Frontinus desires to afford Hannibal only an evanescent advantage in the text; or – to look at it another way – Hannibal's greatest victory finds itself eclipsed immediately by the final Roman victory. As Alice König observes, Frontinus sometimes describes the successful deployment of a stratagem by, and then against, a general, which may inadvertently undermine the authority of the text itself.[18] Frontinus may be using this seemingly counterintuitive arrangement of his examples to make his point: Hannibal was successful against the Romans in the first – and second, third, and fourth (!) – instances, but eventually the Romans learned from Hannibal and used his stratagems against him. The near side-by-side placement of these examples amplifies the message that the Romans learn *very* quickly from their defeats; the author simply (re)assures the reader from the first time that the Romans lose to Hannibal that they will win. That being said, it is important to reflect that Zama is the subject of only one direct stratagem in Frontinus' text – though it is a very long passage at two paragraphs; and he does not mention the battle by name, simply referring to 'Hannibal against Scipio in Africa', as if not to close off Hannibal as an example of a successful general by naming the specific place of the defeat that ended the war.

The next *exemplum* features Hannibal utilising his Numidian cavalry in a deception, pretending to surrender to the Romans, only to attack them with concealed weapons (2.5.27).[19] This stratagem appears in the chapter 'on ambushes' (*de insidiis*), a substantial collection of forty-seven examples.[20] On the one hand, it is not surprising that Hannibal appears frequently in this chapter (in fact, he appears more than any individual with seven *exempla*, five of which appear consecutively, §§21–25), since it aligns broadly with Roman thought on Carthaginians generally and Hannibal specifically: Punic treachery (*Punica fraus*), the national characteristic that was ubiquitous in representations of Carthaginians in Roman texts – though, it is important to note, the traditional words to indicate treachery or deception do not appear.[21] He probably intends for Hannibal to serve as an exemplar of cleverness (*sollertia*).[22]

While Frontinus appears to follow broadly the Roman historical tradition on the Carthaginians, he does not acquit – in fact, he would appear to endorse – the Romans' use of deception: in fact, the first example in the section features Romulus at Fidenae (2.5.1); the next three examples also feature Romans luring the enemy into an ambush instead of offering open battle.[23] The fourth example features Licinius Metellus doing so *against* the Carthaginians during the Sicilian (First Punic) War, though Frontinus perhaps thinks he justifies this by mentioning the Carthaginian general Hasdrubal's superior tactical position due to the very large number of elephants under this command (2.5.4). Chronologically and textually, the Romans through Romulus appear to invent such a tactic, and the Carthaginians learn this means of battle *from* the Romans.

Reflection upon the textual arrangement of the whole chapter is informative: a stratagem that features Epaminondas appears in the midst of the Hannibal examples (2.5.26).[24] This is illustrious company for Hannibal, to be sure. However, the arrangement is informative: Hannibal's greatest victory does not read as a natural continuation of a stratagem perfected elsewhere, including, ironically, Trebia (2.5.23) and Trasimene (2.5.24). The author tightly tethers these five stratagems with a recurring *idem* in the opening clause beginning with the second example: *Hannibal . . . idem Hannibal . . . idem Hannibal . . . idem ad Trasumennum . . . idem Hannibal*; beginning at 2.5.27, *Hannibal* perhaps prompts the reader to expect a new sequence, only to find that this stratagem appears in isolation; a textual *ex post facto* ambush perhaps. An *exemplum* that features Epaminondas disrupts abruptly, like a wedge, the flow of Hannibal's success.

The insertion of the *exemplum* of Epaminondas against the Spartans may have additional contextual significance, serving as a thematic tease: the reader is of course aware that Epaminondas' longer-term campaign is a successful one.[25] Hannibal's deployment of deception, however, is in each case successful, but only in the limited temporal context of the specific situation. He still loses the war: there is no cumulative benefit from this tactic, just as there is no cumulative narrative on Hannibal's exercising the stratagem.

Moreover, a temporally appropriate, and arguably deliberately positioned, Roman *exemplum* appears a few sections after: Scipio Africanus employs an ambush against Syphax and the Carthaginians (2.5.29).[26] In textual terms, Scipio's counteraction is almost immediate: it appears after an interval of only one stratagem, and one that Frontinus summarises *very* briefly, in a single sentence – that of the Iapydes against Romans under Publius Licinius (2.5.28). Moreover, the abrupt temporal shift after the stratagem of Scipio – to the first century BCE – suggests that Scipio, at least for a while, delays and negates deception as a valid tactic against Romans: the next *exemplum* features Mithridates failing to employ treachery against Lucullus (2.5.30).

The author's discussion of Cannae under three interrelated topics of battle indicates its perceived relevance to understanding strategy more broadly, and the nearly millennium span and geographic expanse of the stratagems results in Cannae being interwoven into a broader, transhistorical and transcultural overview of warfare. To Frontinus, then, Cannae is an important link in a longer chain that provides a thorough perspective on the ways and means of war. However, the *brevitas* of Frontinus' style denies the reader the chance to dwell in the *hic* and *nunc* of each *exemplum*: he must consider the broader perspective as he proceeds from stratagem to stratagem to stratagem. Therefore, structure of the text is itself a weapon by which the author dilutes the uniqueness of Hannibal's stratagems.[27]

II. How to lose – Roman style

Five *exempla* that relate to Cannae appear in book 4.[28] The transition from the practical tactics to the theoretical – as the author confesses in the preface to the book: 'those which are examples more of military science than stratagems' ~ *quae parum apte videbantur et errant exempla potius strategicon quam stratagemata* – demonstrates the utility of Cannae as an opportunity to understand the phenomenon of defeat in wider terms.

The first chapter focuses on the topic of discipline (*de disciplina*); more precisely, it concerns deserters. The vastness of this chapter – forty-six examples – points to its centrality in Frontinus' evaluation of war. Here Frontinus shifts how he writes about Cannae, recalibrating from the Carthaginian to the Roman – that is, from the winner's to the loser's – perspective. It surely raises the stakes in the author's approach, since he must presumably undertake a subtle yet significant shift from undermining Hannibal's victory to assuaging Rome's defeat. The first *exemplum* concerns the soldiers taking the *ius iurandum* – that is, it provides a glimpse of the Roman army *before* battle (4.1.4).[29] Specific reference to Cannae is absent, as if the name itself assigns defeat; instead, the names of the consuls confirm the battle's identity. In this instance, it confirms what the Romans will do in battle: not take flight. What Frontinus grants the Romans in one passage he appears to take away in another: he later discusses the Romans' refusal to allow the surviving soldiers from Cannae (*ex pugna Cannensi*) to return to Italy (4.1.44).[30] This creates – briefly – the impression of fairness in how Frontinus writes about Hannibal and the Romans, if one recalls the strategy the author deploys in book 2 as discussed earlier. However, in the third passage, the author eschews the pejorative nomenclature of *desertores*, using instead *milites*, as if to facilitate, if not perhaps to initiate, their restoration to the Roman army.

These *exempla* effectively interweave Cannae and the Second Punic War with the larger argument on Roman discipline. Romans dominate the discussion: they are the subject of forty-one of the forty-six examples. This is a much higher Roman content than the book as a whole, one that is more Roman-inclined than the previous three.[31] Therefore, Cannae participates in Frontinus' implied argument on the Romans as superior exemplars of discipline, which in turn compensates partially the Romans' defeat in the battle.

Three consecutive examples from Cannae appear in the chapter 'on determination' (*de constantia*), all of which concern the Romans in the denouement of the battle and its immediate aftermath.[32] The first features the consul L. Aemilius Paulus, who, wounded, receives an offer of a horse by which he can make his escape, which he declines, dying instead at the enemy's hands (4.5.5).[33] The second features Paulus' colleague, C. Terentius Varro, post-battle receiving the Senate's thanks for not giving up, and his self-deprecating retort (4.5.6).[34] Finally, a small group of soldiers resolve to fight on under their tribunes, making their way to safety (4.5.7). This third *exemplum* comes *before* the second in temporal terms: thus Frontinus features the dying and surviving consuls side-by-side, collapsing the temporal and textual space from the (near-)end of defeat to the (pre-) beginning of victory. In textual terms, Varro's *constantina* appears to have an immediate positive – and retroactive – effect: the author notes that the general's behaviour affects positively the sub-commanders and soldiers under his leadership. Frontinus also pushes the narrative forward – way forward – noting Varro's service to Rome post-Cannae, which he extends beyond Rome's final victory in the war: 'but he proved for the rest of his lifetime that he had remained alive not from desire for life, but because of his love for his country'.[35] Interestingly, Frontinus provides a rare narrator's injection here, stating that Varro demonstrates 'even greater resolution' (*vel maiore constantia*).

Here, then, in the antepenultimate section of the treatise, Frontinus features prominently the Roman character persevering through the immediate crisis that Cannae represents, and planting the seeds of recovery. A survey of the whole chapter validates this: many of the *exempla* pertain to defeat in battle that was only a temporary reverse in the war. The reference to Leonidas at Thermopylae is surely an example *par excellence* of this (4.5.13). Moreover, the absence of direct reference to Hannibal in these examples allows for an unobstructed focus on the Romans; Carthaginian victory does not contaminate Roman recovery. Understanding defeat in battle on its own terms, without direct reference to the agent of that defeat, creates an environment to contemplate the eventual victory in the war that will result.

The final *exemplum* appears in the final section of the *Stratagems*, which features examples that the author feels do not fall in any of the earlier covered topics (4.7, *de variis consiliis*). It is here as the text closes that the author aligns with Livy's language for describing the battle: *post Cannensem cladem* (4.7.39) – this is the only time Frontinus uses *clades* with reference to Cannae.[36] In this instance, Frontinus does so only because it suits the context of the example: he describes the despair of the defeated Romans as particularly acute – they even consider the abandonment of Italy. However, Frontinus inserts Publius Scipio, whose efforts, almost Camillus-like, lead to those present into swearing an oath to carry on; Zama, Scipio's victory and Hannibal's eventual defeat, therefore, lurks in the subtext of this example.[37]

One reference to Cannae is incidental: it concerns Lucius Bantius, who fought at Cannae. Hannibal shows him kindness by freeing him (3.16.1).[38] That the Roman general, Claudius Marcellus, shows a similar kindness to Bantius represents a possible point of inversion, coming as it does at the middle-point in Frontinus' Cannae sub-narrative whereby the Romans begin to assume formally the advantage in the conflict.[39] That is, one can perhaps imagine a dividing line in book 3 with respect to the Cannae *exempla*: the stratagems that appear in book 2 notionally present Hannibal as the ascendant general, potentially heading towards victory; in book 4, however, there are (very) early indicators of Roman recovery from Cannae.

It is not, *stricto sensu*, an *exemplum* of desertion or treachery (*qua ratione proditoribus et transfugis occurratur*), the topic of the chapter; Bantius only plans – a thought-crime, if you will – to betray his town, Nola, to Hannibal. Frontinus follows this *exemplum* with three that pertain to Carthaginians dealing with deserters, one each from Hamilcar, Hanno, and Hannibal (3.16.2–4); Roman kindness (*benignitate*) brings into clearer focus Carthaginian cruelty in these subsequent examples (implied by the verb used to describe Hannibal's action at §4: *ultus est*); and this in turn reveals Roman treatment of their 'deserters' in book 4 to be rather placid. The first two references to Carthaginians and deserters, interestingly, refer to the First Punic War, and therefore are not directly relevant to the narrow(er) context of this specific war; Hannibal's cruel trick, therefore, stands in sole balance against Marcellus' counter-kindness. In fact, Hannibal is not harsh, since all he does is secretly expose traitors; it is the Romans who cut off the hands of the deserters.

III. Hannibal and the military manual

The analysis heretofore argues that the references to Cannae do not exist in a vacuum, devoid of connection to Frontinus' treatise as an organic whole. In fact, Hannibal claims the laurel as the most prominent character, Roman and non-Roman, in the treatise: he appears no fewer than forty-five times, scoring other awe-inspiring tactical victories over the Romans, including Trebia and Trasimene.[40] In a text that features nearly six hundred stratagems, Hannibal features in approximately eight percent or one in thirteen (as a point of comparison, Valerius Maximus provides thirty-five *exempla* that feature Hannibal across a work that is at least twice as long). Hannibal dwarfs the next two most prominent non-Romans, Alexander and Epaminondas, who feature thirteen and twelve times, respectively.[41] Unlike Hannibal, however, neither of these men were a threat to Rome; therefore there is less at stake historically in discussing, and by so doing frequently accentuating, their reputations as brilliant tacticians.[42] Frontinus' lack of authorial comment, and lack of reference, direct or indirect, to his own military experiences, would appear to assign the greatest authority as a practitioner of stratagems to he who appears in the text in the most: Hannibal.[43] The approach to Cannae, therefore, indicates that Frontinus intends for the reader to learn from Hannibal from both the perspective of how he produces victory and how to compensate for the same. One might argue that Frontinus misses an opportunity in not also presenting Hannibal as a *learner* in the field of warfare: Vegetius correctly notes that Hannibal on his invasion availed himself of a tactician, Sosylus, who probably provided with military treatises and perhaps composed one himself while observing Hannibal in action.[44] Frontinus does not mention his presence, which acquires significance when one notes that in another case he mentions a tactician: Xanthippus the Spartan, advisor to the Carthaginians in the First Punic War, which reveals Carthaginian military culture as one in which learning is a continuous process, even ocurring in the midst of a war.[45]

Hannibal also features in the treatise of Polyaenus, Frontinus' near-immediate successor in the genre of the military treatise, though not nearly as prominently.[46] Polyaenus agrees with his predecessor on the basic facts of Hannibal's tactical brilliance: his deception with the retreating of the centre of the battle-line (6.38.3), and his use of the wind (6.38.4). In Polyaenus, Hannibal appears alongside Hamilcar, Hasdrubal, and the Carthaginians generally, but their collective position is very much one of lesser prominence: Polyaenus provides only ten stratagems for Hannibal in a treatise that is more than twice the length as Frontinus'. This may reflect a *communis opinio* that Hannibal and his compatriots not pertinent to Lucius Verus' Parthian campaign: Polyaenus includes only those *exempla* that may prove *directly* useful. The stratagem on using the wind to one's advantage, for instance, would seem singularly relevant to a campaign in Parthia: one might imagine that Polyaenus wishes to ensure that Verus does not forget that wind and a sandstorm could help – or hinder – the Romans in battle. If true, Polyaenus' text may expose what initially appears to be a flaw in Frontinus' treatise of which Hannibal is but the most obvious symptom: that by allocating such a prominent position to Hannibal, whose invasion represents a kind of military situation which Rome conceivably would never face again, Frontinus inadvertently provides a text that is of antiquarian interest only, not directly useful to his contemporary reader – Domitian or Trajan, for instance.[47]

On the one hand, Polyaenus' structure, clustering *exempla* into national groups in the first instance, and then (where applicable) an individual's examples together, allows Hannibal's success to exist without dilution from Roman stratagems – or those of other famous foreign generals: Epaminondas or Agesilaus, for instance – before or after each.[48] On the other hand, Polyaenus also arranges his material to lessen Hannibal's tactical advantage: within the sixth book, which contains Carthaginians and 'other' Greeks, Polyaenus isolates Hannibal from his fellow Carthaginians: descriptions of tactics employed by Thessalians (6.39) and

Masinissa (6.40) appear between those on Hannibal (6.38) and Hamilcar (6.41) and Hasdrubal (6.42); and all of these appear separate from stratagems concerning the Carthaginians prior to conflict with the Romans (6.16).[49]

Moreover, one might point out that Polyaenus' text proves restrictive in that it does not easily allow for the more organic exploration of warfare that Frontinus provides. While Polyaenus allows the antiquarian to acquire an overview of a particular general or community, Frontinus insists upon evaluation of each general in the wider, supranational context. Through near-seamless shifting from examples that feature Romans, Greeks, Carthaginians, and others, Frontinus' treatise allows the reader or reader-general to acquire the broadest, that is, transcultural and transhistorical, and therefore surely more nuanced and ultimately utilitarian, perspective of war: the reader compares side-by-side-by-side similar tactics employed by different generals across the near-totality of time.[50] To be blunt, to study *exempla* clustered around situations rather than individuals and nations surely affords a more efficient means to visualise the ways and means to victory, assuming the reader-general prefers to direct his attention to a part of a text that contains information directly relevant to the situation he faces at that particular moment.[51]

While some might read Frontinus' structure as 'destabilising in a general way', that could precisely be the point.[52] War *is* destabilising: thus there is a latent irony in a text that seeks to make a general-reader better equipped to wage a successful war and at the same time acutely aware that war is an inherently chaotic and resistant to the imposition of order. The shift in time and location as the author proceeds stratagem by stratagem by stratagem is a consciously discombobulating act. Simply put, Frontinus' medium is a – if not *the* – message about war. The sub-microstructure does not detract from the broader thematic message; in fact, it helps the reader appreciate it: Frontinus' creation of order from chaos draws attention to the fact that chaos exists in the first place.

IV. Defeat as text

> [I]n the Second Punic War, [the Romans] were not able to counter Hannibal. So it was that after so many generals, so many armies lost, they only finally achieved victory when they had been able to learn military science and training.[53]

From Vegetius' perspective, Hannibal was the reason for the Romans to begin the study of military science, which, one may speculate, includes the writing and reading of military manuals.[54] After all, Hannibal himself studied tactics, as Vegetius notes (3 Preface 7): surely the Romans cannot fail to benefit from understanding Hannibal and the situations that enabled their victory over him. That the Romans admired Hannibal as a strategist is clear from the fact that Pliny the Elder records that in his – and therefore probably also Frontinus' – time there were no fewer than three statues in Rome of her greatest rival.[55]

In her recent study of Roman military defeat in the Middle Republic, Jessica Clarke observes that the Romans lost battles about as often as they won them. How the Romans contextualised an individual defeat could mitigate its impact and possibly frame it as a victory.[56] This chapter suggests that perhaps the treatise reflects, and at the same time validates, this cultural phenomenon. But whereas the Romans of the Middle Republic probably evaluated their defeats in a narrow(er) temporal and geographical (that is, exclusively Roman) frame, Frontinus coerces his reader to reflect upon the near-totality of history, Roman and non-Roman. Writing about Cannae and the other Roman defeats inserts the reality and potentiality of defeat into the didactic frame of the military treatise. In blunt terms, Frontinus does this so as to emphasise that the reader-general needs to study the widest possible range of examples of warfare – victory and defeat, ancient and modern, in order to calculate his route to victory. Cannae, one of the great set-piece battles of military history (still studied step-by-step-by step in military colleges, not to mention by students of Roman history), remained firmly in the Roman historical consciousness when Frontinus composed his text. Frontinus participates in the elevation of Cannae to the *Weltgeist* of war.

Frontinus' willingness to discuss Roman defeat has implications for the consideration of the generic frame of the military manual. As a text composed by a Roman author for Roman readers (presumably both experts and laypersons), which features prominently, thought admittedly not exclusively, Romans, what would be defined by any metric as a Roman defeat might appear counterintuitive.[57] To be sure, one might suspect that the topic of defeat falls outside the scope of the military treatise if one defines the genre narrowly: to provide guidance on how to *win*. To include defeat might seem counterintuitive, some might argue, potentially normalising it as a facet of the Roman war experience. One might instead recognise that the study of war necessitates coming to terms with the reality that in any battle someone must lose: one side benefits from the deployment of a stratagem, and the other side loses as a result. In so doing, Frontinus expands the generic parameters of the military treatise. Moreover, the intratextual relationships between the Hannibal and Cannae *exempla* and their respective chapters surely demonstrate the sophistication of the treatise and the sophisticated ways in which one could contemplate the universality of the phenomenon of war.

Notes

1 *si verum est, quod nemo dubitat, ut populus Romanus omnes gentes virtute super-arit, non est infitiandum Hannibalem tanto praestitisse ceteros imperatores prudentia, quanto populus Romanus antecedat fortitudine cunctas nationes. nam quotienscumque cum eo congressus est in Italia, semper discessit superior.* Unless otherwise stated, all dates are BCE. Translations (modified) come from the Loeb Classical Library editions.

2 John F. Lazenby, 'Cannae', *OCD* 4e. Clark (2014) 63 is more cautious: 'utterly unprecedented figures for a single day of fighting before the advent of mechanized warfare, and still almost unparalleled'; see also her comment 63 n.34. Owens (2017) 710: 'the defeat at Cannae was an unprecedented disaster for the Romans and the slaughter was

unmatched in Roman history'. On the battle, see Cornelius (1932); Lazenby (1978) 77–86; Walbank (1957) 435–441; Goldsworthy (2001); Hanson (2002) 99–132; Daly (2007); Owens (2017) 704–711; and Sears (2019) 203–205.

3 Rosenstein (1990); Sordi (1990); Rich (2012); Östenberg (2013); Clark (2014); and Turner and Clark (2018) are important on the subject of defeat, but it is, arguably, a woefully understudied topic.

4 On Cannae in Livy, see Jaeger (1997) 94–107 and Chaplin (2000) 62–70. On defeat in Livy, see Levene (2010) 261–316.

5 Allia: 2.6.1, 3.13.1, 3.15.1. Caudium: 1.5.16. Teutoburgwald: 2.9.4, 4.7.8. Frontinus does not mention Carrhae; there are, however, references to the Parthians: 1.1.6, 2.2.5, 2.3.15, 2.5.35, 2.5.36, 2.5.37, 2.13.7, 4.2.3. That the Romans thought about a defeat in comparative terms with other defeats, note Livy's narrative aside at 22.50.1–3, where he compares Cannae to the Romans' defeat at the hands of the Gauls.

6 Cf. König (2018) 155.

7 These first three examples in Frontinus that feature Cannae appear in a single passage of Valerius Maximus: 7.4 ext.2, which arguably restricts Hannibal's victory into a single, isolated textual space (see also later, 191–192, on Hannibal in Polyaenus). Appian, *Hannibalic War* 26 notes that at Cannae, Hannibal deploys four stratagems in 'his rare and splendid victory' (νίκην ἀρίστην τε καὶ σπάνιον): the first two appear as the first and second discussed by Frontinus.

8 Cf. Livy 22.43.10–11; Florus 1.22.16; Plutarch, *Fabius* 16.1; Polyaenus 6.38.4; Seneca, *Quest. Nat.* 5.16.4; Machiavelli, *Art of War* 4.23. Frontinus incorrectly identifies the river here as the Volturnus, when he means the Aufidus.

9 *bonum ducem convenit nosse magnam partem victoriae ipsum locum, in quo dimicandum est, possidere. elabora ergo, ut conserturus manum primum auxilium captes ex loco.* Laederich (1999) 110 n.27 notes the possible connection between these passages. Cf. Vegetius 3.20.24–26; 3.26.11, 'terrain is often more value than bravery' (*amplius prodest locus saepe quam virtus*); and Onasander 21.3–4, on which see Petrocelli (2008) 230–232. Vegetius mentions Hannibal thrice: 1.28.8–9; 3 Preface 7; 3.24.6.

10 König (2017) 162–163 notes Frontinus' habit of presenting generals as the subject and using a strong verb of 'action' as they deploy the stratagem.

11 Östenberg (2018), who discusses Cannae in Livy (250–251). She notes that Polybius does not mention the dust, but Hannibal notes the advantage afforded him by the terrain (3.111.2–3).

12 König (2017) 159.

13 Livy 27.2.4–12; Plutarch, *Marcellus* 24.4–5. See Levene (2010) 203 n.90. Frontinus possibly plays a counterfactualist turn in writing about Numistro, since, at least as far as Livy and Plutarch represent the battle, it was indecisive. Livy 27.2.6: 'for a long time the battle favoured neither side' (*diu pugna neutro inclinata stetit*).

14 In Livy, Marcellus refers to his victory against Hannibal after Cannae (27.2.2, *qui post Cannensum pugnam victoria Hannibalem contuderit*); this connects formally Cannae to Numistro, which may have influenced Frontinus' placement of the *exempla*.

15 On Marcellus, see also later, 190.

16 Livy 22.47; Polybius 3.115; Plutarch, *Fabius* 16.2–3. See König (2018) 167–172 on the interplay between examples of Romans fighting Carthage and the Roman civil war.

17 This is how Livy presents it through the structure of the third decade: Cannae is the final narrative episode of book 22, and Metaurus of book 27. In Hannibal's reaction to learning about Hasdrubal's defeat at Metaurus, which comes upon delivery of the latter's head, Livy records Hannibal's acknowledgment that he would lose the war (27.51.11–12): see Jaeger (1997) 94–99, who argues specifically how Metaurus is a response to Cannae.

18 König (2017) 168–169. The analysis of this chapter supports her point, but whereas she reads Frontinus' arrangement of *exempla* as 'disorientating' by the constant temporal and geographical shifts, this chapter argues that Frontinus insists that the reader to

acquire a transcultural and transhistorical understanding of war, and that a stratagem from one general in one war frames the interpretation of a similar stratagem from a general of a different culture and time.

19 Livy 22.48; Valerius Maximus 7.4. ext.2; Appian, *Hannibalic War* 20; Polyaenus 6.38.3.

20 On *insidiae*, see Wheeler (1988b) *passim*.

21 According to Livy, Hannibal (infamously) possesses 'treachery greater than Punic' (*perfidia plus quam Punica*, 21.4.9). Frontinus does not use *perfidia* to apply to Hannibal. Cf. Valerius Maximus 7.4. ext 2: 'such was Punic bravery, equipped with tricks and treacheries and deceit. That is now the surest excuse for our inveigled valour, since we were deceived rather than vanquished' ~ *haec fuit Punica fortitudo, dolis et insidiis et fallacia instructa. quae nunc certissima circumveniae virtutis nostrae excusatio est, quoniam decepti magis quam victi sumus.* On trickery more broadly, including in Frontinus, see Wheeler (1988a) 8–9, 14 and Wheeler (1988b) 50–92.

22 Wheeler (1988b) 56–58, who notes that Frontinus intends for *ars* to indicate innate cleverness. See also König (2017).

23 Polyaenus also mentions this stratagem (8.3.2).

24 Frontinus discusses Hannibal earlier in this section at 2.5.13.

25 Cf. Polyaenus 2.3.9.

26 Frontinus mentions Syphax at 1.1.3, 1.2.1, and 2.7.4. The first two suggest Syphax and his people as reasonably easily deceived.

27 One might compare the approach of Polyaenus: see later, 191–192.

28 This chapter does not intend to address the question of the authenticity of this book, since arguments against which pertain primarily to style. Scholars now seem willing to accept it as Frontinus' work, if not a direct continuation of books 1–3. See König (2017) 174 n.54.

29 Cf. Livy 22.38.5.

30 Livy 25.5–7; Valerius Maximus 2.7.15; Plutarch, *Marcellus* 13.2–5.

31 As Campbell (1987) 15 n.11 notes, in the first three books Roman *exempla* only account for 49 percent; Hannibal is therefore the *de facto* leader of the non-Roman 51 percent. There is a significant reversal in book 4: 72 percent of *exempla* concern Romans. This provides a subtle subtext for undermining Hannibal as a successful general; he finds himself demoted in this book as Frontinus reimposes Roman control in the text.

32 In his Loeb translation, Charles E. Bennett inserts the qualifying phrase 'the will to victory'. Indeed some of the stratagems that appear before the three related to Cannae demonstrate Romans (all are Roman-centred) grabbing eventual victory from the jaws of defeat. Especially noteworthy is the *exemplum* immediately before that which features Paulus, where Claudius Marcellus not only avoids defeat at the hands of the Gauls, but also secures the *spolia opima* (4.5.4): thus Frontinus pre-negates Paulus' defeat.

33 Livy 22.49.6–12; Plutarch, *Fabius* 16.5–8.

34 Livy 22.51.14–15; cf. Valerius Maximus 3.4.4 and 4.5.2.

35 *non autem vitae cupiditate, sed rei publicae amore se superfuisse reliquo aetatis suae tempore approbavit.*

36 Livy consistently labels Cannae as *clades* up to the point where Roman victory in the war is assured: 22.50.1, 23.4.6, 23.30.11, 23.30.19, 23.35.1, 24.18.3, 24.45.2, 25.22.3, 26.41.13, 27.1.4, 27.2.2. Frontinus uses the word once to refer to Trasimene: 4.7.25; he uses it with reference to Caudium (1.5.16) and Teutobergwald (4.7.8).

37 Cf. Valerius Maximus 5.6.7. It is surprising that Frontinus does not mention Zama as a Roman victory. He twice discusses Scipio moving the Roman forces to Africa (1.3.8, 1.12.1), and Hannibal's and then his arrangement of his troops at the battle (2.3.16–17). Polyaenus only discusses Zama in the context of Scipio's treatment of Hannibal's spies before the battle (8.16.8).

38 On the treatment of prisoners in the text, see Wheeler (1988a) 21–22 and 28–29 nn.66–71.

39 Machiavelli refers to this incident in his *Art of War*, where he notes Marcellus 'used so much humanity and liberality toward him' (7.116).

40 Trebia: 2.5.23; Trasimene: 2.5.24, 2.6.4, 4.7.25. Frontinus does not mention Ticinus. Clark (2014) 56–57 notes that Ticinius and possibly Trebia do not feature in the Roman collective historical consciousness as defeats: this could explain why Frontinus does not afford them more prominent places in the text. Had Frontinus done so, Hannibal's prominence would be even more conspicuous.

41 On Epaminondas, see also earlier, 187–188. Frontinus mentions Alexander seventeen times in total in the text, but in some instances he uses Alexander's name only as a point of reference (e.g., 2.4.11).

42 It is important to note a complicating aspect of the text: Frontinus features stratagems employed by Romans against other Romans in civil war. On one level, this expands the scope of the text by its offering a truly totalising perspective on war.

43 As König (2017) 161 nicely puts it, 'having set himself up as an authority in the preface, Frontinus departs the arena and leaves it to the generals who populate each section to provide the instruction'.

44 3 Preface 4. Vegetius does not name Sosylus specifically, instead noting a *doctor* present with Hannibal. The phrase *tot consules tantasque legiones* surely implies Cannae.

45 2.2.11; 2.3.10.

46 Whitehead (2008) 143 connects the two authors as working within a newly emerged genre (or subgenre?) of stratagem-collections. This might suggest that Frontinus would be in a position of influence over Polyaenus in terms of content, if not in structure. See also Wheeler (2010) 19–38.

47 One could accuse Polyaenus of proffering an out-of-date (or stale) text: see Chapter 6; cf. also Chapter 11. While Frontinus is an author with not insignificant military experience writing about war, which may presume that he writes for a professional audience, the prevalence of Hannibal in the text may betray Frontinus' zeal to appeal to the non-specialist audience. As Chapter 3 suggests, some manuals appear specifically geared towards the casual reader.

    However, as Turner (2007) 430–431 observes, Frontinus' wording for describing Domitian in one section (2.11.7) appears to echo an example that describes Quintus Fabius Maximus (1.3.3). The reader surely could reasonably infer, though possibly at a bit of a stretch, that Domitian could campaign not wholly unsuccessfully against a Hannibal-like figure: avoiding a scenario like Cannae would seem the advice here. On Polyaenus' and Frontinus' broad differences of approach, see Turner (2007) 435–437 and Pretzler (2010) 89. To be sure, Polyaenus too can face the charge of antiquarianism, given that he does not cover in a substantial way the enemies against whom they Romans have more recently fought, or are probably about to fight; and the most recent (from Polyaenus' perspective) Roman exemplar is Augustus.

48 See Wheeler (2010) 48–54 on Polyaenus' 'military ethnography': Carthage exists as a 'civilised' nation. The prominence of Hannibal in Frontinus' text, and the broadly positive representation of him, mean that Frontinus also holds Carthage in the same regard.

49 Cf. Pretzler (2010) 98–104. The chapter on Hasdrubal is missing from the text.

50 Here one should observe that Polyaenus can lay claim to the greater temporal scope, including as he does mythological figures – Dionysus, Heracles, and Theseus – who predate the earliest figures whom Frontinus features (for example, Romulus; see earlier, 187).

51 Cf. the discussion of Chapter 11.

52 Cf. König (2017) 169.

53 Vegetius, 1.28.8–9: *ut secundo Punico bello Hannibali pares esse non possent. tot itaque consulibus, tot ducibus, tot exercitibus amissis, tunc demum ad victoriam pervenerunt, cum usum exercitiumque militare condiscere potuerunt.*

54 Note the observation of Chapter 9 that military defeat may have, at least in the short term, increased interest in the study of military treatise (in this instance, Crassus' defeat at Carrhae generated consumer demand for military texts in the late 50s (172)).

55 *Natural History* 34.32: 'in three places in this city there were indeed statues of Hannibal, he alone of our enemies threw a spear in over the walls of Rome' ~ *Hannibalis etiam statuae tribus locis visantur in ea urbe, cuius intra muros solus hostium emisit hastam.*
56 Clark (2014) 16–49. See also Kerreman (2018) and Beek (2020), especially 105–109.
57 The political context in which Frontinus wrote, the Flavian regime, would seem not wholly conducive to exploring defeat. Rome was not facing any serious external threat, though the pain of the last iteration of civil war left fresh wounds. On Frontinus and the Flavians, see Turner (2007); Malloch (2015); and König (2018) 173–175.

Bibliography

Beek, A. L. 2020. 'Campaigning against Pirate Mercenaries: A Very Roman Strategy?', in R. Evans and M. De Marre, eds., *Piracy, Pillage and Plunder in Antiquity*. London: Routledge. 97–114.

Bennett, C. E. 1925. *Frontinus: Stratagems and Aqueducts of Rome*. Cambridge, MA: Harvard University Press.

Campbell, B. 1987. 'Teach Yourself How to Be a General'. *JRS* 77. 13–29.

Chaplin, J. D. 2000. *Livy's Exemplary History*. Oxford: Oxford University Press.

Chlup, J. T. 2014. 'Just War in Onasander's *Strategikos*'. *JAH* 2. 37–63.

Clark, J. H. 2014. *Triumph in Defeat: Military Loss and the Roman Republic*. Oxford: Oxford University Press.

Cornelius, F. 1932. *Cannae. Das militärische und das literarische Problem*. Leipzig: Dieterich.

Cusmano, G. 2005. *Sesto Guilio Frontino: Il IV libro degli "Strategemata"*. Rome: Aracne.

Daly, G. 2007. *Cannae*. London: Routledge.

Gallia, A. B. 2012. *Remembering the Roman Republic: Culture, Politics, and History under the Principate*. Cambridge: Cambridge University Press.

Gilliver, C. M. 2007. 'Battle', in P. Sabin, H. van Wees, and M. Whitby, eds., *The Cambridge History of Greek and Roman Warfare Volume II: Rome from the Late Republic to the Late Empire*. Cambridge: Cambridge University Press. 122–157.

Goldsworthy, A. 2001. *Cannae: Hannibal's Greatest Victory*. London: Cassell & Co.

Hanson, V. D. 2002. *Carnage and Culture: Landmark Battles in the Rise to Western Power*. New York: Doubleday.

Jaeger, M. 1997. *Livy's Written Rome*. Ann Arbor: University of Michigan Press.

Kerremans, B. 2018. 'A Real Roman Defeat: Memory, Collective Trauma and the *clades Lolliana*'. *AC* 61. 69–98.

König, A. 2017. 'Conflicting Models of Authority and Expertise in Frontinus' *Strategemata*', in J. König and G. Woolf, eds., *Authority and Expertise in Ancient Scientific Culture*. Cambridge: Cambridge University Press. 153–181.

——— 2018. 'Reading Civil War in Frontinus' *Strategemata*: A Case-Study for Flavian Literary Studies', in L. D. Ginsburg and D. A. Krasne, eds., *After 96 CE: Writing Civil War in Flavian Rome*. Berlin: De Gruyter. 145–177.

Kuchma, V. V. 1984. 'Some Disputed Questions Concerning the *Stratagemata* of Frontinus'. *VDI* 4. 45–55.

Laederich, P. 1999. *Frontin: Les Stratagèmes*. Paris: Economica.

Levene, D. S. 2010. *Livy on the Hannibalic War*. Oxford: Oxford University Press.

Lazenby, J. F. 1978. *Hannibal's War: A Military History of the Second Punic War*. Warminster: Aris and Phillips.

Malloch, S. J. V. 2015. 'Frontinus and Domitian: The Politics of the "Strategemata"'. *Chiron* 45. 77–100.

Milner, N. P. 1993. *Vegetius: Epitome of Military Science*. Liverpool: Liverpool University Press.

Östenberg, I. 2013. 'War and Remembrance. Memories of Defeat in Ancient Rome', in B. Alroth and C. Scheffer, eds., *Attitudes towards the Past in Antiquity: Creating Identities*. Stockholm: Acta Universitatis Stockholmiensis. 255–265.

——— 2018. 'Defeat by the Forest, the Pass, the Wind: Nature as an Enemy of Rome', in J. H. Clark and B. Turner, eds., *Brill's Companion to Military Defeat in Ancient Mediterranean Society*. Leiden: Brill. 240–261.

Owens, E. 2017. 'The Second Punic War', in M. Whitby and H. Sidebottom, eds., *The Encyclopedia of Ancient Battles*. Malden, MA: Wiley Blackwell. 668–796.

Petrocelli, C. 2008. *Onasandro: Il generale*. Bari: Edizioni Dedalo.

Pretzler, M. 2010. 'Polyainos the Historian? Stratagems and the Use of the Past in the Second Sophistic', in K. Brodersen, ed., *Polyainos: neue Studien-Polyaenus: New Studies*. Berlin: Verlag Antike. 85–107.

Rich, J. 2012. 'Roman Attitudes to Defeat in Battle under the Republic', in F. Marco Simón, F. Pina Polo, and José Remesal Rodríguez, eds., *Vae Victis! Perdedores en el Mundo Antiguo*. Barcelona: Col·lecció Instrumenta. 83–111.

Rodgers, R. H. 2004. *Fontinus: De aquaeductu urbis Romae*. Cambridge: Cambridge University Press.

Rosenstein, N. 1990. *Imperatores Victi: Military Defeat and Aristocratic Competition in the Middle and Late Roman Republic*. Berkeley: University of California Press.

Sears, M. A. 2019. *Understanding Roman Warfare*. London: Routledge.

Schettino, M. T. 1998. *Introduzione a Polieno*. Pisa: Edizioni ETS.

Sordi, M., ed. 1990. *Dulce et decorum est pro patria mori: la morte in combattimento nell'antichità*. Milan: Vita e Pensiero.

Turner, A. 2007. 'Frontinus and Domitian: Laus principis in the *Strategemata*'. *HSPh* 103. 423–449.

Turner, B. and Clark, J. H. 2018. 'Thinking about Military Defeat in Ancient Mediterranean Society', in *idem*, eds., *Brill's Companion to Military Defeat in Ancient Mediterranean Society*. Leiden: Brill. 3–22.

Walbank, F. W. 1957. *A Historical Commentary on Polybius*. Vol. I. Oxford: Oxford University Press.

Wheeler, E. L. 1988a. 'The Modern Legality of Frontinus' Stratagems'. *MGM* 43. 7–29.

——— 1988b. *Stratagem and the Vocabulary of Military Trickery*. Leiden: Brill.

——— 2010. 'Polyaenus: *Scriptor Militaris*', in K. Brodersen, ed., *Polyainos: neue Studien ~ Polyaenus: New Studies*. Berlin: Verlag Antike. 7–54.

Whitehead, D. 2008. 'Fact and Fantasy in Greek Military Writers'. *AAHung* 48. 139–155.

11 Vegetius' *regulae bellorum generales*

Jonathan Warner

The third book of Vegetius' *Epitome of Military Science* (*Epitoma rei militaris*) concludes with a list of thirty-three pithy and memorable precepts entitled 'general rules of war' (*regulae bellorum generales*, 3.26.1–33).[1] These precepts (*regulae*) restate and summarise Vegetius' most important arguments in the first three books, emphasising the prime importance of training and discipline, the virtues of careful planning, and the benefits of avoiding direct confrontation. Despite their immense popularity in the medieval period, the authenticity of these *regulae* has been questioned by some modern scholars. The main objections are the similarity of Vegetius' rules to later collections of precepts, inconsistencies between the *regulae* and the *Epitome* as a whole, and a paucity of internal references to the *regulae*.[2]

This chapter examines and rebuts the main arguments that the *regulae* were interpolated. The manuscript evidence makes clear that the rules were penned at an early date, and nothing opposes Vegetius himself having composed them. By noting literary parallels in other military handbooks, we can see that Vegetius' summarising list is consistent with the conventions of other Greek and Latin military handbooks. Whether or not such texts constitute a formally distinct genre, many exhibit similar strategies of list-making that distill military knowledge down to select *exempla* and general principles, a literary approach that maintains a conceit of practical value entirely separate from actual military usefulness.[3] Seen in this light, Vegetius' *regulae* partake in a long literary tradition and serve an important function within his *Epitome*, lending the treatise an air of pragmatism and presenting the text as a complete guidebook and programme of reform. I conclude the chapter by suggesting that Vegetius' precepts were also significant in their specific historical context. Both literary conventions and contemporary circumstances elucidate the rhetorical purposes underlying Vegetius' *regulae*.

I. Objections to the *regulae*

Since the nineteenth century, scholars have questioned the authenticity of the *regulae bellorum generales*. In both his editions of the *Epitome*, Carl Lang bracketed the passage. In his *apparatus criticus*, he notes the 'extraordinary agreement' between the first *regula* and the first rule in the *Codex Laurentianus*' so-called

Precepts of Military Matters (*Praecepta de re militari*), a list of Greek maxims in a tenth-century Byzantine collection of military writings.[4] Grammatically, there are some important differences between the two passages, but Lang was right to note a relationship. Evidently, he believed that Vegetius' *regulae* were an interpolated list from another source, part of which eventually found its way into the *Codex Laurentianus*. Still, he could not show whether both originated from a common source or the Greek translated Vegetius. As Lang acknowledged, the other rules in the section 'differ very much from our *regulae*', so we should be suspicious that this is evidence of interpolation, especially since we have no idea of the link between the text of Vegetius and these Byzantine *Praecepta*.

Another objection is the purported lack of unity between the *regulae* and the rest of the work. Dankfrid Schenk pointed to four particular rules which he argued did not derive from any other section of the *Epitome*, showing that neither Vegetius nor an interpolator created the *regulae*.[5] Instead, Schenk argued that Vegetius drew from Frontinus' no-longer extant *On Military Matters* (*De re militari*) and clumsily inserted Frontinus' list, including the rules summarising sections which do not appear elsewhere in Vegetius' text.[6] This hypothesis may seem attractive, but it is founded on the assumption that Vegetius was willing to copy irrelevant rules from another source but could not devise such rules on his own. Likewise, Schenk ignored the possibility that an interpolator could have written the rules. We should take a much more cautious approach to Vegetius than Schenk's rigid *Quellenforschung* that assumes a single source for each book.[7]

Upon closer examination, Schenk's examples fail to show that the *regulae* are out of place in the *Epitome*. Two of the *regulae* cited by Schenk, 7 and 30, clearly refer to other parts of the text.[8] Rule 7 – 'In attracting and taking in the enemy, it is a great benefit if they come in good faith, since deserters harm the enemy more than if they were killed' – restates a point made from earlier in the book (3.6.33).[9] Rule 30 – 'in camp, fear and punishments correct soldiers; on the march, hope of rewards makes them better' – summarises Vegetius' longer discussion of discipline (3.4.3–6).[10]

The other two *regulae* cited by Schenk, 27 and 28, do not militate against Vegetian authorship either. Rule 27 calls for the general to send all his soldiers into their tents to apprehend enemy scouts, and rule 28 encourages the general to change his plans when they are discovered. Although these maxims have no explicitly corresponding passages in the rest of the work, they are consistent with some of the *Epitome's* recurring themes. The stratagem described in rule 27 – sending all soldiers to their tents in order to catch enemy spies in the open – may have no direct parallel, but it has a slight echo in Vegetius' prescribed remedy for enemy ambushes (3.26.27, 3.6.25–28).[11] Moreover, his previously stated maxim that what helps the general harms the enemy and its corollary would suggest that the general should seek to apprehend enemy spies as assiduously as he directs his own (3.26.1 and 3.6). A general is his own undoing if his plans are discovered, so Vegetius advises the commander to keep his next move secret, even from his own men (3.6.8–12). It is only a short step to *regula* 28: 'when you discover that your plan has been betrayed to the enemy, you should change your strategy'

(3.26.28).[12] It is invalid to assume that every single *regula* should have a directly correlated passage in the rest of the work. Some are sufficiently aphoristic to refer not just to specific sections but to recurring themes and principles.[13]

Another argument marshalled against Vegetius' *regulae* is the lack of references to the rules elsewhere in the *Epitome*. The heading, 'GENERAL RULES FOR WAR' (*REGULAE BELLORUM GENERALES*), if authentic, is the only indication that a list of rules is coming. In all but one of the earliest manuscripts, the *regulae* are listed one after another with no spacing or indentation, an arrangement which might be jarring to a reader, especially due to the lack of connective particles between the rules.[14] In fact, this confusion may have led manuscripts descended from the ε-hyparchetype to mistake the text at 3.26.2 for a title.[15] In Michael Reeve's opinion, the disjointedness of the rules is the strongest argument in favour of the interpolation of the *regulae*.[16]

Much of this question hinges upon whether the tables and headings in Vegetius are authentic or whether they are themselves the result of later marginalia.[17] On the topic, Vegetius himself writes, 'I attempt to present ancient practice regarding the recruitment and training of soldiers in certain steps and *tituli*' (1.1.5).[18] Some translators take *tituli* here to mean 'titles' or 'headings'.[19] N. P. Milner writes that Vegetius' use of *tituli* here 'indicates that the rubrics are the author's; their style is homogeneous with the text, and they share in the same systems of *variatio*'.[20] Erik Andersson, through an analysis of both Vegetius' extant works, argues that the language and syntax of the *tituli* in general, and the *regulae* specifically, are consistent with Vegetius' general style.[21] Nevertheless, Lang argues that the word *tituli* refers to sections of the work and not the headings in and of themselves, an attested use of the word.[22] Michael Reeve adds that Vegetius often links across from section to section using connective particles, making headings superfluous.[23] But this does not prove that headings were not included, and Reeve ultimately can only state that 'the answer . . . remains *non liquet*'.[24]

What does this mean for the *regulae*? Given Vegetius' own statements about *tituli*, it is certainly plausible that the heading *regulae bellorum generales* is authentic. But even if it is not, the lack of connective particles joining the rules to one another and the rest of the work is not a fatal argument. After all, there are no connective particles before the epilogue which follows the *regulae* (3.26.35–38), so unless an interpolator excised such connectors after adding the rules, the *regulae* are no less connected to the text than Vegetius' conclusion to book three, a passage everyone agrees is genuine. Moreover, syntactic parallelism between *regulae* renders transition from rule to rule less abrupt. Relative clauses introduce twelve of the thirty-three, most of which are arranged successively, and the appearance of the nominative subject at the outset of two-thirds of the *regulae* emphatically marks out each maxim.[25]

Finally, it has been argued that the vocabulary of the rules is not sufficiently consistent with the rest of the text. Everett Wheeler has pointed out that the title is the only occurrence of the word *generalis* in the *Epitome*, but this should not be read into too far, especially in a work as brief as Vegetius'.[26] The choice of the word *regulae* to describe rules is also noteworthy. Besides the heading *regulae bellorum generales*, the word only occurs twice in the *Epitome*, at 3.10.20 and 3.26.38. Wheeler has suggested that this latter mention of a *regula proeliandi* 'may have suggested insertion of the *regulae*' by an interpolator, but this argument can, of course, cut both ways. Vegetius' emphatic reference to *regula proeliandi*, a 'rule-book of battle', could tie in an otherwise unconnected section of the work, namely a set of practical maxims.[27] Likewise, the expression 'these are the summaries' (*digesta sunt*) with which the epilogue begins could be a reference not only to the preceding three books, but more proximately to the preceding *regulae*.[28] Immediately following the *regulae*, Vegetius declines to comment on cavalry, instead writing that there are already 'many precepts' (*multa praecepta*) on the subject.[29] *Praecepta* has a broad range of meanings, but could it not also allude to the *regulae* immediately above?[30] It is worth noting that Lang did not bracket Vegetius' remark on existing precepts on horsemanship. If we are to excise 3.26.1–33, but not 34, that would leave a rather abrupt transition from Vegetius' discussion of retreats (3.25) to his mention of *praecepta de equitatu*. Although the *regulae* may seem unattractive to some critics, the alternative leaves the reader with an abrupt throwaway line on the principles of horsemanship.

The arguments against Vegetian authorship of the *regulae* are unconvincing, and the manuscript evidence gives some positive evidence. All authoritative MSS include the *regulae*, and this fact alone is a strong point in favour of Vegetian authorship. A closer look at the MSS suggests that the *regulae* were included in the text at an early date. MSS descended from the ε-hyparchetype append to book four a subscription by an otherwise unknown Flavius Eutropius:

> I, Flavius Eutropius, emended this without an exemplar at Constantinople in the seventh consulship of the emperor Valentinian and of Abienus.[31]

This subscription gives a *terminus ante quem* of 450 CE for the entire work.[32] Because ε includes both the *regulae* and this subscription, and since the other authoritative MSS all attest the same *rules* as well, 3.26.1–33 could at the latest have been an addition by Eutropius.[33] Wheeler suggests that this man omitted the name of the dedicatee and inserted the *regulae*, but this is speculative.[34] Because the MSS of the families δ, β, and φ lack this subscription but include the *regulae*, it is unlikely that the *regulae* are Eutropius'.[35] The study of other recensions in the fourth and fifth centuries suggests that the interpolation of summarising lists was not a common feature of *emendatio*.[36] Moreover, the inclusion of the phrase *sine exemplario* in the subscription suggests that Eutropius did not edit or contaminate the text but rather corrected it, what Michael B. Charles sees as 'by far the most likely possibility'.[37]

Whatever the case, the *regulae* became part of the text by 450 CE at the latest. And although none of the evidence here proves that Vegetius himself wrote the *regulae*, there are not strong arguments against his authorship either. The fact that the rules appear at the end of book three and not book four could be significant. A later interpolator would presumably have appended a summarising list to the end of the entire work after book four, as Eutropius did with his subscription. The earlier position of the rules, before the remarks in praise of the emperor at the end of book three, could belong to the author's original scheme. The concluding *regulae* and *encomium* perhaps capped off a self-contained set of three books before Vegetius later added a fourth on siegecraft and nautical matters as 'a complement to the work' (4 Preface 8).[38]

II. List-making in a literary context

If we turn our attention to other ancient military handbooks, we can begin to see the ways in which Vegetius' *regulae* accord with established generic norms. Given the lack of explicit ancient theorising about a genre of military manuals, it is difficult to decide whether the corpus constitutes a single genre or a 'a cluster of related genres'.[39] For the sake of argument, I consider all ancient treatises that deal with purely military matters a loose category of analysis, even if such an approach elides distinctions between subgenres.[40] For all their various aims, such works tend to strive for selectivity and concision and frequently profess a desire for pragmatic use. List-making was a way for writers of handbooks to pursue these literary aims, allowing a presentation of *exempla* and general principles which could theoretically be recalled and referenced quickly, even if we can only speculate on their practical value.[41] For Vegetius and his predecessors, the chief value of the list was rhetorical. It structured and condensed knowledge, giving the author the privilege of claiming mastery or holistic understanding of a field.[42]

The earliest extant military handbook, Aeneas Tacticus' fourth-century-BCE *How to Survive under Siege*, probably included no summarising list or table of contents, but several portions of the text give some of the earliest signs of list-making in military writing.[43] The most prominent such list is the proclamations to be made by the general (10.3–5). This section presents sample orders as imperatival infinitives, one after another, to be reproduced by the general at the appropriate time. Unlike the other recommendations that follow (10.5 ff.), these (or something like them, τοιάδε) are meant to be reproduced and announced by the general.[44] As with other lists in Aeneas' text, the discrete items are articulated with minimal syntactic connectors.[45]

This paratactic and grammatically simple list would be theoretically easy to use, for the general can modify and expand it as needed, but the regulations also have a didactic aim, assimilating the role of the reader with that of the author. The composition of lists is the task of both Aeneas and the general, a fact that becomes explicit a few sentences later: 'write down (ἀπογράφεσθαι) individuals who have more than one weapon'.[46] Aeneas also urges the keeping of a catalogue of resident aliens present after all the riffraff have been deported.[47] Just as Aeneas presents his reader with a list of announcements, so too must the general be prepared to make his own lists and impose order on a city under threat. That the list of announcements is presented as oral and public while the administrative catalogues are to be written and confidential highlights the level of expertise and literacy assumed by the author.[48] The consumer of the text is to self-identify as a commander with access to specialised knowledge, a 'technician-warrior' who can relate Aeneas' commands to the less knowledgeable public.[49]

The Hellenic *Tacticae* of Asclepiodotus, Arrian, and Aelian represent a shift in military writing in a more philosophical and systematic direction. The addition of tables of contents, diagrams, and headings shows organisational development in list-making. Although the manuscript history of such lists renders their origins uncertain for Arrian and Asclepiodotus,[50] Aelian's preface confirms the authenticity of his table of contents:

> On account of the pressure of your business, I have prefaced the work with subject-headings, so that you can, without reading the book as a whole, determine its contents from a few words and without spending much time easily find the places you want to look up.
>
> (Preface 7)[51]

This lays out the reasons for an introductory list; it gives structure to the text and aids the busy reader. A different list ends each *Tactica*, effectively framing each treatise with paratactic inventory.[52] All three authors give a sequence of model commands not joined by connectors but listed one after another. This asyndeton between items is a suggestive parallel for Vegetius' *regulae*, underscoring the inclination in list-making to set aside normal grammatical constraints in favor of a more excerptable layout. It is not clear whether the lists in these *Tacticae* would have been of much practical use, but they meaningfully highlight the author's desire to present his text as eminently usable.

In tandem with this tradition, the genre, or subgenre, of stratagem collections developed, lists of *exempla* organised thematically, chronologically, ethnographically, or by character.[53] As it stands, there are only two extant stratagem collections from antiquity: those by Frontinus and Polyaenus.[54] In his preface, Frontinus explains the value of a list of historical anecdotes:

> I think an attention to brevity is owed to busy men. Indeed, it is a lengthy task to pursue examples one at a time and strewn about through the massive body of histories. And those writers who have selected noteworthy examples have bewildered the reader just as if under a large heap.[55]

The stratagem collection, then, more than any other sort of military work, found its value specifically in the abridgement and organisation of information. These texts were more than exhaustive catalogues; they ideally sought to distill a vast heap of anecdotes down to quintessential *exempla*. In the fourth book, probably by Frontinus himself, we see a shift away from anecdotes to even more generic advice, culminating in a list of maxims without reference to specific events.[56] Their aphoristic character and final position offer a suggestive parallel to Vegetius' *regulae*. Each book was sorted into categories such as 'on concealing plans'

(*De occultandis consiliis*) or 'on quelling mutinies' (*De seditione militum compe-scenda*). Frontinus boasted that this arrangement would allow quick reference, and he listed the *tituli* at the beginning of each book.[57] Given the exigencies of ancient warfare, we might doubt the actual value of such a text in the field, but there was a literary point to Frontinus' sorted lists. His collection conveyed its own practical value and the idea of selection by an expert.

Whereas Frontinus' lists were organised by strategic and tactical situation, Polyaenus' *Stratagems* was sorted ethnographically, chronologically, and proso-pographically, making it extremely impractical to use.[58] Nevertheless, Polyaenus still presented his work as having pragmatic value, a testament to the preoccupa-tion of military handbooks with the *topos* of use. Ostensibly unable to join the Parthian campaign due to old age, Polyaenus presented his work to Marcus Aure-lius and Lucius Verus as a substitute for actual service:

> I offer these supplies of strategic insight, as stratagems of past generals, a great compilation of past deeds both for yourselves and for those men sent by you – polemarchs, generals, legates, tribunes, prefects, or any other com-manders – who teach the prowess and skills of ancient successes.
>
> (1 Preface 2)[59]

For the ancient reader, stratagem collections seem to have drawn their appeal from a belief in the general applicability of historical *exempla* and the ostensible conve-nience that such a list could give.[60] A reader could hope for personal improvement by studying an exemplary catalogue, for the author had culled the most essential episodes from the dross of history.

In a similar vein, Onasander produced a treatise on generalship outlining the best qualities of a commander. Instead of using historical *exempla* to illustrate his points, he began by listing and describing the moral and personal qualities neces-sary in a general – self-restraint, vigilance, and so forth (1.1–18). As with Veg-etius' *regulae*, it is unclear how Onasander's gnomic statements on generalship were set apart from the rest of the text. At least two manuscripts have numbers in the margin, and one manuscript puts the first letter of each new 'quality' in red.[61] Whatever the case, we see in Onasander's treatise a list that goes beyond a simple selection of commands or examples. By beginning with a summarising list, Ona-sander unambiguously displays the principles which will govern his treatise. In effect, he gives an abridgement of his work which 'not only arrayed guiding prin-ciples of generalship, but also endeavoured to make out the art of the general and the practical wisdom in these things' (Preface 3).[62]

Authors of military handbooks simultaneously sought to instruct, to entertain, and to display learning.[63] One may judge lists of commands, stratagems, virtues, or maxims to be trivial or common-sense catalogues, but this assessment misun-derstands the generic purpose of these works. Approached on their own terms, the lists found in ancient military handbooks reflect concerns of concision, selec-tivity, and practical use. The curated presentation made the author an authority, someone who had sifted through vast amounts of material and mined the best examples and principles. The list's abbreviated format gave military handbooks a veneer of practicality.

As part of this tradition of list-making, Vegetius' *regulae* make more sense. Just as Onasander lists the main topics of his book on generalship, Vegetius restates his main points from the first three books of the *Epitome*.[64] The recurrence of jussive subjunctives in Vegetius' *regulae* is also similar to the lists of commands in Aeneas Tacticus and the Hellenic *Tacticae*, a discourse which calls to mind the general's role in issuing commands. Finally, the extraction of general rules from historical examples reflects an intellectual project like that of the stratagem collections.[65] Far from an uncharacteristic interpolation, the *regulae* are exactly what we might expect from an author steeped in Greek and Roman military literature.[66]

Vegetius' *regulae* also differ from this tradition in some significant ways. Like other paratextual elements, the *regulae*'s references are to aid and guide the reader, but Vegetius' list differs from other military tables of contents in two respects.[67] First, it appears at the end of the work, so it more explicitly refers to what has already been read rather than what will be read.[68] Second, and more importantly, the rules do not point back to specific sections of the text in any easily discernible pattern.[69] Rather, they jump back and forth between general aphorisms and specific stratagems. Thus, the *regulae* convey general knowledge and cap off the work with a reminder of basic themes without proceeding systematically through the *Epitome*'s structure.

In addition to this structural difference, his choice of the word *regula* is unique. The word originally referred to a ruler or measuring rod, derived from the same root as such words as *regere*.[70] By applying the transferred meaning, 'rule' or 'principle', to describe his summary, Vegetius suggests that military expertise could provide a standard for correct decisions, just as a ruler measures out lines. Indeed, Vegetius presents his whole work as a 'a ruler for battle (*regula proeliandi*), nay an art of victory (*vincendi artificium*)', both of which evoke the idea of the warrior as a craftsman (3.26.28).[71] One could even usefully compare Vegetius' second-person rules to instructions in medical and mathematical texts, but the mingling of more general rules in the third-person suggests we consider as a parallel a different kind of *regulae*, legal rules that bore the same title.[72] The Justinianic *Digest*, in its concluding section of legal *regulae*, quotes the jurist Paul's definition: 'a *regula* is that which briefly describes something. The law (*ius*) does not proceed from the *regula*, but the *regula* comes from the law (*ius*) that exists'.[73] If Vegetius had such juristic parallels in mind, he may have thought of his *regulae* as having general applicability, rules of thumb that could substitute for historical examples or systematic strategic knowledge. The *regulae* put Vegetius in the position of master jurist or craftsman, handing down basic tools with which to achieve victory.

III. Vegetius' *regulae* in historical context

As we have seen, Vegetius' summarising rules were consistent with literary conventions that emphasised pragmatic concision and selectivity. It should come as no surprise, then, that his *regulae* – a distillation of the author's most incisive observations and prescriptions, a summary of a summary – were poised to become a very popular section of his work, often commented upon in marginalia and independently excerpted.[74] Subsequent reproductions, translations, and adaptations of Vegetius' rules are numerous, and the afterlife of the *regulae* underscores the longevity of the author's precepts and the enduring search for universal principles in war.[75]

In the more immediate historical context, however, there could be other reasons why Vegetius' rules might have been well received. Latin persisted as the 'supreme' language of military command and civil administration as late as the seventh century, forming a 'significant integrative structure' even in the largely Greek-speaking eastern empire where Vegetius' work was known.[76] On the ground, however, the linguistic makeup of the empire's army was far from homogenous.[77] Maurice's *Strategikon* (c. 600 CE) informs us that rules and regulations were to be read out by officers to their respective units.[78] In such circumstances, a select list of thirty-three general precepts – easier to reference or memorise than a longer collection or treatise – could prove useful for training a diverse army whose Latin abilities surely varied.[79] To be sure, the *topos* of practical utility pervaded ancient handbooks and should not always be taken seriously, as I have noted, but Vegetius also wrote with an eye to contemporary circumstances, as his well-crafted dedications to the emperor attest.[80] He presented his rules as suited to the empire's needs and refused to give advice where 'present doctrine suffices' (3.26.34).[81] Against the backdrop of a diverse, polyglot army, a problem at the fore of Vegetius' mind, the *regulae* were a set of precepts that could constitute a shared doctrine, standardised in brief, if sometimes banal, Latin sentences.[82]

We cannot know how much influence Vegetius had. To judge from several MS subscriptions, he was a highly situated official in the imperial service, bearing the rank of 'illustrious man' (*vir illustris*) and 'count' (*comes*).[83] Familiar with the empire's administration, he may have thought his *regulae* would be well received within a bureaucracy built around schedules of fees, hierarchies of offices, and lists of regulations.[84] An analogy could again be made to legal *regulae*, which may have been used by lower government officials and were compiled in a list at the end of the Justinianic *Digest*.[85] We have precious few indications of how Vegetius' own *regulae* were received, but our *testimonia* to the *Epitome* suggestively cluster around the eastern capital. Eutropius emended the text there, and, about a century later, John Lydus, a bureaucrat within the Praetorian Prefecture, knew of Vegetius as a military writer.[86] Priscian, a grammarian at Constantinople, quoted from Vegetius, and the *Strategikon* included translations of many of the *regulae* alongside other aphorisms.[87] Based on these connections to the imperial center, the *Epitome* could plausibly have had an administrative audience, for whom Vegetius' list of *regulae* may have been attractive.

IV. Conclusion

Arguments for the interpolation of the *regulae bellorum generales* are unpersuasive. The manuscript evidence points to their presence at an early date, probably due to Vegetius himself. Although other lists in Greek and Latin military hand-books differed substantially in their specific forms and aims, they all shared strategies of concision, selectivity, and the appearance of practical value. The *regulae* were consistent with this long tradition, offering a succinct summary that tied together the work and maintained the conceit of pragmatism. In the immediate context of the fourth or fifth century, Vegetius' *regulae* could have addressed present circumstances in addition to this wider literary heritage. The Latin rules fit Vegetius' project of regularising a diverse army, and they may have even been crafted to find appeal within the imperial service. In the end, the 'general rules of war' stand out not as a peculiar addition to the text of the *Epitome* but as an integral part of the literary whole, speaking to the expectations of genre as well as present circumstances and audience.

Notes

1 Aguilar (2006) 57 only counts thirty-two *regulae*, presumably because he does not count the first, longer maxim (3.26.1). *Regula* 33 is dislocated in the MSS and actually belongs after 26 or 27. My text and enumeration follow Reeve (2004) 116–119. All translations, unless otherwise noted, are my own. On the date, see later, 209 n.32.

2 Lang (1885); Schenk (1930); and Wheeler (2012).

3 On the problems of ancient manuals as historical sources with practical value, see Chapters 2 and 3.

4 Lang (1885) 120–121. Vegetius 3.26.1. *Mediceo-Laurentianus gr.* 55, 4, folio 131. Part of Constantine VII Porphyrogenitus' encyclopaedic project, the codex contained ancient and contemporary military texts, a catalogue of which may be found in Bandini (1768) II, cols. 218–238. On the *Praecepta*, which come from the codex's version of Maurice's *Strategikon.* 8.2, see Dain and de Foucault (1967) 362. For an edited text of Maurice 8.2, which may be an interpolation, see Dennis and Gamillscheg (1981), with the discussion of textual history at 28–42.

5 Schenk (1930) 58–59, citing 3.26.7, 27, 28, 30.

6 Schenk (1930) 59.

7 Milner (2011) xvi–xvii and Janniard (2008) 20 believe that many modern commentators skirt close to this tradition in interpretations of Vegetius that stress the patchwork of his sources rather than his own creativity. See Pretzler (2010) 92 for a discussion of the flaws in traditional *Quellenforschung*.

8 Müller (1997) 299 overlooks Schenk's error.

9 3.26.7, *In sollicitandis suscipiendisque hostibus, si cum fide veniant, magna fiducia est, quia adversarium amplius frangunt transfugae quam perempti.*

10 3.26.30, *Milites timor et poena in sedibus corrigit, in expeditione spes ac praemia faciunt meliores.*

11 This stratagem is by no means unusual and can be found in Maurice 9.5.20. Spaulding (1937) 110 notes its use by Napoleon on the island of Lobau during the Wagram campaign.

12 *Cum consilium tuum cognoveris adversariis proditum, dispositionem mutare te convenit.*

13 E.g., 3.26.12 – *paucos viros fortes natura procreat, bona institutione plures reddit industria* – could summarise much of books one and two.

14 Reeve (2000) 277; cf. (2004) xxx. Z (Vat. Lat. 5957) writes out each *regula* on a separate line. In the former, Reeve notes 'many chapters are linked to their predecessor across the title by *sed/vero/autem/tamen, ergo/igitur, quoque/etiam/praeterea,* or more rarely *enim* or a retrospective pronoun, and nothing in a text picks up anything in a title'.

15 Reeve (2000) 276. I have followed Reeve's tripartite stemma, which presumes three different lost hyparchetypes: ε, β, and δ (as well as φ, a descendant of δ which provides independent testimony to reconstitute the end of the text (4.39.1 ff.) which dropped from δ).

16 Reeve (2004) xxxviii.

17 Reeve (2000) 276–277. Cf. Schröder (1999) 144–145.

18 *De dilectu igitur atque exercitatione tironum per quosdam gradus et titulos antiquam consuetudinem conamur ostendere.*

19 Milner (2011) 2: 'by a number of stages and headings'; Müller (1997) 31: 'über verschiedene Stufen und Titel'. Stelten (1990) 9 erroneously translates *tituli* as 'ranks', and Phillips (2011) completely avoids translating *tituli* and instead takes *per quosdam gradus et titulos* as 'in some order'.

20 Milner (2011) 2 n.2.

21 Andersson (1938), 42–48.

22 *OLD* s.v. *titulus*, 3b; Blaise (1967) s.v. *titulus*, 4. Lang (1885) xv and Reeve (2000) 276.

23 Reeve (2004) xxxvii.

24 Reeve (2004) xxxviii.

25 '*In bello qui . . .* ' 3.26.2, '*Qui . . .* ' 3.26.16–26. Exceptions to the pattern of nominative priority (ignoring adverbs and conjunctions) are 3.26.1, 2, 7, 10, 11, 12, 14, 27, 33, 28, 29, many of which begin with prepositions, subordinate clauses, or indirect questions that arguably also distinguish them from preceding maxims.

26 Wheeler (1994) 138. The *TLL* cites the use of the adjective *generalis* by Vegetius' contemporaries in historiographic (e.g., Ammianus 14.11.25, 28.5.14), ecclesiastical (e.g., Leo M., *sermo* 38.3), legal (e.g., *Cod. Theod.* 1.1.4, 10.3.7), and philosophical (e.g., Augustine, *City of God* 4.20, Macr. *Somn.* 1.3.11) contexts. Handbooks from other periods use the adjective, cf. Cels. 5.28.11 and Var. *R.* 2.1.12, '*una quaeque [scientia] in se generalis partis habet minimum novenas*'. Palladius, roughly a contemporary of Vegetius, links *generalis* with synonyms of *regula*, cf. Pallad. *de Agric.* 1.42.4, '*Expletis his, quae pertinent ad generale praeceptum . . .* ' Seneca also writes of *generalia praecepta* (*Epist.* 94.32).

27 Wheeler (2012). Milner's translation of *regula proeliandi* (2011, 119).

28 Vegetius 3.26.35, *Digesta sunt, imperator invicte, quae nobilissimi auctores . . . prodiderunt.*

29 Vegetius 3.26.34, *De equitatu sunt multa praecepta; sed cum haec pars militiae usu exercitii, armorum genere, equorum nobilitate profecerit, ex libris nihil arbitror colligendum, cum praesens doctrina sufficiat.*

30 Cf. *OLD* s.v. *praeceptum*, esp. 3, 'A principle, rule'. s.v. *regula*, 2, 'A basic principle, rule, standard, or sim'.

31 *Fls. Eutropius emendavi sine exemplario Constantinopolim consuls Valentiniano August. VII et Abieni.* Reeve (2000) 246, M (Munich Bayer. Staatsbibl. Clm 6368). Other variants have minor differences. See Önnerfors (1995) 260–61.

32 A *terminus post quem* of 383 is provided by the mention of *divus Gratianus* at 1.20. Within this range, opinion remains divided over Vegetius' unnamed emperor. A majority of scholars side with Theodosius I (Barnes (1979); Richardot (1998a); Milner (2011) xxxvii–xli; and Lenski (1997) 147–148), but there is also a strong case for Valentinian III: Charles (2007) and Goffart (1977). See also Chapter 1 for a late date on the basis of references to the Huns, and Chapter 12 for an argument for an earlier date based on the nautical material.

33 Reeve (2004) xviii–xix and 116–120 and Önnerfors (1995) 187–192. Simple corruptions of spelling consistent throughout B (Bern Burgerbibl. 280) and M suggest that the ε-hyparchetype postdates Eutropius' fifth-century recension.

34 Wheeler (2012).

35 Reeve (2004) xvii, xxxiii–iv. δ lacks the end of book four, so its lack of the subscription does not indicate much.

36　Emendation was a popular activity of leisure and study in the fourth and fifth centuries, but notes tended to be limited to haphazard marginalia, see Zetzel (1980) 38–39. Emendation was neither a purely pagan activity nor serious scholarship Cameron (2011) 421–497. A sizable market for forged manuscripts was especially vibrant in the second century, but these forgeries tended to present alternate readings of antiquarian and grammatical interest, not summarising lists: Zetzel (1973) 225–243.

37　Charles (2007) 37. Editing is also a possibility, but Charles imagines it as a reorganisation or joining of the different books.

38　*Ad complementum igitur operis maiestatis vestrae praeceptione sucepti rationes . . . in ordinem digeram.* But the wording is not specific, and, as Charles (2007) 34 n.75 warns, attempts to reconstruct a chronology of composition are suspect. Still, the introduction to book three (which looks back to the first two books and presents itself as 'sounding the *classicum*' (3.1.1)), the grand epilogue in praise of the emperor (3.26.35–38), and the lack of references to a fourth book suggest that the first three books were meant to form a coherent whole, of which the *regulae* offered a fitting summary.

39　Single genre: Whately (2015) 252; 'cluster of related genres': Whitehead (2008) 141.

40　On these distinctions, see the other contributions to this volume, especially Chapter 1. Chapter 3 draws a distinction between the works of armchair tacticians, like Vegetius and Asclepiodotus, and military writers, such as Frontinus.

41　Campbell (1987) 27 and Pretzler (2010) 105–107. Bachrach (1995) 7–11 identifies a shift to institutionalised use of handbooks in late antiquity, but some of Bachrach's evidence for the use of handbooks by the officer corps is circumstantial.

42　There are many examples of ancient lists, from the banal and administrative (e.g., *Notitia Dignitatum*) to the literary and complex. Some lists convey the impression of infiniteness (as Eco 2009 argues is the case with Homer's catalogue of ships in *Iliad* 2), others display organisational mastery (e.g., the table of contents in Pliny the Elder's *Natural History*), and still others are meant purely for show (e.g. Apuleius' listing of fish in *Apology* 38). A list's simple and paratactic presentation of information can imply a complete transmission of knowledge from author to reader. For a discussion of tables of contents and their ideological and literary dimension, see Riggsby (2007).

43　The chapter headings and divisions are presented in the manuscript tradition in such an inconsistent way that they likely never existed in the original text (Whitehead 2008: 47). A summarising list of authoritative maxims could of course have been in a no longer extant portion of Aeneas' work, but it is also possible that Aeneas, who exhorts the reader to think for himself (2.8), did not deign to write a list like Vegetius' *regulae*. Cf. the concluding summary at Xenophon, *Agesilaus* 11, which includes pithy sentences and longer paragraphs.

44　Aeneas 10.3, 'Then, make announcements such as these after some time to frighten and deter plotters . . . ' ~ Ἔπειτα κηρύγματα ποιεῖσθαι τοιάδε διά τινος χρόνου, φόβου καὶ ἀποτροπῆς τῶν ἐπιβουλευόντων ἕνεκεν . . .

45　. . . τε . . . τε . . . μηδὲ . . . μηδὲ . . . cf. 8.2–5

46　Aeneas 10.7, ὅπλα οἷς ἐστιν ἑνὸς πλείω ἀπογράφεσθαι. Cf. 10.9. See König (2017) 160–162 for a similar argument about the significance of writing in the first *exemplum* of Frontinus.

47　Aeneas 10.8.

48　The level of literary skill required to compose such a list – what Thomas (2011) 40 has called 'officials' literacy' in the context of Athenian democracy – was of course much different than that required to compose a literary work such as Aeneas', but the text ignores such differences in favor of a closer identification between author and reader.

49　Cuomo (2007) 75.

50　All but one of Asclepiodotus' MSS include a table of contents (Poznanski (1992) 1), but the text's schematic format leaves open the possibility of it having been redacted (Oldfather (1923) 232). For Arrian, the MS is mutilated at the beginning, rendering a reconstruction impossible (Roos (1907) v. 2).

51 διὰ μέντοι τὰς ἀσχολίας προέγραψα τὰ κεφάλαια τῶν ἀποδεικνυμένων, ἵνα πρὸ τῆς
ἀναγνώσεως τοῦ βιβλίου τὸ ἐπάγγελμα τοῦ συγγράμματος δι' ὀλίγων κατανοήσῃς καὶ
οὓς ἂν ἐπιζητήσῃς ἀναγνωσθῆναι τόπους ῥᾳδίως εὑρίσκων τοὺς χρόνους μὴ τρίβῃς.
Devine (1989) 44, trans. Cf. Aulus Gellius, *Attic Nights* preface 25.

52 Arrian, *Tactics*, τὰ παραγγέλματα δὲ ἔστω τοιάδε . . . In Asclepiodotus' case, a putative
lacuna limits our understanding of how the list was articulated (Asclepiodotus 12.11;
Oldfather (1923) 332–333, n. 8). Some MSS of Aelian include similar expressions
(Aelian 40–43; Köchly and Rustow (1853) 549), and Devine conjectures such an addition in his translation (Devine (1989) 58).

53 Cf. Frontinus, preface: *sollertia ducum facta, quae a Graecis una* στρατηγημάτων
appellatione comprehensa sunt ('The clever deeds of generals that the Greeks express
with the single term *strategemata*').

54 Wheeler (1988) 1, 19–20. Demetrius of Phalerum may have written stratagems (Dio-
genes Laertius 5.80; Plutarch, *Sayings of the Kings and Commanders* 189D), and
Hermogenes, son of Charidemus, wrote two lost books of stratagems. Clement of
Alexandria wrote of Moses' 'stratagems', and Julius Africanus listed stratagems in
Kestoi 7.

55 *Sed, ut opinor, occupatis velocitate consuli debet. Longum est enim, singula et sparsa
per immensum corpus historiarum persequi; et hi, qui notabilia excerpserunt, ipso
velut acervo rerum confuderunt legentemn thi.* On this passage, see also earlier, 6.

56 For the consensus on Frontinus' authorship of book four, see Chapter 10 (specifically
195 n.28.), as well as references in König (2017) 174, n.54.

57 Frontinus, preface: *species eorum, quae instruant ducem in his, quae ante proelium
gerenda sunt.*

58 Morton (2010) 119. On this point, see also Chapters 6 and 10 (specifically 121–122 and
191–192) in this volume.

59 ἀλλὰ τῆς στρατηγικῆς ἐπιστήμης ἐφόδια ταυτὶ προσφέρω, ὅσα τῶν πάλαι γέγονε
στρατηγήματα, ὑμῖν τε αὐτοῖς πολλὴν ἐμπειρίαν παλαιῶν ἔργων, τοῖς τε ὑπὸ ὑμῶν
πεμπομένοις πολεμάρχοις ἢ στρατηγοῖς ἢ μυριάρχοις ἢ χιλιάρχοις ἢ ἑξακοσιάρχοις ἢ
ὅσαι ἄλλαι ὅπλων ἀρχαί, διδασκομένοις ἀρχαίων κατορθωμάτων ἀρετὰς καὶ τέχνας.

60 This attitude was not unique to antiquity. In the eighteenth century, Shepherd dedicated
a translation of Polyaenus to Lord Cornwallis to help his campaigns in India (Shepherd
1796: xix–xx).

61 Oldfather (1923) 374–375.

62 μὴ μόνον στρατηγικὰς συνεταξάμην ὑφηγήσεις, ἀλλὰ καὶ στρατηγικῆς ἐστοχασάμην
καὶ τῆς ἐν αὐτοῖς φρονήσεως.

63 Campbell (1987) 27.

64 Milner (2011) xxvi: 'These rules were intended to provide an aide-mémoire of the main
principles of field strategy and tactics; such recapitulation was a valued technique of
late antique didactic writers'. Milner notes that Palladius uses the same strategy when
he lists his *sententiae* at 1.6.1ff.

65 Formisano (2017) 150–151 argues that this relationship between *exemplum* and *norm*
is one of the primary contributions of Vegetius to military thought.

66 We know that Vegetius at least had read Cato, Celsus, Paternus, and Frontinus' *On
Military Manuals* (*De re militari*) – as he clearly indicates at 1.8, 1.13, 1.15, and 2.3,
respectively.

67 For a general discussion of ancient lists of chapters, see Fruyt (1997) 28–29, where
all the examples precede the texts just as modern English tables of contents today. On
'paratext', see Genette (1997), esp. 316–8 on tables of contents. For Roman paratexts,
see the contributions in Jansen (2014), esp. Wibier's chapter on the common topog-
raphies of legal texts (58–72) and the browsing and 'horizontal reading' which this
facilitated, even without tables of contents (68–72).

68 Columella (11.3.65) is one of the only ancient authors to include a list of *capitula* at the
end of his text: Petitmengin (1997) 497.

69 One exception to this is the seven 'modes' of battle, given in the same order as in the text: 3.26.18–24; cf. 3.20.

70 *OLD*, s.v. *regula*, 1.

71 *regula proeliandi, immo vincendi artificium*

72 See Asper (2007) 198, who calls this *Rezeptstil*. *Regulae* with the second person: 3.26.1, 5, 14, 33, 28, 29. For a discussion of *libri regularum,* see Ferrary (1997) 248–249.

73 Digest 50.17.1, *Regula est, quae rem quae est breviter enarrat. Non ex regula ius sumatur, sed ex iure quod est regula fiat.*

74 Allmand (2011) 3, 39–41 and Löfstedt (1985) 498–499.

75 E.g., Maur. *Strat.* 8.2; Peter the Deacon's additional rules (Reeve (2000) 272–273); the *Chapel des fleur de lis* of Philippe de Vitry (Piaget 1898); Macchiavelli's *Dell'arte della guerra,* 7 (Formisano (2002) 113–114). For medieval reception, see Richardot (1998b) and Allmand (2011). For early-modern reception, see Richardot (2002).

76 I use 'supreme' rather than 'official' following Adams (2003) 608, who notes that Greek was often used. Quotation from Millar (2006) 92. For the army in the later period, see Rance (2010) 63–64.

77 For evidence of a polyglot army from the early to later empire, see Adams (2003) 20, n.61 and Haynes (2013) 301–311.

78 See Maurice 1.7–8, where the author orders a list of punishments be read in Latin and Greek and *Strat.* 3.5, where the author or a later copyist transliterates into Greek a Latin harangue (cf. *Strat.* pr.).

79 Maurice *Preface.* For evidence of soldiers attempting to learn Latin, see Adams (2003) 617–623.

80 Dedications to the emperor: 1.preface, 1.28, 2.preface, 3.preface, 3.26.35–38, 4.preface, 4.31.

81 *Ex libris nihil arbitror colligendum, cum praesens doctrina sufficiat.* Cf. 4.46.9.

82 Milner (2011) xxix, 'it is evident that the driving-force behind Vegetius' reforms was the desire to reduce and down-grade the rôle of all non-Roman ethnic forces'. See especially 1.2–3, for recommendation of recruits from temperate climates, and 2.2.5–6, where the united *antiqua legio* is compared to auxiliaries who 'coming from diverse places and units (*ex diversis locis ex diversis numeris*), are not suited to one another in discipline, knowledge, or affection (*nec disciplina inter se nec notitia nec affectione consentiunt*). Their training (*instituta*) is different, their use of arms differs'. Linguistic standardisation, although not explicitly part of Vegetius' project, would fit his call for uniformity.

83 On Vegetius' position, perhaps *comes stabuli*, see the discussion in Milner (2011) xxxv–xxxvi with references. For the responsibilities of the *comes stabuli,* see Jones (1964) v. 2, 625–626; Scharf (1990).

84 Notable examples of such administrative lists are the *Notitia Dignitatum*, the inscribed *Album* at Timgad (Chastagnol (1978), the list of regulations at Qasr el-Hallabat and elsewhere (Kelly (2004) 256, n.50), and the laws in *Cod. Theod.* and *Cod. Iust.* Such lists give 'a striking blueprint of a tightly regimented officialdom, its impressive concern for hierarchy and position beautifully illustrated in meticulously detailed lists or, in some cases, more permanently displayed on an inscribed wall' (Kelly (2004) 43). Cf. MacMullen (1962) 376 for a different aspect of Vegetius' characteristically bureaucratic style.

85 Stein (1966) 80–82. For a different interpretation, see Ferrary (1997) 248–249.

86 Lydus, *On Magistracies* 1.47.

87 Priscian, *Grammatical Institutes* 3.21; Maurice 8.2.

Bibliography

Adams, J. N. 2003. *Bilingualism and the Latin Language*. Cambridge: Cambridge University Press.

Aguilar, D. P. 2006. *Compendio de técnica militar*. Madrid: Cátedra Letras Universales.

Allmand, C. T. 2011. *The De Re Militari of Vegetius: The Reception, Transmission and Legacy of a Roman Text in the Middle Ages*. Cambridge: Cambridge University Press.

Andersson, E. A. 1938. "Studia Vegetiana." Upsaliae.

Asper, M. 2007. *Griechische Wissenschaftstexte: Formen, Funktionen, Differenzierungsgeschichten*. Stuttgart: Franz Steiner.

Bachrach, B. S. 1995. 'Some Observations Concerning the Education of the "Officer Corps" in the Fifth and Sixth Centuries', in F. Vallet and M. Kazanski, eds., *La Noblesse romaine et les chefs barbares du IIIe au VIIe siècle*. Saint-Germain-en-Laye: Musée des antiquités nationales. 7–13.

Bandini, A. M. 1768. *Catalogus codicum manuscriptorum Bibliothecae Mediceae Laurentianae varia continens opera graecorum patrum*. Florence: Praesidibus Adnuentibus.

Barnes, T. D. 1979. 'The Date of Vegetius'. *Phoenix* 33. 254–257.

Blaise, A. 1967. *Dictionnaire latin-français des auteurs chrétiens*. Turnhout: Brepols.

Cameron, A. 2011. *The Last Pagans of Rome*. New York: Oxford University Press.

Campbell, B. 1987. 'Teach Yourself How to Be a General'. *JRS* 77. 13–29.

Charles, M. B. 2007. *Vegetius in Context: Establishing the Date of the Epitoma Rei Militaris*. Stuttgart: Franz Steiner.

Chastagnol, A. 1978. *L'album municipal de Timgad*. Bonn: R. Habelt.

Cuomo, S. 2007. *Technology and Culture in Greek and Roman Antiquity*. Key Themes in Ancient History. Cambridge: Cambridge University Press.

Dain, A. and de Foucault, J. A. 1967. 'Les Stratégistes byzantins'. *Travaux et Mémoires* 2. 317–392.

Dennis, G. T., and Gamillscheg, E., eds. 1981. *Das Strategikon des Maurikios*. Vienna: Österreichischen Akademie der Wissenschaften.

Devine, A. M. 1989. 'Aelian's Manual of Hellenistic Military Tactics: A New Translation from the Greek with an Introduction'. *Ancient World* 19. 31–64.

Eco, U. 2009. *The Infinity of Lists*. New York: Random House.

Ferrary, J.-L. 1997. 'Les titres des textes juridiques', in J. C. Fredouille, M.-O. Goulet-Cazé, P. Hoffmann, and P. Petitmengin, eds., *Titres et articulations du texte dans les œuvres antiques: actes du Colloque international de Chantilly, 13–15 décembre 1994*. Paris: Institut d'études Augustiniennes. 233–253.

Formisano, M. 2002. 'Strategie da manuale: l'arte della guerra, Vegezio e Machiavelli'. *Quaderni di Storia* 55. 99–127.

——— 2017. 'Fragile Expertise and the Authority of the Past: The "Roman Art of War"', in J. König and Greg Woolf, eds., *Authority and Expertise in Ancient Scientific Culture*. Cambridge. 129–152.

Fruyt, M. 1997. 'Sémantique et syntaxe des titres en latin', in J. C. Fredouille, M.-O. Goulet-Cazé, P. Hoffmann, and P. Petitmengin, eds., *Titres et articulations du texte dans les œuvres antiques: actes du Colloque international de Chantilly, 13–15 décembre 1994*. Paris: Institut d'études Augustiniennes. 9–34.

Genette, G. 1997. *Paratexts: Thresholds of Interpretation*. Translated by J. E. Lewin. Cambridge: Cambridge University Press.

Goffart, W. 1977. 'The Date and Purpose of Vegetius' "De Re Militari"'. *Traditio* 33. 65–100.

Haynes, I. 2013. *Blood of the Provinces: The Roman Auxilia and the Making of Provincial Society from Augustus to the Severans*. Oxford: Oxford University Press.

Janniard, S. 2008. 'Végèce et les transformations de l'art de la guerre aux IVe et Ve siècles'. *Antiquité Tardive* 16. 19–36.

Jansen, L., ed. 2014. *The Roman Paratext: Frame, Texts, Readers*. Cambridge: Cambridge University Press.

Jones, A. H. M. 1964. *The Later Roman Empire, 284–602: A Social, Economic and Administrative Survey*. Oxford: Oxford University Press.

Kelly, C. 2004. *Ruling the Later Roman Empire*. Cambridge, MA: Harvard University Press.

Köchly, H. A. T. and Rustow, W. 1853. *Griechische Kriegsschriftsteller: Griechisch und deutsch, mit kritischen und erklärenden Anmerkungen*. Leipzig: Engelmann.

König, A. 2017. 'Conflicting Models of Authority and Expertise in Frontinus' *Strategemata*', in J. König and G. Woolf, eds., *Authority and Expertise in Ancient Scientific Culture*. Cambridge. 153–181.

Lang, C., ed. 1885. *Epitoma rei militaris*. Second edition. Leipzig: Teubner.

Lenski, N. 1997. 'Initium Mali Romano Imperio: Contemporary Reactions to the Battle of Adrianople'. *TAPA* 127. 129–168.

Löfstedt, L. 1985. 'Végèce en Moyen Âge: motifs et modifications des traducteur et des copistes', in A. M. Cano González, ed., *Homenaje a Alvaro Galmés de Fuentes*. Vol. 1. Madrid: Universidad de Oviedo. 493–500.

MacMullen, R. 1962. 'Roman Bureaucratese'. *Traditio* 18. 364–378.

Millar, F. 2006. *A Greek Roman Empire: Power and Belief under Theodosius II (408–450)*. Berkeley: University of California Press.

Milner, N. P. 2011. *Vegetius: Epitome of Military Science*. Second edition. Liverpool.

Morton, J. 2010. 'Polyaenus in Context: The Strategica and Greek Identity in the Second Sophistic Age', in K. Brodersen, ed., *Polyainos: Neue Studien = Polyaenus: New Studies*. Berlin: Verlag Antike. 108–132.

Müller, F. L., ed. 1997. *Abriß des Militärwesens*. Stuttgart: Franz Steiner.

Oldfather, W. A. 1923. *Aeneas Tacticus, Asclepiodotus, Onasander*. Cambridge, MA: Harvard University Press.

Önnerfors, A. 1995. *P. Flavii Vegeti Epitoma rei militaris*. Stuttgart: Teubner.

Petitmengin, P. 1997. 'Capitula païens et chrétiens', in J. C. Fredouille, M.-O. Goulet-Cazé, P. Hoffmann, and P. Petitmengin, eds., *Titres et articulations du texte dans les œuvres antiques: actes du Colloque international de Chantilly, 13–15 décembre 1994*. Paris: Institut d'etudes augustiniennes. 491–509.

Phillips, T. R. 2011. *The Military Institutions of the Romans (De Re Militari)*. Translated by J. Clarke. Mansfield Centre, CT: Martino Fine Books.

Piaget, A. 1898. 'Le Chapel des fleurs de lys, par Philippe de Vitri'. *Romania* 27.105. 55–92.

Poznanski, L. 1992. *Traité de tactique*. Paris: Les Belles Lettres.

Pretzler, M. 2010. 'Polyainos the Historian? Stratagems and the Use of the Past in the Second Sophistic', in K. Brodersen, ed., *Polyainos: Neue Studien-Polyaenus: New Studies*. Berlin: Verlag Antike. 85–107.

Rance, P. 2010. 'The *De Militari Scientia* or Müller Fragment as a Philological Resource: Latin in the East Roman Army and Two New Loanwords in Greek: *Palmarium* and **recala*'. *Glotta* 86. 63–92.

Reeve, M. D. 2000. 'The Transmission of Vegetius's "Epitoma Rei Militaris"'. *Aevum* 74. 243–354.

——— 2004. *Epitoma rei militaris*. Oxford: Oxford University Press.

Richardot, P. 1998a. 'La datation du 'De re militari' de Végèce'. *Latomus* 57. 136–147.

——— 1998b. *Végèce et la culture militaire au Moyen Age*. Paris: Economica.

——— 2002. 'La tradition moderne du De re militari de Vegece (XVe-XVIIIe siecles)', in P. Defosse, ed., *Hommages à Carl Deroux V: Christianisme et Moyen Âge, Néo-latin et survivance de la latinité*. Brussels. 537–544.

Riggsby, A. M. 2007. 'Guides to the Wor(l)d', in T. Whitmarsh and J. König, eds., *Ordering Knowledge in the Roman Empire*. Cambridge: Cambridge University Press. 88–107.

Roos, A. G. 1907. *Flavii Arriani Quae Exstant Omnia*. Leipzig: Teubner.

Scharf, R. 1990. 'Der Comes Sacri Stabuli in der Spätantike'. *Tyche* 5. 135–147.

Schenk, D. 1930. *Flavius Vegetius Renatus: Die Quellen der Epitoma Rei Militaris*. Klio, Beiheft; XXII (Neue Folge, Hft. IX). Leipzig: Teubner.

Schröder, B.-J. 1999. *Titel und Text: zur Entwicklung lateinischer Gedichtüberschriften, mit Untersuchungen zu lateinischen Buchtiteln, Inhaltsverzeichnissen und anderen Gliederungsmitteln*. Berlin: De Gruyter.

Shepherd, R. 1796. *Polyænus's Stratagems of War: Translated from the Original Greek, by Dr. Shepherd, F.R.S.* Second edition. London: Printed for George Nicol.

Spaulding, O. L. 1937. *Pen and Sword in Greece and Rome*. Princeton: Princeton University Press.

Stein, P. 1966. *Regulae Iuris: From Juristic Rules to Legal Maxims*. Edinburgh: Edinburgh University Press.

Stelten, L. F. 1990. *Epitoma Rei Militaris*. New York: P. Lang.

Thomas, R. 2011. 'Writing, Reading, Public and Private "Literacies": Functional Literacy and Democratic Literacy in Greece', in W. A. Johnson and H. N. Parker, eds., *Ancient Literacies: The Culture of Reading in Greece and Rome*. Oxford: Oxford University Press. 13–42.

Whately, C. 2015. 'The Genre and Purpose of Military Manuals in Late Antiquity', in G. Greatrex, H. Elton, and L. McMahon, eds., *Shifting Genres in Late Antiquity*. Farnham: Ashgate. 249–262.

Wheeler, E. L. 1988. *Stratagem and the Vocabulary of Military Trickery*. Leiden: Brill.

———— 1994. 'Review of *Vegetius: Epitome of Military Science*, by N. P. Milner'. *The Journal of Military History* 58. 136–138.

———— 2012. 'Review of *The De Re Militari of Vegetius: The Reception, Transmission and Legacy of a Roman Text in the Middle Ages*, by Christopher Allmand'. *Review in History* 1293 (www.history.ac.uk/reviews/review/1293).

Whitehead, D. 2008. 'Fact and Fantasy in Greek Military Writers'. *AAHung* 48. 139–155.

Wibier, M. 2014. 'The Topography of the Law Book: Common Structures and Modes of Reading', in L. Jansen, ed., *The Roman Paratext: Frame, Texts, Readers*. Cambridge: Cambridge University Press. 58–72.

Zetzel, J. E. G. 1973. 'Emendavi Ad Tironem: Some Notes on Scholarship in the Second Century A. D.'. *HSCPh* 77. 225–243.

———— 1980. 'The Subscriptions in the Manuscripts of Livy and Fronto and the Meaning of Emendatio'. *CPh* 75. 38–59.

12 Vegetius' naval appendix and the Battle of the Hellespont (324 CE)[1]

Craig H. Caldwell

Vegetius devotes the final section of his late antique *Epitome of Military Science* (*Epitoma Rei Militaris*) to Roman naval warfare, including the nomenclature of warships, shipbuilding techniques, navigation, and tactics. While other parts of the *Epitome* have inspired debates over the author's antiquarianism or the 'barbarisation' of the Roman army, Vegetius' consideration of ships has generally remained on an analytical island, only occasionally relevant to a historian of the late Roman navy.[2] Fresh attention to *Epitome* 4.31–46 can connect a neglected part of this military manual with late antique military history, specifically the naval engagement between the rival emperors Constantine and Licinius in 324. That battle itself has often 'lacked present-day interest', but deciphering its influence upon Vegetius can reveal an underappreciated aspect of the Epitome: analysis of recent events.[3] Reconstructing the Battle of the Hellespont provides critical context for Vegetius' discussion of the navy, and the author emerges as a historian of his own time at the intersection of actual events with his military manual.[4]

I

To begin, one should consider the nature of Vegetius' work. It falls within a genre of military manuals intended to teach audiences of emperors and civilian elites.[5] The question of the *Epitome*'s probable addressee will arise at the conclusion of the chapter, but Vegetius clearly intends to advance an imperial agenda, and his methods reinforced his status as an authority on military matters. Encyclopaedism and antiquarianism distinguish the style of the *Epitome*. To craft an encyclopaedic work, Vegetius broadens the definition of 'precepts of naval warfare' to include shipbuilding and navigation in order to present a complete view of his subject. This expansive scope regarding the elements of sea power mirrors his attitude toward war on land, where he includes recruitment and choosing of battlefields along with typical attention to types of soldiers and tactics. His comprehensive erudition emphasises the usefulness of the work: the *Epitome* offers the hope of continuing Roman military success to the emperor who follows its instructions.[6] Vegetius' discussion of these two additional subjects avoids the repetition that characterises most of the *Epitome*, indicating the wealth of available source material, which included Varro's 'naval books' that are now lost.[7] Shipbuilding and matters of winds and currents are also important clues regarding the recent history examined in the following. Vegetius' persuasiveness also depended upon his ability to ground his principles in the victorious Roman past, and thus his naval advice resurrects terminology from centuries earlier.[8] This antiquarian tendency is generally late antique rather than unique to Vegetius; Zosimus shares the *Epitome*'s distant memory of giant warships and the origins of the liburnian type.[9] Diverging from Zosimus' references to the clashes of ships in the Punic Wars, Vegetius invokes Actium on two separate occasions: to introduce the liburnian, what he calls the standard imperial warship after Augustus (4.33); and to impress his readers with the size of warships in the past (4.37).[10]

The late fourth century was apparently awash in the kind of technical vocabulary of naval warfare that Vegetius presents at the end of his *Epitome*; it even appears in hagiography. For instance, Jerome begins his *Life of Malchus* with a description of naval training that would not be out of place in a military manual:

> Those who intend to fight in a naval battle first in a harbour and on a quiet sea ply the rudders, pull up the oars, prepare the iron hands and hooks, and accustom the marines lined up on deck with halting step and uncertain footing to stand firm, so that they may not fear in a real context what they have learned in a mock battle.[11]

Jerome might not have read Vegetius as he wrote this introduction to his saint's life in 391 or 392, but his choice of words indicates the diffusion of naval concepts among educated civilians.[12] One reason advanced for Vegetius' own precepts regarding warships is the rise of Vandalic naval power in the fifth century, but interest in sea battles clearly predates the Vandals or even the naval expeditions of the general Gildo in 398 or the naval battle fought by the Gothic commander Gainas in 400.[13] The overlooked context for Vegetius' advice regarding ships was what Chester Starr called 'the only real sea battle in the history of the Roman Empire', the clash between the fleets of Constantine and Licinius in 324.[14]

II

Some initial remarks about the name and date of the battle of the Hellespont are necessary. The current scholarly consensus favours two separate naval battles in 324: one off the coast of Elaius (near the modern village of Seddülbahir in Turkey) and another near Callipolis (now Gallipoli or Gelibolu), more than 30 nautical miles up the straits.[15] The historical problem here is that the two battles appear in two different ancient sources, as discussed later, and the way that these two accounts can be combined to produce a single narrative is not self-evident. Yet, the sources agree that the focus of the campaign was the control of the channel between Europe and Asia, specifically the Hellespont, now called the Dardanelles or Çanakkale Boğazı. In the interest of accuracy, in recognition of the discrete clashes of fleets within the battle, and following the ancient understanding of the objective of the opposing navies, this chapter refers to these related engagements as the 'Battle of the Hellespont'.[16] Concerning the date of this battle, the *terminus post quem* should be July 3, which several chronicles record as the date of the Battle of Adrianople, which prompted the naval campaign. The Battle of the Hellespont must also have occurred prior to the Battle of Chalcedon or Chrysopolis, which all but ended the civil war on September 18.[17] Since the *terminus ante quem* includes an amphibious invasion of Asia Minor by tens of thousands of soldiers and the associated logistical requirements for that event, we may safely assign the naval campaign that cleared the straits for Constantine to July or August of 324.

The fact that the Battle of the Hellespont occurred is not controversial. Vague references to one or more naval engagements during the civil war are relatively common, but specific details are rare. This comment by the historian Socrates of Constantinople in the early fifth century is typical: 'Not long afterwards [the rival emperors] took up arms against each other as declared enemies. And after several engagements both by sea and land, Licinius was at last utterly defeated'.[18] If one wants more particulars than that, one has to turn to two very different ancient accounts. The first source is an anonymous history usually called the *Origo Constantini*, or the *Origin of Constantine*, which seems to have been written around the time of that emperor's death in 337. The *Origo* focuses on political and military details, and because relatively few years separated it from the events it describes, historians have come to trust its accuracy.[19] Since the *Origo* preserves the complex chronology of the early fourth century in ways that one can confirm from contemporary sources, but that later authors began to confuse, one may presume that those details unique to the *Origo* are reliable. The second source, the *New History* of Zosimus, differs from the *Origo* in date, style, and precision. Writing around the year 500, Zosimus presented the decline of the Roman Empire as a *fait accompli* of its Christianisation, which began with the emperor Constantine. Since Licinius was a defender of traditional religion, he is a more sympathetic figure and a more capable opponent in Zosimus than in the accounts of the Christian historians. Furthermore, for the part of his history that is germane to this discussion, Zosimus seems to have copied from or rephrased the *Histories* of Eunapius of Sardis, a fourth-century philosopher noted for his hostility to the new imperial religion of Christianity.[20] The advantage of Zosimus' dependence upon Eunapius, whose work has only survived in fragments, is that this earlier history considered events within a century of the battle, and a native of Asia Minor composed it relatively near to the Hellespont. Eunapius and Zosimus had a different religious vantage point into the early fourth century, and they thus collected evidence that their Christian counterparts passed over. But Zosimus never completed the final revision of his work, and the draft transmitted to us includes factual and chronological mistakes that were probably less apparent in the sixth century than they would have been at the time of the events. Readers of Zosimus also discover that he and his source Eunapius were not annalists who reported events in strict order according to years; their histories were content with regnal dating, placing events such as the Battle of the Hellespont within a particular emperor's reign.[21] One special nightmare for the modern reader of Zosimus is his frequent repetition of specific people or places, and so discerning an actual recurring event from careless scribal 'doubling' requires careful historical judgment.[22]

Narrative accounts of the naval aspect of the civil war between Constantine and Licinius are not new. Almost every modern biography of Constantine mentions the clash of navies in 324, though Constantine's first biographer, the bishop Eusebius of Caesarea, elided the details of the Battle of the Hellespont in favour of Constantine's miraculous victories on land against Licinius 'the God-hater' (θεομισής).[23] Edward Gibbon provided a detailed description over two centuries ago in his famous *Decline and Fall of the Roman Empire*, and more recent books recount the battle within larger narratives of the evolution or decline of ancient navies.[24] But scholars have either tended to blend ancient sources without critical attention to their authors' perspectives, or they have ignored the geographic setting for the battle. The last scholar to tackle both challenges was Edwin Pears, who discussed the battle in detail in a 1909 article with the evocative title 'The Campaign against Paganism A. D. 324'. One can improve on Pears with close attention to both the *Origin of Constantine* (*Origo Imperatoris Constantini*) and Zosimus' *New History*, and one may also set aside the struggle between religions that captivated Pears as 'one of the epoch-making events of history' as no longer the overriding historiographical concern, allowing us to examine the battle as a subject in itself.[25]

Turning to the basic outline of events in the *Origin of Constantine*, of which one can be reasonably confident for the reasons mentioned earlier, Constantine began this phase of the civil war in 324 by mounting parallel invasions of Licinius' domain by both land and sea. As Constantine set out eastward across the Balkans with his army, he dispatched his son Crispus, whom he had named Caesar, or junior emperor, to command 'a huge fleet to take Asia'.[26] Licinius opposed this threat with his own fleet under Amandus, who commanded a comparable force stationed in the Hellespont. According to the *Origin*, Crispus sailed up the strait to face Amandus at Callipolis (modern Gallipoli), where the Constantinian fleet overwhelmed the Licinian ships, destroying and capturing all of them. Amandus himself only escaped 'with help from those who had stayed on shore'.[27]

In contrast to the *Origin of Constantine*, Zosimus describes two naval engagements in his *New History*. One battle occurred at the mouth of the Hellespont near Elaius, where Crispus deployed a smaller number of ships that outmanoeuvred the larger fleet of Amandus (here called Abantus) in the narrow confines of the strait. In the second encounter, a sudden shift of winds wrecked almost the entire remainder of the Licinian fleet on the Asian shore of the Hellespont. As in the *Origin of Constantine*, Amandus escaped with a remnant of his original force. To harmonise these narratives, some historians have cast the second battle in Zosimus as identical to the decisive engagement at Callipolis in the *Origo*, though Zosimus actually claims that catastrophe befell Amandus' ships near the base of the Constantinian fleet at Elaius.[28] Yet, Zosimus does hint that the chokepoint at Callipolis was important following the destruction of the Licinian navy, for he says Licinius sent part of his army to Lampsacus (near modern Lapseki), across the Hellespont from Callipolis, to block a possible invasion of Asia.[29] Both sources agree that once Constantine and Crispus had control of the sea, they used transports to cross the Bosphorus and force Licinius to make a final stand in Asia Minor.

One can draw several conclusions about the characteristics of the battle. First, the entire series of engagements revolved around the objectives and requirements of the two armies. The catalyst for the Battle of the Hellespont was Licinius' defeat at the Battle of Adrianople. His retreat to Byzantium, to which Constantine then laid siege, placed a premium on control of the passage between Europe and Asia. As the *Origin of Constantine* says, Licinius 'felt secure to seaward', and he could stymie Constantine's advance so long as Amandus' ships protected his route of supply and communications.[30] Zosimus writes that Licinius had 'hurried through Thrace to reach his fleet', which indicates its importance in bringing food and equipment to continue the fight.[31] The unusual promotion to co-emperor of Licinius' *magister officiorum* Martinianus is another indication of the interdependence of ships and armies.[32] Licinius already had a deputy emperor or Caesar, his young son Licinius II, and coinage reveals that Martinianus was actually named as Augustus, or senior co-emperor. The *Origin of Constantine* dismisses this delegation of power as 'foolishness' (*vanitas*), but Zosimus can help navigate around the *Origin*'s scorn.[33] His history includes two relevant details: Licinius sent part of his army and navy away from Byzantium to Asia due to overcrowding during the siege, and Martinianus' task was to defend Lampsacus and prevent a Constantinian army from crossing the Hellespont there.[34] After his initial naval defeat, Licinius did not fear that Crispus might emulate Alexander the Great by marching inland from a landing at Abydos, so Constantine's rival chose instead to block the strait further north. Based on where Licinius' mints produced coins in honour of Martinianus, one may conclude that he led an army along the shore of the Propontis (now the Sea of Marmara) from the capital Nicomedia (now İzmit) through Cyzicus (near modern Erdek).[35] If one follows the *Origin of Constantine*'s chronology of events, it is possible that Martinianus' reinforcements constituted 'those on shore' who enabled Amandus' escape from the destruction of his fleet near Callipolis. Amandus might also have intended to join forces with Martinianus and gain the safety of a shoreline protected by a friendly army, but the wind surprised him while his ships were too close to shore.[36] Even the victorious ships ended up being incidental to Constantine, who needed to procure 'galleys and swift-sailing skiffs' of shallower draft to make his amphibious invasion across the Bosphorus.[37] Unlike the battles of Salamis and Actium, where the fleets themselves were the focus, at the Hellespont the navies were somewhere between the cavalry and the baggage train of the opposing armies. Or – to put it another way – Licinius named Martinianus as his co-emperor to command an army that he could not afford to lose as it guarded the Hellespont; he did not make his admiral Amandus into a second Agrippa, who had been the indispensable general and admiral of the emperor Augustus.

III

An initial objection to the relevance of the Battle of the Hellespont to the *Epitome* is that the clash of fleets was a regression from an earlier golden age of the Roman navy.[38] The Battle of Actium, at least as Vegetius remembers it, set a high bar for the size of the opposing warships and the decisive deployment of a new type of ship. Insofar as the Constantinian and Licinian ships set out to fight each other rather than simply to transport soldiers or to blockade a city, their collision merits Starr's description of it as *the* Roman imperial naval battle. But in keeping with the uneven comparison of Martinianus with Agrippa, the Battle of the Hellespont did not directly involve the two principal leaders, nor did it decide the civil war as Actium did.[39] Setting Vegetius' naval precepts alongside important elements of the battle in 324 reveals how he attempts to derive general principles from the most recent naval engagement. The *Epitome* therefore intends to improve the current state of Roman military affairs, so it uses the context of the Hellespont as an inspiration for restoration and reform. Taken with Vegetius' references to military practices and individuals from the distant past, the *Epitome* emerges as a unique text that blends ancient (from the perspective of a Roman in the late fourth or early fifth century) and 'modern' or near-contemporary history to provide a comprehensive overview of Roman warfare.[40]

Beyond tactics, Vegetius' inclusion of supplemental information such as ship-building (4.34–37) and navigation (4.38–43) derive from the realities of relatively recent naval warfare. Crispus constructed his entire fleet in the new military harbour at Thessalonica, so his efforts formed the background to the *Epitome*'s advice on choosing trees for cutting and not constructing ships hastily with green wood. Naturally, Vegetius describes winds and currents as essential parts of the naval battlefield. With the general advice that 'he is rarely shipwrecked who makes a thorough study of the winds' (4.38) and 'he who is going to fight a naval battle ought to find out the characteristics of the sea and locality before any encounter' (4.42), the *Epitome* draws conclusions from the treacherous 'battlefield' of the Hellespont. Here the winds and the currents take the place of the high ground that Vegetius praises as a 'large part of a victory' on land (3.13).

As with topography on land, naval geography was fundamental to what transpired in 324. From a military standpoint, the Hellespont comprises three major bottlenecks: the southern entrance near Elaius, a middle-zone between Madytus and Abydos, and the northern entrance between Callipolis and Lampsacus. Not coincidentally, the ancient sources place the two engagements of the Battle of the Hellespont at the outer chokepoints of the strait. The width of the channel is not the only tactical consideration here: the current in the Hellespont demanded the attention of every ship's captain. In general, the southwesterly current of roughly one-and-a-half knots near Elaius should have favoured Amandus' attack, but it may have pushed his ships too precipitously against Crispus' line. Conversely, it may have been as simple as Zosimus writes, 'the place was narrow and not suited to a vast number', and like the Persians at Salamis or the Ottoman warships in the Battle of Elli in 1912, the Licinian fleet could not bring its numbers to bear quickly enough to avoid defeat.[41] Following that engagement, the Constantinian ships spent the night at Elaius, while Amandus retreated to the harbour at Rhoiteion.[42] In general, the advantage of a landing and launching point on the Asian side is a reduced current, but not in this portion of the Hellespont, where the current is stronger.[43] The Licinian ships had to row out through the current to face their enemy again. While Zosimus has Amandus hesitate to attack, it is not unreasonable that his fleet actually withdrew up the strait when it became clear that Crispus had been reinforced by additional ships. Amandus' ships were carrying or escorting thousands of Licinian soldiers who could still impede the Constantinian advance from a position further up the Hellespont.

Crispus' inability to force a second engagement was a result of the geography of the strait, and Vegetius's remark about how 'the force of a tide is not overcome by the force of oars' is relevant here (4.42). Toward the ancient cities of Madytus, Abydos, and Sestus, in what is now called 'the Narrows' of the Dardanelles, the current is the strongest of any point in the Hellespont, and it runs at the same speed across the channel.[44] (The maximum current of four knots equals the steady cruising speed of the reconstructed trireme *Olympias*, so this part of the journey up the strait would have favoured the faster Licinian ships, even as it required the effort of the oarsmen against the current).[45] Sails would not have been helpful, as Zosimus tells us that Amandus sailed out into a strong north wind.[46] Vegetius notes that 'even the wind yields to [the current] on occasion' (4.42). This breeze was the etesians (called the *meltem* in Turkish), the strong, dry Aegean winds that blow for the better part of the year, mostly from the northeast in the Hellespont.[47] The sudden shift of winds that befell the Licinian fleet was unusual; in the nineteenth century, the Hydrographic Office of the Admiralty advised sailors that southwest and southeast winds are rarely encountered during the sailing season from March until September.[48] Strictly speaking, a southerly wind should have assisted Amandus in evading Crispus, and it should have been very difficult for the Licinian ships to be driven aground on the Asian side. Again, in general the Asian side is shallower but less suddenly shoaly. One place that this disaster could have occurred is the channel toward Sestus and Abydos (the Narrows), but if Amandus was able to sail toward Callipolis, he may well have been blown onto what is now the beach of Lapseki, near the ancient harbour of Lampsacus. The critical problem was the surprise. The strait here is nearly two miles wide, but Amandus may have stayed close to the Asian side for protection from a northeastern wind along with a reduced current. The abrupt shift in the wind, following hours of rowing against the current of the Hellespont to escape Crispus' ships, overwhelmed Amandus' oarsmen and completed the Constantinian victory. 'If enemy sailors are weary from lengthy rowing, if pressed by head-winds, if the tide is flowing against the ships' beaks', the *Epitome* advises an immediate attack against them, though the wind actually gained Crispus' victory before he could press the advantage (4.45). The obvious lesson from the demise of the Licinian fleet is summed up by Vegetius: 'it is advantageous for your fleet always to use the deep, open water, while the enemy's is pushed inshore' (4.46). This precept is not unique, as Polybius had drawn the same conclusion from the defeat of a Roman navy off the coast of Sicily in the First Punic War, but Vegetius combines this general instruction with the principles of sailing.[49] The *Epitome* asserts that battles take place on calm seas, but it notes the many catastrophic perils of the setting and the need for knowledge to avoid them. Instead of depending upon the nature of the battlefield for a decision, however, Vegetius preferred that the navy be reformed and improved so that ships determine their own fates with their mounted weapons, which never occurred in the Battle of the Hellespont.

IV

The naval portion of the *Epitome* may further help us to answer the contentious question of its date – one of the most discussed topics with respect to Vegetius' treatise. One scholarly approach links its position at the end of the work to its sudden and immediate relevance, and thus to the rise of Vandal naval power in the fifth century.[50] That argument fails to account for the typical position of naval precepts after siege tactics, however, in Aeneas Tacticus (now lost) and Philo of Byzantium, for instance.[51] A better *terminus ante quem* could be the naval expeditions of Gildo in 398 followed by Heraclianus' invasion of Italy by sea in 413. Vegetius offers no advice on transporting armies by sea as a prerequisite for warfare on land, though it had proven decisive in the civil wars of Julius Caesar, Septimius Severus, and Constantine.[52] As naval affairs came to revolve around delivering armies at the close of the fourth century, the *Epitome* either ignores this trend in favour of decisive naval battles, or more likely took shape prior to the dominance of naval transport. Vegetius expected that his addressee the 'invincible emperor' (4.31, *imperatore invicte*) would fight any future naval battles in the manner of Actium rather than the prelude to Pharsalus. Constantine's response to his son's victory at the Hellespont is relevant here, too: he regressed from the *Epitome*'s model of Augustus and only gained the surrender of his enemy Licinius by building a new fleet of transports. For a fourth century that had seen a lesser civil war at sea, Vegetius thus offered a programme to restore the Roman naval prowess through which *iam dudum pacato mari* ~ 'the sea has long been pacified' (4.31).

At least a century after Vegetius, Peter the Patrician remembered the civil war of the summer of 324 and made the defeated Licinius recite Homer: 'Old man, for certain it is that young warriors distress you, and your strength has been destroyed, and dire senility pursues you' (*Iliad* 8.102–103).[53] The 'momentous achievements of Crispus, Constantine's son', including the victory at the Hellespont, were still noteworthy for Peter, and Vegetius' naval appendix also reflects the importance of that battle.[54] Since the *Epitome* contains antiquarian elements deployed with an encyclopaedic purpose, it has frustrated historians who have sought to contextualise it. It omits mention of military events that modern textbooks cannot ignore, such as the catastrophic Battle of Adrianople in 378. Focusing on a specialised section of the text, however, shows a current that is less apparent in the 'sea' of the larger work: Vegetius wrote in the decades following a major naval battle, the first of its kind in centuries, and his advice on naval matters prepares his imperial addressee to be like Crispus.

Notes

1 All passages from Vegetius follow the Oxford Text of Reeve (2004), with translations from Milner (1993). All dates are CE unless otherwise indicated. In writing this chapter I must acknowledge my debt to my student Christian Moncelle, who served as an insightful interlocutor on Vegetian matters. His Undergraduate Research Assistantship with me in 2014 was supported by the Office of Student Research at Appalachian State University.
2 Lammert (1940); Emanuele (1974); Baatz and Bockius (1997); and Charles (2007).
3 Rodgers (1939) 6, blaming 'scanty accounts' of the campaign for its obscurity.

4 An exception to the conclusion of Chapter 2 that a layman's work on military affairs is mostly irrelevant to recent events, since this chapter argues that exploring the relationship between a historical battle and a specific section of a military treatise can inform our approach to the latter.

5 Whately (2015) 253–255. On civilian readers of military manuals (including Vegetius), see Chapter 3.

6 Formisano (2013) 209 and Formisano (2017).

7 *Epitome* 4.41, along with a reference to Vergil, *Georgics*, especially to 1.351–460. Milner (1996) xxvii notes the lack of repetitions.

8 On the selective editing of military treatises for Roman imperial audiences to omit or abbreviate naval theory and practice, see Murray (2012) 284–285. Vegetius depends upon a Roman past preserved in Latin rather than descriptions of Hellenistic navies in Greek.

9 Zosimus 5.20.4.

10 See Charles (2005a) on the evolution of the term 'liburnian'. Similarly to Vegetius, Zosimus 5.20.4 cites a missing passage from Polybius describing the huge 'sixes' used in the Roman and Carthaginian navies.

11 *Life of Malchus*, trans. Gray (2015) 79, with commentary at 96–102: *qui navali praelio dimicaturi sunt, ante in portu et in tranquillo mari flectunt gubernacula, remos trahunt, ferreas manus, et uncos praeparant, dispositumque per tabulata militem, pendente gradu, et labente vestigio stare firmiter assuescunt, ut quod in simulacro pugnae didicerint, in vero certamine non pertimiscant.*

12 Gray (2015) 40–41.

13 See Charles (2007) 174–180 on the proposed connection between Vegetius and the Vandals, criticised by Wheeler (2008). On naval affairs at the close of the fourth century, see Charles (2005b).

14 Starr (1989) 112 contra Rankov (1995) 78, who consider Actium to be the last of the great naval battles.

15 Lenski (2012) 76.

16 As does Sarantis (2013) 199.

17 Adrianople in the *Codex-Calendar of 354*: Salzman (1990) 141; Chalcedon in the *Consularia Constantinopolitana*: Burgess (1993) 236.

18 Socrates, *Church History* 1.4 (translation is by Zenos (1890)).

19 See the introduction in Lieu and Montserrat (1996) 39–43, and the critical edition and commentary of König (1987).

20 Photius, *Library* 98.

21 Ridley (1972) 289.

22 Ridley (1972) 290. On doublets, see Kraus (1998).

23 Eusebius, *Life of Constantine* 2.17–18.

24 Pitassi (2009) 292–293, but with the battle misdated to 323. See also Courtois (1939); Kienast (1966); Reddé (1986) 331–369; Hocker (1995); and Aiello (2002).

25 Pears (1909) 16. Against the idea of a religious war: Cameron (2011), but note the prominent opposition of Ratti (2012). The reality of the ideological struggle between Constantine and 'pagan' Rome was championed by Alföldi (1948) and now Odahl (2010) 177–182, who calls this stage of the civil war 'the eastern crusade'.

26 *Origin of Constantine* 23.

27 *Origin of Constantine* 26.

28 Lenski (2012) 76.

29 Zosimus 2.25.2.

30 *Origin of Constantine* 25.

31 Zosimus 2.22.7, trans. Ridley (1982): ἤλαυνεν διὰ τῆς Θρᾴκης, ὡς ἂν τὸ ναυτικὸν καταλάβοι.

32 Jones, Martindale, and Morris (1971) 563, s.v. 'Martinianus 2.'

33 *Origin of Constantine* 25.

34 Zosimus 2.25.2.

35 Bruun (1966) 69, n.3.
36 *Origin of Constantine* 26.
37 Zosimus 2.26.1.
38 Emanuele (1974) 90.
39 Although Lazenby (1987) contends that ancient naval battles were seldom if ever strategically significant, the importance of morale in civil wars could define exceptions.
40 From the distant past, Vegetius invokes Hannibal (1.28) and war elephants (3.24). According to the *Epitome*, Hannibal exploited the Romans' lack of continuing military preparation following the First Punic War, so the Romans should preserve esoteric knowledge such as how to defeat elephants in battle. The naval precepts in 4.31–46 comprise a similar font of specialised knowledge from centuries earlier, principally from Actium.
41 Zosimus 2.23.3, trans. Ridley (1982): οἷα τοῦ τόπου διὰ τὴν στενότητα πλήθει νεῶν οὐκ ὄντος ἐπιτηδείου.
42 Zosimus 2.23.4.
43 Hydrographic Office (1877) 38.
44 Hydrographic Office (1877) 37. This same current frustrated the British minesweepers in the Gallipoli campaign with devastating consequences for the Allied battleships that attempted to force the Straits in 1915.
45 Rankov (2012).
46 Zosimus 2.24.1.
47 Lionello (2012).
48 Hydrographic Office (1877) 39.
49 See Polybius 1.51 on the Battle of Drepana in 249 BCE, with commentary at Walbank (1957) 113–114. The infamous incident of P. Claudius Pulcher and the sacred chickens prior to this engagement occurs neither in Polybius nor Vegetius.
50 Charles (2007) 180.
51 Wheeler (2008). On Aeneas Tacticus, see Lammert (1940); on Philo, see Chapter 5.
52 See, for example, Caesar, *Civil War* 3 (on Caesar); Herodian 3.2 (on Septimius); Zosimus 2.26 (on Constantine).
53 Fragment 209 in Banchich (2015) 143; cf. Sozomen, *Church History* 1.7.
54 Fragment 209 in Banchich (2015) 143.

Bibliography

Aiello, V. 2002. 'Il controllo militare del Mediterraneo in età tetrarchica e costantiniana', in M. Khanoussi, P. Ruggeri, and C. Vismara, eds., *L'Africa romana: lo spazio marittimo del Mediterraneo occidentale, geografia storica ed economia: atti del XIV Convegno di studio, Sassari, 7–10 dicembre 2000*. Rome: Carocci editore,. 201–220.
Alföldi, A. 1948. *The Conversion of Constantine and Pagan Rome*. Oxford: Oxford University Press.
Baatz, D. and Bockius, R. 1997. *Vegetius und die römische Flotte*. Bonn: Verlag des Römisch-Germanischen Zentralmuseums.
Banchich, T. M. 2015. *The Lost History of Peter the Patrician: An Account of Rome's Imperial Past from the Age of Justinian*. Abingdon: Routledge.
Bruun, P. M. 1966. *The Roman Imperial Coinage*, Vol. 7, *Constantine and Licinius, A.D. 313–337*. London: Spink.
Burgess, R. W. 1993. *The Chronicle of Hydatius and the Consularia Constantinopolitana: Two Contemporary Accounts of the Final Years of the Roman Empire*. Oxford: Oxford University Press.
Cameron, A. 2011. *The Last Pagans of Rome*. Oxford: Oxford University Press.
Charles, M. B. 2005a. 'Vegetius on Liburnae: Naval Terminology in the Late Roman Period'. *SCI* 24. 181–193.

———— 2005b. 'Transporting the Troops in Late Antiquity: Naves Onerariae, Claudian and the Gildonic War'. *CJ* 100. 275–299.

———— 2007. *Vegetius in Context: Establishing the Date of the Epitoma Rei Militaris*. Stuttgart: Franz Steiner.

Courtois, C. 1939. 'Les politiques navales de l'empire romain'. *Revue historique* 186. 225–259.

Emanuele, P. D. 1974. *Vegetius on the Roman Navy: Translation and Commentary, Book Four, 31–46*. Master's thesis. University of British Columbia.

Formisano, M. 2013. 'Late Latin Encyclopaedism: Towards a New Paradigm for Practical Knowledge', in J. König and G. Woolf, eds., *Encyclopaedism from Antiquity to the Renaissance*. Cambridge: Cambridge University Press. 197–218.

———— 2017. 'Fragile Expertise and the Authority of the Past: The "Roman Art of War"', in J. J. König and G. Woolf, eds., *Authority and Expertise in Ancient Scientific Culture*. Cambridge: Cambridge University Press. 129–152.

Gray, C. 2015. *Jerome, Vita Malchi: Introduction, Text, Translation, and Commentary*. Oxford: Oxford University Press.

Hocker, F. M. 1995. 'Late Roman, Byzantine, and Islamic Galleys and Fleets', in R. Gardiner and J. S. Morrison, eds., *The Age of the Galley*. London: Conway Maritime Press. 86–100.

Hydrographic Office, Admiralty, Great Britain. 1877. *Sailing Directions for the Dardanelles, Sea of Marmara, and the Bosporus*. Second edition. London: Hydrographic Press.

Jones, A. H. M., Martindale, J. R., and Morris, J. 1971. *The Prosopography of the Later Roman Empire*, Vol. 1, *A.D. 260–395*. Cambridge: Cambridge University Press.

Kienast, D. 1966. *Untersuchungen zu den Kriegsflotten der römischen Kaiserzeit*. Bonn: Habelt.

König, I. 1987. *Origo Constantini: Anonymus Valesianus, Teil 1; Text und Kommentar*. Trier: Verlag Trierer Histor.

Kraus, C. S. 1998. 'Repetition and Empire in the *Ab urbe condita*', in C. Foss and P. Knox, eds., *Style and Tradition: Studies in Honor of Wendell Clausen*. Stuttgart: Franz Steiner. 264–283.

Lammert, F. 1940. 'Die älteste erhaltene Schrift über Seetaktik und ihre Beziehung zum Anonymus Byzantinus des 6. Jahrhunderts, zu Vegetius und zu Aineias'Strategika'. *Klio* 33. 271–288.

Lazenby, J. F. 1987. 'Naval Warfare in the Ancient World: Myths and Realities'. *International History Review* 9. 438–455.

Lenski, N. I. 2012. 'The Reign of Constantine', in N. I. Lenski, ed., *The Cambridge Companion to the Age of Constantine*. Revised edition. Cambridge: Cambridge University Press. 59–90.

Lieu, S. N. C. and Montserrat, D. 1996. *From Constantine to Julian: Pagan and Byzantine Views, a Source History*. London: Routledge.

Lionello, P. 2012. *The Climate of the Mediterranean Region: From the Past to the Future*. Amsterdam: Elsevier.

Milner, N. P. 1993. *Vegetius: Epitome of Military Science*. Second edition. Liverpool: Liverpool University Press.

Murray, W. M. 2012. *The Age of Titans: The Rise and Fall of the Great Hellenistic Navies*. Oxford: Oxford University Press.

Odahl, C. M. 2010. *Constantine and the Christian Empire*. Second edition. London: Routledge.

Pears, E. 1909. 'The Campaign against Paganism, A.D. 324'. *English Historical Review* 24. 1–17.

Pitassi, M. 2009. *The Navies of Rome*. Woodbridge: Boydell and Brewer.

Pohlsander, H. A. 1984. 'Crispus: Brilliant Career and Tragic End'. *Historia* 33. 79–106.

Rankov, B. 1995. 'Fleets of the Early Roman Empire, 31 BC – AD 324', in R. Gardiner and J. S. Morrison, eds., *The Age of the Galley*. London: Conway Maritime Press. 78–85.

——— 2012. 'On the Speed of Ancient Oared Ships: The Crossing of L. Aemilius Paullus from Brindisi to Corfu in 168 BC', in B. Rankov, ed., *Trireme Olympias: The Final Report*. Oxford: Oxbow Books. 145–151.

Ratti, S. 2012. *Polémiques entre païens et chrétiens*. Paris: Les Belles Lettres.

Reddé, M. 1986. *Mare Nostrum: Les infrastructures, le dispositif et l'histoire de la marine militaire sous l'empire romain*. Rome: Bibliothèque des écoles françaises de Rome.

Reeve, M. D. 2004. *Vegetius: Epitoma rei militaris*. Oxford: Oxford University Press.

Ridley, R. T. 1972. 'Zosimus the Historian'. *ByZ* 65. 277–302.

——— 1982. *Zosimus: New History*. Sydney: Australian Association for Byzantine Studies.

Rodgers, W. L. 1939. *Naval Warfare under Oars, 4th to 16th Centuries: A Study of Strategy, Tactics and Ship Design*. Annapolis: United States Naval Institute.

Salzman, M. R. 1990. *On Roman Time: The Codex-Calendar of 354 and the Rhythms of Urban Life in Late Antiquity*. Berkeley: University of California Press.

Sarantis, A. 2013. 'Tactics: A Bibliographic Essay', in A. Sarantis and N. Christie, eds., *War and Warfare in Late Antiquity*. Leiden: Brill. 177–207.

Starr, C. G. 1989. *The Influence of Sea Power on Ancient History*. Oxford: Oxford University Press.

Walbank, F. W. 1957. *A Historical Commentary on Polybius*. Vol. 1. Oxford: Oxford University Press.

Whately, C. 2015. 'The Genre and Purpose of Military Manuals in Late Antiquity', in G. Greatrex and H. Elton, eds., *Shifting Genres in Late Antiquity*. Farnham, UK: Ashgate. 249–261.

Wheeler, E. L. 2008. 'Review of Charles 2007'. *Bryn Mawr Classical Review* 2008.06.42.

Zenos, A. C. 1890. 'The Ecclesiastical History by Socrates Scholasticus', in P. Schaff and H. Wace, eds., *Nicene and Post-Nicene Fathers*, Second Series. New York: Hendrickson Publishers.

13 Justinian's warfare as role model for Byzantine warfare?

The evidence of the military manuals

Clemens Koehn

In a seminal paper on Byzantine military strategy published more than thirty years ago, Walter E. Kaegi argued for a detailed study of how the Byzantines handled strategy, operations, and stratagems.[1] He did so in the aftermath of John Keegan's *Face of Battle*, a book which led to a revival of military history studies – but which did not deal with ancient or Byzantine warfare at all.[2] Kaegi's main argumentation consists in stating that a close study of Byzantine strategy can only be done in connection to a study of Greek and Roman military thought in general, since Byzantine strategy was firmly based on the latter. Since then, not only has a large amount of textual research and critical editing of core sources for Byzantine military thinking been done, but also, a wider range of thematic studies devoted to that topic has been published, culminating in Edward Luttwak's *The Grand Strategy of the Byzantine Empire*.[3] However, those studies are much less focused on the continuities in military theory and thinking than Kaegi's initiating paper was; instead, they deal with a particular 'Byzantine way of war'. Permanently exposed to hostile pressure from outside and suffering endemic shortages of resources, the Byzantines followed a military strategy which was focused on avoiding direct confrontations, using all diplomatic options, and adapting to the military cultures of their enemies.

I

The French scholar Gilbert Dagron analysed the representation of foreign peoples in Byzantine military manuals, concluding that with the so-called *Strategicon* of Maurice a completely new manner of perceiving enemies who were fighting in very different ways was introduced in Byzantine military thinking. Instead of taking any military superiority for granted, as the ancient Greeks and Romans did, the late-sixth-century Byzantines understood well that they had constantly to adapt to the fighting techniques of their enemies (ἁρμόζεσθαι πρὸς τοὺς πολεμίους). That is why they theorised about the way to adapt properly in their military manuals, something that is absent from any ancient military literature, since Greeks and Romans felt superior and did not need to consider any possible advantages their enemies might have had.[4]

Special attitudes of the Byzantines to war and warfare are also stressed by John Haldon in his seminal book on warfare and the state in Byzantium from the

late sixth to the twelfth century. Haldon analyses at length a society which para-
doxically was rather focused on peace and philanthropy though being constantly
engaged in wars. One key element of his argumentation is the often-stressed point
that the Byzantines avoided direct confrontations and preferred militarily indirect
approaches in coping with hostile threats. Although he discussed different modes
of warfare including offensive operations, Haldon focused on the guerrilla-like
actions thematised in Byzantine military literature.[5] Central for this approach is
the treatise on 'skirmishing' from the tenth century and dealing with frontier warfare against raiding Muslims.

This particular work, which is entitled *peri paradromēs* (literally: running
alongside the enemy), is also the base for the detailed argumentation put forward
in a series of articles by the Italian scholar Gastone Breccia.[6] According to Breccia, the Byzantine way of war was completely different from the classical, Roman
tradition. Whereas the war-longing Romans epitomised 'the Western way of war'
by waging wars for defeating and consequently conquering the enemy, and therefore seeking the military decision in conventional pitched battles, the Byzantines
were rather unwilling of cultivating any war-like ethos, and therefore preferred
modes of guerrilla warfare.[7] Thus, the Byzantines were exhibiting alternative
approaches by developing a theory for this non-traditional way of war. Although
the Romans had to cope with many situations in which they had to fight more of
a guerrilla-type of war than a war filled with open battles, they never developed
theories about this other approach to warfare. Breccia argues that this was due
to the Roman concept of an honourable Just War (*bellum iustum*) which did not
allow any alternative approach.[8]

The most rigorous presentation of this scholarly tendency to see Byzantine
warfare as completely distinctive from its classical predecessors is the already-
mentioned book by Edward Luttwak, which was conceived as a sequel to his
much-discussed book on the grand strategy of the Roman Empire. Luttwak argues
that the Byzantines accepted the fact that they no longer possessed the resources
for the type of warfare and strategy their Roman predecessors could afford. Surrounded by a world of enemies whom they could hardly fight successfully alto-
gether simultaneously, the Byzantines used a wide range of different political and
military options to protect the empire. Diplomacy was central for gaining allies or
splitting hostile alliances. Militarily, their warfare aimed at sparing the precious
resources of the empire. Small yet well-equipped and trained forces were not
engaged in wars of annihilation, but rather in wars of containing the threats posed
by the many enemies, since a weaker enemy was better than a destroyed formerly
strong enemy, who might follow an even stronger enemy, whom the Byzantines
then had to fight again from scratch.[9]

In contrast to Kaegi's call for an interconnected analysis, recent scholarship
tends rather to deny the long-range continuities of Greco-Roman and Byzantine
strategies. However, any of the scholars just mentioned stick to Kaegi's argument that the Byzantine strategy was consistent over many centuries in avoiding
direct confrontation and pitched battles and in preferring wars of attrition rather
than those of annihilation – wars characterised by the intensive use of stratagems,
ambushes, trickery, and guerrilla-like modes of warfare. Already Kaegi claimed

not the invention but the exemplarity of this kind of warfare for the wars waged by Justinian I in the sixth century: 'The mode of warfare that already was evident in the reign of Justinian was to be the dominant form of Byzantine warfare for more than five hundred years.'[10] Toeing the same line, Breccia wrote: 'The decisive moment for the transformation of Byzantine military theory was certainly the sixth century: Belisarius is already its advocate.'[11] And again, it is put more or less in the same words by Luttwak: 'It was with Belisarios that [successful stratagems] first became a Byzantine speciality, to so remain for centuries to come'.[12]

This paper will discuss the theorem of a clearly discernible Justinianic pattern in Byzantine warfare by exploring to what extent the extant military literature compiled by later generations of Byzantines shows any influence of the way Justinian's generals and armies fought their way back through the former Western provinces of the Roman Empire. Earlier in this volume, Hans Michael Schellenberg has articulated substantial doubts about the value of the ancient military literature as a source for understanding warfare historically. While these doubts are generally fully justified, the case of Byzantine 'manuals' is in some important aspects different. Firstly, they are usually written by persons having a more or less strong connection to the military and are in one way or the other practitioners of warfare (emperors, generals, officers); and secondly, they are claiming two important features mostly absent from ancient manuals: they are explicitly designed to fix a military practice in use for some time in written form (as Nikephorus' *paradrome* tactic or Constantine's campaign order) and they are regarded not as an 'adequate substitute for actual practice' (Schellenberg) but rather as its complimentary (as Mauricius' and Leo's manuals). Thus, they can be considered as valuable sources, even if they provide only little actual information on historical military events.[13]

II

The so-called age of Justinian is a pivotal period in the transition from antiquity to the medieval age. His large-scale attempt to restore politically, religiously, and legally the lost splendours of the *Imperium Romanum* failed in the end.[14] But his long reign represents a culmination of the old imperial traditions and at the same time their transformation. So, to what extent are his wars already exemplary for Byzantine warfare?

In fact, the campaigns in the West to conquer the lost provinces occupied by Germanic realms such as the Vandal and Ostrogothic kingdoms were fought by relatively small armies which were often outnumbered by their adversaries. Nevertheless, 'the Byzantines succeeded in exploiting some importantly perceived asymmetries in weaponry and fighting techniques between themselves and their adversaries', as Walter Kaegi aptly remarked.[15] Crucial for this kind of 'lean warfare' were highly mobile cavalry units capable of fighting both in close combat and on distance. And Belisarius can certainly be regarded as the exponent of a warfare of trickery and attrition, 'a protracted struggle by means of wiles and clever contrivances', to quote the eye-witness of those events, Procopius of Caesarea.[16] Furthermore, it has to be admitted that most of the Byzantine military action – especially on the Eastern front during the seventh and eighth century – was exactly that kind of guerrilla-like warfare postulated in modern scholarship. Many campaigns were rather small-scale events targeting not the destruction of the enemy in open field battles but on shadowing and repelling raiding Arab invaders who were trying to get away quickly with their booty. And this warfare is described by contemporary Byzantine military manuals such as the one explaining the *paradrome* tactics.

So far, those are the obvious similarities. But can we really say that this is the heritage of Justinian's wars? What was really left of its strategic implementations? Certainly, we find a direct influence on military developments in the time after Justinian. The *Strategicon* of Maurice, the most significant and influential Byzantine military manual we have, deals at length with cavalry warfare which still bears the marks of the fighting experiences gained by Justinian's army. There is good ground for the argument put forward by the Russian scholar Petr Shuvalov that one of the central manoeuvres described in the *Strategicon*, the surprise attack of a hidden second wing, goes back to tactical ideas evolved in leading military circles around Justinian's cousin Germanus.[17] But other than that, the influence seems rather minimal. Belisarius, the obvious exponent of the new Byzantine way of war, is mentioned only by one manual, the *peri strategikes* which nowadays is attributed to the ninth century author Syrianus Magister. Interestingly, he gets mentioned as favouring a certain stratagem (to attack the enemy when he is forced to split his forces due to supply shortages), which does not fit with what we know from Procopius' narrative about Belisarius' stratagems.[18]

It is different when we look at the narratives of warfare and fighting in Byzantine historiographical sources. Here, we find reminiscences of themes well known from the Justinianic context. When Leo Phocas addresses his troops on the eve of a confrontation with the Arabs in 960 CE, reminding them that wars are not necessarily won in open battles but by clever planning, that one should not rush into danger but consider his actions carefully, and that the best way to defeat the enemy is to take him by surprise, then this speech echoes some of Belisarius' stratagems. But does this mean that Leo Phocas is showing his awareness of finding himself in a long-standing military tradition which can be traced back to the time of Justinian's wars?[19] Hardly likely; it is rather that the historian narrating these events, Leo Diaconus, is moulding this speech in the patterns he found in his famous historiographical predecessor. So, Leo Phocas is not following the role model of Belisarius exhibiting strategies of battle avoidance and clever contrivances, but Leo Diaconus is following the role model of Procopius in composing adhortative speeches addressed to the soldiers.[20] Equally, when the chronicler Georgios Cedrenos in the eleventh century is referencing Procopius' history of Justinian's wars as 'the stratagems of Belisarius' (τὰ τοῦ Βελισαρίου στρατηγήματα), does that show a distinctive reception of Justinian's wars as a role model for contemporary Byzantine warfare (1)?[21] Certainly not; it rather shows the success Procopius had in focussing on his real hero, who was not Justinian but his boss Belisarius. Already the church historian Evagrius in the late sixth century cites the work as τὰ κατὰ Βελισάριον (4.12).[22]

Despite obvious similarities and a wider reception of the master narrative of Justinian's wars as written by Procopius, there are more discontinuities than continuities between Justinianic warfare and later Byzantine warfare. Certainly, the biggest difference consists in the weaponry branches used for combat actions. Justinianic warfare is famously characterised by highly mobile and versatile cavalry units consisting of mounted archers, who occasionally are called *hippotoxotai* by our sources. The fighting capabilities of these *hippotoxotai* are described in a much-discussed passage from the preface of Procopius' *Persian War*:

> The bowmen of the present time go into battle wearing corselets and fitted out with greaves which extend up to the knee. From the right side hang their arrows, from the other the sword. And there are some who have a lance also attached to them and, at the shoulders, a sort of small shield without a grip, such as to cover the region of the face and neck. They are expert horsemen, and are able without difficulty to direct their bows to either side while riding at full speed, and to shoot an opponent whether in pursuit or in flight. They draw the bowstring along by the forehead about opposite the right ear, thereby charging the arrow with such an impetus as to kill whoever stands in the way, shield and corselet alike having no power to check its force.
>
> (1.1.12–15; translation by H. Dewing)[23]

We cannot go here into recent discussions about the function of this lengthy representation for the historiographical aims and purposes of Procopius.[24] Of course, we have to be aware that only a small part of the cavalry units of Justinian's armies consisted of this type of hybrid warriors. Their availability might have been one of the reasons why the expeditionary forces sent by Justinian were relatively small. And we have to be aware that this combined type of weaponry branch – armoured horse archers fighting both long-range and close-range – did not have a long tradition in the Roman army. Surely, *hippotoxotai* were used since the early imperial days, first as auxiliary units, later also as regular units: Arrian mentions in his *Tactics* that Hadrian gave orders to adopt fighting techniques of Parthian and Armenian *hippotoxotai* for the Roman cavalry (44.1). But they were essentially a light-armed troop designed for skirmishing and long-range encounters.

In regard to the combined capabilities of Justinian's horse archers to fight both in close combat and on distance, we can therefore safely assume that this was the product of a rather recent development which was triggered by the encounter that the Romans had with Hunnic warfare.[25] This becomes clear also from the fact that this weaponry branch underwent some significant changes during Justinian's reign. As Procopius states in his preface (written around 540 CE), some of them had lances additionally to bow and sword (εἰσὶ δὲ οἷς καὶ δόρυ προσαποκρέμαται). Agathias gives for a later period of Justinian's wars an albeit shorter description according to which the horsemen deployed by Narses in Italy were 'carrying spears, and shields, while a bow and arrows and a sword hung at their sides; a few of them also held lances' (2.8.1; tr. J.D. Fendo).[26] According to the *Strategicon* of Maurice from the end of the sixth century, those cavalry men were all equipped with lances designed both for thrusting and throwing.[27] The fact that there is a development going from horse archers partially equipped with lances (as in Procopius), to horse archers still partially equipped with lances but now regularly carrying spears (as in Agathias), to cavalrymen carrying bow and lance and thus finally performing the roles both of archers and lancers (as in the *Strategicon*) implies that the tactical role of Justinian's horse archers was not yet as fixed as it was later in the sixth century.[28]

Modern scholars are tempted to see the introduction of this type of highly versatile cavalry as a crucial phase for the development of Byzantine small-war tactics.[29] But when we have a closer look on their role in the military manuals, then we will see that they are almost exclusively a specialty of Justinianic warfare. In the following centuries, the military tasks of the cavalry were again diverging.[30] In the same way that the Justinianic horse archers were obviously an amalgam of different types of cavalry in order to combine different tactical roles, this amalgam dissolved again into its component parts. The military manuals illustrate this process well. The *Strategicon* of Maurice in particular represents clearly the culmination of this process of amalgamation. The cavalrymen were able to fight both as lancers and archers, a capability only partially gained by Justinian's wonder fighters. While the Byzantine military literature is heavily influenced by the *Strategicon*, it becomes nonetheless evident that the changes within the weaponry branches are quite significant.

Regarding mounted archery, the *Tactics* of Leo VI largely copy the prescriptions of the *Strategicon*.[31] But there are fine differences. In the *Strategicon*, there is an exemption provided for non-Roman recruits who are not able to shoot properly with the bow and who are allowed to have lance and shield only; in the *Tactica*, this exemption is granted to all recruits.[32] However, all Roman recruits (up to forty years old) regardless of their shooting capabilities are urged to have bow and quiver.[33] Leo gives an explanation: 'The fact that archery has been completely neglected and fallen into disuse among the Romans has caused a great deal of harm nowadays'.[34] Even if we assume that in the sixth and early seven century cavalrymen with wide-ranging capabilities were more an ideal than a reality, Leo's remark shows the difficulties in maintaining former standards.[35] Apparently, the Byzantines were no longer able to raise larger units of versatile *hippotoxotai* of the sixth-century type.[36]

In this context, it is interesting to see Leo supplementing the information he found in the *Strategicon* with older tactical literature:

> I do not think it unprofitable to call to mind, even briefly, the ancient armament of the infantry and the cavalry, as Aelian and the other authors on *tactica* have divided it. Among the ancients, the commanders divided the cavalry force according to two different kinds of armament, the one called heavy armed and the other not heavy armed.
>
> (6.25; translation by G. T. Dennis)[37]

He then speaks about *kataphraktoi* (heavy-armed cavalry) and *akrobolistai* (light-armed cavalry), who consisted of *hippakontistai*, mounted javelin men, and *hippotoxotai*, mounted archers. This is exactly what Aelian and Asclepiodotus are describing.[38] On first glance, this seems to be an odd reference. But on second glance, it becomes clear that the information of the older literature seems to serve – at least to a certain extent – the necessities of contemporary warfare much better than the highly demanding conception of the *Strategicon*. 'This much, then', concludes Leo, 'about the armament of each individual soldier we have read in the ancient and more recent tacticians. We have organized it and defined it so that, possessing this knowledge, you may choose what is beneficial' (6.35).[39] Leo, thus, is leaving it to his officers to look for the best option to follow.[40]

The process of divergence is further reflected in the tenth-century military literature. The anonymous *Sylloge Tacticorum* focuses on very heavily armed cavalry, *kataphraktoi*, who are equipped with lance, shield, and two swords. Additionally, they have a bow and quiver. Next to the *kataphraktoi*, the *Sylloge Tacticorum* mentions *doryphoroi*, with less heavy armour and two lances, swords, and axes. Finally, there are two types of light-armed cavalry: *akontistai*, who apart from the lance and shield use two or three javelins, and *hippotoxotai*, who use bow and sword (31). The portfolio of arms is interesting. In principle, this manual seems to give evidence that the Byzantines of that period still stick to the concept of multi-role cavalry by equipping the main heavy-armed cavalry force with bow and quiver. So, apparently, this is still in line with Leo's chapters copied from Maurice's *Strategicon*. However, given the heavy armour and armament, it seems not very plausible to see those cavalrymen engaged according to the same tactical premises as the *hippotoxotai* of the sixth century. If they used their bows, then certainly only in those occasions when they dismounted and fought on foot. The main task of the heavy-armed *kataphraktoi* was to break through the enemy battle line.[41] The distance fighting was assigned again to light-armed cavalrymen, who were used only for that purpose. The tenth-century *hippotoxotai* are again the kind of cavalry they had been before the tactical revolution of the late fifth and early sixth century, when light-armed horse archers were transformed into heavier-armed multipurpose troopers by equipping them with proper armour, shields, and lances.[42]

This development is taken further in the short military treatise of Nicephorus Phocas, a treatise influenced by the *Sylloge Tacticorum*. Here again, the main weaponry branch of the cavalry is comprised of heavy, armed *kataphraktoi*, whose principal weapon, apart from lances and various types of swords, were maces; they do not have the bow and quiver anymore. Next to them, the *Praecepta* prescribe *hippotoxotai* who are armoured (of course, to a lesser extent than the *kataphraktoi*) and armed with bow and sword. Additionally, there are also *akontistai* armed with javelins, shields and swords.[43] Although the *hippotoxotai* here seem to be more heavily armoured than those described in the *Sylloge Tacticorum*, and although they are prescribed to operate in the same tactical formation as the *kataphraktoi*, they only have one main tactical role: to fight from a distance. From the disposition given in the *Praecepta*, it is clear that they had to fend off any arrow barrage of the enemy, and not to attack in close combat, which was the task of the *kataphraktoi*.

As we can see, the high time of Justinian's wonder fighter so much praised by Procopius was a rather short one. Instead of continuing the use of amalgamated versatile cavalry, which reached its peak in Maurice's *Strategicon*, the later Byzantines split again the tactical roles and used horse archers for distance fighting only. Given the porous state of our sources, it is hard to determine what the stages of this process have been. At the crushing defeat at the battle of Dazion in 838, the horse archers seem already to be in decline or underperforming.[44] However, the notion of an ideal warrior, which is connected with that type of weaponry branch, did indeed have a long-lasting influence, but much more in the area of art and literature than in the military realities of Byzantine warfare.[45] As suitable as those multi-role cavalrymen seem for the *paradrome* tactics so favoured by modern scholars, they were only one type of weaponry branch among others the Byzantines could use to gain their objectives. For example, infantry plays a prominent role in the *paradrome* tactics, a concept developed to cope with raiding armies which were usually hard to force into a direct confrontation on the battlefield.[46] On the other side, when possible, the commanders using the *paradrome* tactics had to fight in the battle line, if they had enough troops at their disposal.[47] The word *parataxis*, battle order, is as present in this manual as it is in Byzantine manuals in general, which were primarily in its core parts conceived as battle instructions.[48] The later Byzantines did not cultivate anymore a military monoculture as the early Byzantines did by trying to homogenise the cavalry in being a truly all-rounder weaponry branch.[49]

III

So far, by drawing upon information provided by Byzantine military manuals on particular weaponry and tactics, we have discussed what separates Justinianic warfare from later Byzantine warfare. To conclude, we should come back to Walter Kaegi's advice to look on Byzantine strategy as firmly rooted in the Greco-Roman tradition, advice largely neglected by recent scholarship. Three aspects are stressed by modern scholarship when trying to separate the Byzantine from the Greco-Roman tradition, in order to demonstrate the 'Byzantine way of war' as opposed to the 'Western way of war': firstly, war seen as an evil which has to be avoided if possible; secondly, the advice to use indirect methods such as ambushes and guerrilla-type actions as opposed to direct methods, such as confronting the enemy in pitched battles; and thirdly, consideration of the enemy's strengths and adaptation to them, the ἁρμόζεσθαι πρὸς τοὺς πολεμίους.[50] Without getting into a deeper discussion, here only two points shall be discussed in order to demonstrate that there is more continuity than disrupture in strategic thinking from antiquity to Byzantine times.

First, let us have a short look on one particular Classical author, who is a fine military writer and represents both the Greek and the Roman tradition: Polybius. For all three statements mentioned previously and identified as signature features of Byzantine strategy by modern scholars, we can find a corresponding quotation in Polybius. War is a terrible thing, he writes in book four (in order to criticize the Messenians to avoid it at all costs, something which a Byzantine would have seconded without doubt).[51] His long digression on the usefulness of mathematical knowledge, which he ends by pointing to a larger discussion of it in his own treatise on tactics (now lost), is introduced by stating 'that in military operations what is achieved openly and by force is much less than what is done by stratagem and the use of opportunity, can easily be learned from the history of former wars' (tr. W. Paton).[52] For the third point, the adoption of enemy fighting techniques, there are plenty of examples in Polybius: the adaption of Greek cavalry equipment and Spanish swords by the Romans, the manipular tactic employed by Hannibal, and so on.[53] For the Romans themselves, adaptability became an internalised myth as narrated by the *Ineditum Vaticanum* and other texts.[54] The Roman concept of *disciplina* in total is based on the understanding that it is necessary to create a constant learning and adaptation process in order to make the soldiers able to cope with all eventualities. It is not by chance that the first Roman military manual, entitled *On Military Matters* (*De re militari*) and written by the elder Cato, is often cited as *On Military Discipline* (*de disciplina militari*).[55] In this context, the assumption that the Romans did not develop any theory of small-scale warfare is less likely. The problem is, of course, that, apart from Vegetius and the strategemata literature in the style of Frontinus and Polyaenus, we do not have any extant Roman military manuals. That said, even the scanty fragments of Cato's *On Military Matters* give a hint of fighting techniques which do not rely on open battle tactics.[56]

Second, let us look closer on one of the three statements in particular. As it is often stated, the Byzantines stuck to the idea that it is better to overcome the enemy by clever planning and use of strategy than by attacking him directly in pitched battles, which puts oneself always at risk. This idea is very clearly expressed in Maurice's *Strategicon* as well as in Leo's *Tactics*.[57] If we follow the crowd in considering Justinian's strategic approach as foil for later Byzantine times, we could argue that this is quite obviously part of the heritage of his reign. For the latter seems to provide a striking example: When confronted with an invasion of Hunnic Cutrigurs, Justinian did not – despite initial military success – face them in a direct confrontation but set up another Hunnic group against them by instigating a rivalry between the two groups. As Agathias remarks, the emperor was praised for his good planning which brought a great success without using military resources. Menander the Guardsman makes a similar statement in narrating a potential conflict between Byzantium and the Avars at about the same time late in Justinian's reign, describing the aged emperor as avoiding direct military confrontation and intending to fence off the threat by clever planning.[58] Both authors use the word εὐβουλία as keyword for expressing the opposite of potential military options, while Maurice and Leo speak of βουλή to designate the alternative to direct attack and open battle. So far, the continuity in the approach seems to be clear. Yet this was hardly an innovative idea. The same word, εὐβουλία, is used by Polyaenus in his second-century collection of stratagems, who defines it as virtue (ἀνδρεία) when fighting the enemy in open battle but as good planning (εὐβουλία) when overcoming him by stratagems and trickery, thereby saying that the first wisdom of capable generals consists in gaining the victory without danger, that is, without risking a battle.[59] The Greco-Roman heritage of strategic thinking had, as already Kaegi argued, a much longer life than the recent advocates of a peculiar 'Byzantine way of war' acknowledge. In pursuing the tracks of continuity, one should not only scrutinise those bearing the marks of a continuing use of peculiar tactics or stratagems, but rather concentrate on the general principles of war, which, as Kaegi in alluding to Clausewitz pointed out, are simple but not always easy to implement. These principles are as present in the Byzantine tradition of military literature as they had been in the ancient.[60] The overall observation remains untouched from considering the fact that the latter apparently did not yet contain a strategical approach on any geopolitical level, which is usually regarded as the innovative contribution of Byzantine military theory.[61]

IV

To conclude: Byzantine warfare certainly had its peculiarities, but it is neither moulded on the patterns of Justinian's wars, nor is it disconnected from the Greco-Roman tradition of military thinking. To be sure, the wars of Justinian and those of his successors in later periods seem to have many features in common: the use of stratagems, ambushes, faked retreats, trickery, asymmetrical operations, raiding, fighting from strategical strongholds, and so forth. But the political, strategic, and military conditions were completely different. Justinian exploited the military developments of his age to fight successfully Germanic enemies against whom, at least at the beginning, he had a strategic advantage, whereas the later Byzantines had to cope with a multitude of different enemy cultures who often forced their strategic initiatives upon the Byzantines. However, one should not forget that as soon as the latter could regain the strategic initiative in the tenth century, they switched to a more aggressive military policy and focused on offensive actions and direct confrontation, which led to the Macedonian renaissance in Byzantine power.[62]

Notes

1 Kaegi (1983). Translations from Procopius come from the Loeb Classical Library editions.
2 Keegan (1976).
3 Luttwak (2009).
4 Dagron (1987) 207–232. On 'military literature', see Chapter 2.
5 Haldon (1999).
6 Breccia (2001), (2007), and (2008).
7 Breccia models his argumentation on the extremely influential theory by Hanson (1989), who argues that the seeking of military decisions in open battles is a Greek invention which had a long-lasting impact on Occidental warfare. The whole idea is heavily criticised, among others, by Wheeler (2007) and Lynn (2003).
8 Interestingly, Breccia does not cite Traina (1986–1987), who argues in a similar way in regard of the obvious lack of theorising about guerrilla warfare in Roman sources,

but who rightly concludes 'La retorica letteraria della guerra giusta contraposta alla guerriglia barbara è quindi un artificio illusorio che nasconde la realtà delle operazioni militari' (267). On the just war and the military manual, see Chlup (2014).

9   Luttwak (2009).

10 Kaegi (1983) 255, but pointing to the fact that Justinian's warfare was influenced by the long-standing conflict between Rome and the Sassanids.

11 Breccia (2008) 109 n.113: 'Il momento decisivo per la trasformazione della teoria militare bizantina é certamente il sesto seculo: Belisario ne é già un fautore.'

12 Luttwak (2009) 55. The 'Byzantine way of war' concept has been now challenged by Graff (2016), who by comparing Chinese and Byzantine military strategies argues that there is no such thing as a special 'military culture', but that both states pursued the same military policies since both had to cope with the same kind of enemies (Nomads or tribal societies) and their ways to fight.

13 It is sufficient to point here to the fact that terms like πεῖρα/ἐμπειρία are much more frequent in Byzantine manuals than in the ancient ones. For an extensive discussion of literary vs. practical designs of Byzantine manuals, see now Chatzelis (2019), who stresses their practicability.

14 The old question, whether Justinian's politics were based on strategic considerations or mere situational manoeuvres (cf., as early as, Montesquieu [1748/1890]: 202: 'Ces conquêtes, qui avaient pour cause non la force de l'empire, mais de certaines circonstances particulières') is recently answered again in favour of the latter by Heather (2018); I myself tried to argue a more nuanced interpretation (Koehn 2018), in which the pursuing of strategic considerations and using of situational manoeuvres are seen less as a contrast, but rather as complimentary actions.

15 Kaegi (1983) 7.

16 *Gothic War* 3.8.11: ἐν πεδίῳ γὰρ μᾶλλον ἐκ τοῦ εὐθέος διακρίνεσθαι μάχῃ πρὸς αὐτοὺς ἤθελεν (sc. Totila) ἢ τέχναις τισὶ καὶ σοφίσμασι διαμάχεσθαι.

17 Maurice 3.14; cf. Procopius, *Persian War* 1.13.21 and *Vandal War* 2.17 with Shuvalov (2006). Shuvalov's intriguing but somewhat flawed theory about different stages in the textual evolution of Maurice's *Strategicon* is critiqued by Rance (2017b) 218 n.2 and 221 n.12.

18 Syrianus Magister 33. Belisarius is also mentioned by Cecaumenus in chapter 16 of his little treatise originally conceived for his son and usually cited as 'Strategikon', but the name is miswritten (βουλησάριος). Rance (2007a) 710 n.29 was not aware that already Vassilievsky and Jernstedt in their *editio princeps* of this text (St. Petersburg 1896) in the index (p. 106) had explained βουλησάριος as *Belisarius*, and that G.G. Litavrin in his edition (Sovety i rasskasy Kekavmena, Moskow 1972) had put the emended name into the text (without, however, mentioning his predecessors), arguing (p. 363) that Cecaumenus should have known the correct spelling of Belisarius' name, and therefore βουλησάριος has to be regarded as miswriting of the copyist. Nevertheless, Rance is right in stating that 'strictly speaking Cecaumenus' work is more aristocratic counsel than tactical handbook'; thus, this attestation does not invalidate the argument that Belisarius did not leave many traces in the military literature of Byzantium.

19 Leo Diaconus 2.3; reminiscences to Belisarius' speeches in Procopius: Procopius, *Persian War* 2.16.6–7; 19.10; *Vandal War* 1.15.25; reflecting tactical traditions: Breccia (2008) 109.

20 Alternatively, taking the speech at face value, one could argue that Leo Phocas, as a good general, knew the military handbooks and is referring with his remarks, rather than to Belisarius, to prescriptions such as Maurice 8.1.7, 8.2.4, 8.2.68; Leo, *Tactics* 20.11, 20.51, 20.132.

21 τῷ ς´ καὶ ζ´ ἔτει τὰ τοῦ Βελισαρίου στρατηγήματα κατὰ Λιβύην ἐγένοντο, ἅτινα Προκόπιος ὁ Καισαρεὺς ἐν η´ βιβλίοις συνεγράψατο.

22 γέγραπται δὲ Προκοπίῳ τῷ ῥήτορι τὰ κατὰ Βελισάριον συγγράφοντι.

23 οἱ δέ γε τανῦν τοξόται ἴασι μὲν ἐς μάχην τεθωρακισμένοι τε καὶ κνημῖδας ἐναρμοσάμενοι μέχρι ἐς γόνυ. ἤρτηται δὲ αὐτοῖς ἀπὸ μὲν τῆς δεξιᾶς πλευρᾶς τὰ βέλη, ἀπὸ δὲ τῆς ἑτέρας τὸ ξίφος. εἰσὶ δὲ οἷς καὶ δόρυ προσαποκρέμαται καὶ βραχεῖά τις ἐπὶ τῶν ὤμων ἀσπὶς ὀχάνου χωρὶς, οἵα τά τε ἀμφὶ τὸ πρόσωπον καὶ <τὸν> αὐχένα ἐπικαλύπτειν. ἱππεύονται δὲ ὡς ἄριστα καὶ θέοντος αὐτοῖς ὡς τάχιστα τοῦ ἵππου τὰ τόξα τε οὐ χαλεπῶς ἐντείνειν οἷοί τέ εἰσιν ἐφ' ἑκάτερα καὶ διώκοντάς τε βάλλειν τοὺς πολεμίους καὶ φεύγοντας. ἕλκεται δὲ αὐτοῖς κατὰ τὸ μέτωπον ἡ νευρὰ παρ' αὐτὸ μάλιστα τῶν ὤτων τὸ δεξιὸν, τοσαύτης ἀλκῆς ἐμπιπλᾶσα τὸ βέλος, ὥστε τὸν ἀεὶ παραπίπτοντα κτείνειν, οὔτε ἀσπίδος ἴσως οὔτε θώρακος ἀποκρούεσθαί τι δυναμένου τῆς ῥύμης.

24 Cf. Kaldellis (2004–2005); Petitjean (2014); Whately (2016) 181–187; Kruse (2017); and Basso and Greatrex (2018). See also the lengthy discussion in my book on Justinian and the early Byzantine army, where I argue that Procopius' comparison between the contemporary horse archer and Homeric archers follows models of comparing and contrasting developed by Hellenistic scholars commenting on Homer's works; using such models gave Procopius a proper literary frame to counter the contemporary critique of Justinian's cavalry-based warfare: Koehn (2018) 176–188.

25 See Janniard (2015) and (2018).

26 οἱ μὲν οὖν ἱππεῖς ἑκατέρωθεν ἐπὶ τῶν ἄκρων ἐτετάχατο δοράτια φέροντες καὶ πέλτας τόξα τε καὶ ξίφη παρηωρημένοι· ἦσαν δὲ οἳ καὶ σαρίσας ἐκράτουν.

27 Maurice 1.2.

28 Cf. Koehn (2016).

29 Breccia (2008) 69 and 71 and Alofs (2014) 425.

30 For middle Byzantine warfare, see Kühn (1991); McGeer (2008); and Strässle (2006); Birkenmeier (2002); and the relevant chapters in Haldon (2007). For the weaponry of that period, see Dawson (2002).

31 Leo, *Tactics* 6.1–18. For differences in ideological distinctives of Leo's *Tactics* compared to Maurice, see Chapter 14.

32 Maurice 1.2.21f; Leo, *Tactics* 6.3.

33 Maurice 1.2.28–30; Leo, *Tactics* 6.5. A similar deviation from Maurice's template, which changes significantly the tactical outcome, can be found in 12.36: while Maurice prescribes that the soldiers in the middle of the formation should be fighting as archers without using their shields (2.8), Leo says they should not use their lances; in the former's prescription, those soldiers would have been still able to fight close quarters (though lacking their shields if the shield is not attached somehow, which Leo seems to say for training).

34 Leo, *Tactics* 6.5: τῆς γὰρ τοξείας παντελῶς ἀμεληθείσης καὶ διαπεσούσης ἐν τοῖς Ῥωμαίοις τὰ πολλὰ νῦν εἴωθε σφάλματα γίνεσθαι. Cf. ibid. 11.41, with the remarks by Haldon (2014) 160–161.

35 Cf. Haldon (1999) 215–217, esp. 216: 'The Byzantine composite lancer/horse archer is probably something of a myth'; Rance (2007b) 358: 'It is probable that Maurice's 'composite archer-lancer' was also something of an ideal'. Kaldellis (2004–2005) 190–204 discards the existence of horse archers at once.

36 This does not mean that mounted bowmen disappear completely: cf. Kaegi (1964) 101.

37 οὐκ ἄχρηστον δέ μοι δοκεῖ μνημονεῦσαι, κἂν ἐν μικρῷ, καὶ τῆς ἀρχαίας ὁπλίσεως τῶν πεζῶν καὶ τῶν καβαλλαρίων, καθὼς Αἰλιανός τε καὶ οἱ λοιποὶ τῶν τακτικῶν συγγραφεῖς ὑπηγόρευσαν. παρὰ γὰρ τοῖς ἀρχαίοις τὴν μὲν ἱππικὴν τάξιν εἰς δύο διαφορὰς ὁπλίσεων ἐποίουν οἱ στρατηγοί, μίαν μὲν κατάφρακτον λεγομένην καὶ τὴν ἑτέραν οὐ κατάφρακτον.

38 Asclepidotus 1.2; Aelian 2.11–13. Leo refers with his remark rather to Aelian than to Asclepiodotus. The latter does not figure among his sources which were basically the treatises of Maurice, Aelian, and Onasander; for the latest detailed discussion, see Rance (2017b) 295–297, who argues that Leo used these three texts because they were the ones available at that time; this argument is based on the observation that these

authors are transmitted in several majuscule manuscripts which testify their continuous reading/copying in Late Antique and Early Byzantine times.

39 τοσαῦτα μὲν οὖν καὶ περὶ τῆς καθ' ἕνα ἄνδρα στρατιώτην ὁπλίσεως ἔκ τε τῶν παλαιῶν καὶ τῶν νεωτέρων τακτικῶν ἀναλεξάμενοι διεταξάμεθά τε καὶ διωρισάμεθα, ἵνα ἔχων αὐτῶν τὴν γνῶσιν ἐκλέγῃ τὸ χρήσιμον.

40 It is interesting to see that Leo operates in a similar way in Constitution 18, where he deals with different tactics for different types of enemies. In 18.135, he concludes his long discussion by adding that he gives now additional information for consideration to the general, summarizing the basic principles of standard tactics already given in Constitution 12.

41 For the *kataphraktoi*, see Wojnowski (2012).

42 For this earlier process, cf. Koehn (2016).

43 Nicephorus 3.7–9.

44 Cf. Theophanes Continuatus, *History* 3.31. The Byzantine Army could not counter the barrage of arrows shot by the elite corps of Turkish horse archers and failed in coming to close quarters with the Arab army in order to gain advantage of their numerical superiority; that means their own archers were unable to perform as the enemy could and that the Byzantines had to seek a decision in close-combat action.

45 Cf. Theodorus Prodromus, *Historical Poems* 37.70: 'ironclad mounted archer' (ὅλος σιδηρόφρακτος ἱπποτοξότης, in honour of a Byzantine officer killed in action). For the iconography, see Grotowski (2010).

46 Nicephorus 3, 9, 10, 11, 14.

47 Cf. Nicephorus 19–20.

48 Nicephorus 9, 10, 14, 16, 17, 19.

49 Koehn (2018) argues that Justinian's concept of waging lean or limited wars characterised by small but efficient expeditionary forces was based on the increasing use of multiversal troops such as cavalry able to fight both on long range and close range (see esp. ch. II 3 on the hippotoxotai).

50 These aspects are based on sources such as Syrianus Magister 4 ('war as evil'); Leo, *Tactics* 20.11 ('avoiding direct confrontation/battle and using stratagems instead'); Maurice 7.1 and 11.

51 Polybius 4.31.3: 'I admit, indeed, that war is a terrible thing' ~ ἐγὼ γὰρ φοβερὸν μὲν εἶναί φημι τὸν πόλεμον. This is not Polybius' view alone: cf. Thucydides 4.59.2: 'war is an evil is a proposition so familiar to everyone that it would be tedious to develop it' ~ καὶ περὶ μὲν τοῦ πολεμεῖν ὡς χαλεπὸν τί ἄν τις πᾶν τὸ ἐνὸν ἐκλέγων ἐν εἰδόσι μακρηγοροίη; Euripides, *Trojan Women* 400: 'Whoever is wise would flee from war' ~ φεύγειν μὲν οὖν χρὴ πόλεμον ὅστις εὖ φρονεῖ.

52 Polybius 9.12.2, ὅτι μὲν οὖν ἐστὶ τῶν κατὰ πόλεμον ἔργων ἐλάττω τὰ προδήλως καὶ μετὰ βίας ἐπιτελούμενα τῶν μετὰ δόλου καὶ σὺν καιρῷ πραττομένων, εὐχερὲς τῷ βουλομένῳ καταμαθεῖν ἐκ τῶν ἤδη γεγονότων. His treatise on *Tactics* gets mentioned at 9.20.4. Cf. Polyaenus 1 Preface 3: ἔστι πρώτη δεινῶν στρατηγῶν σοφία κτᾶσθαι τὴν νίκην ἀκίνδυνον. The Greco-Roman 'linguistics' of stratagems and trickery is analysed Wheeler (1988); see also Sheldon (2012).

53 Greek cavalry lances: Polybius 6.25.11: ἃ συνιδόντες ἐμιμήσαντο (sc. the Romans) ταχέως· ἀγαθοὶ γάρ, εἰ καί τινες ἕτεροι, μεταλαβεῖν ἔθη καὶ ζηλῶσαι τὸ βέλτιον καὶ Ῥωμαῖοι. Spanish swords: Pol. fr. 179 B.-W.: καὶ τὴν μὲν κατασκευὴν μετέλαβον (sc. the Romans), αὐτὴν δὲ τὴν χρηστότητα τοῦ σιδήρου καὶ τὴν ἄλλην ἐπιμέλειαν οὐδαμῶς δύνανται μιμεῖσθαι.

54 Ined. Vat. FGrHist 839 F 3 (Caeso giving a speech to Carthaginian envoys in which he argues that the Romans would be able to fight the Carthaginians at sea by stressing the point that Rome always overcomes her enemies by adapting their weaponry and tactics); cf. Posidonius 87 F 59; Diodorus 23.3. The theory of Humm (2007), that this idea goes back to Timaeus of Tauromenium, is not convincing, since the speech is composed with the knowledge of the later Roman victory in mind, and Timaeus' work

has hardly covered more than the beginnings of the first Punic war. It is more probable that the idea is a genuinely Roman one (cf. also Polybius' remarks cited in n.53, which make clear that he from a Greek perspective did not think of the Romans as especially more adaptive or more able to adapt than others).

55 For the significance of *disciplina*, cf. Vegetius 1.1 and 1.13; Cicero, *Tusculan Disputations* 1.1–2, *On the Commonwealth* 2.30; for its impact, see the examples offered in Frontinus 4.1–2. For the term (usually understood more in the modern sense of *discipline* instead of *learning process/training*), cf. Mauch (1941) and Phang (2008).

56 fr. 6 JORDAN: 'he sent a part of the cavalry and the ferentarii for looting' ~ *Inde partem equitatus atque ferentarios praedatum misit*. Cf. his speech fr. 13 JORDAN, where he uses the expression *proelia levia* as a term for unconventional warfare; and his adhortatory speech in Liv. 34.13 (based on Cato's own history), which stresses the contrast between unconventional raiding of the enemy territories and conventional fighting in pitched battle; cf. ibid. 16. It is not by chance that Frontinus opens his collection of stratagems with Cato: Frontinus 1.1.1. The Romans adopted and used asymmetrical operational tactics and certainly had also a theoretical understanding of them, *pace* Breccia, who argues that the *bellum iustum* concept prevented any Roman thinking in this direction; but Traina already saw it right (see earlier p. 2f.).

57 Maurice 8.1.7 and 8.2.4; Leo, *Tactics* 20.11, 20.51, 20.136. Of course, Leo is based on Maurice, but Maurice again is based on similar statement found among Vegetius' *regulae generales*, Vegetius 3.26.6: 'It is preferable to subdue an enemy by famine, raids and terror, than in battle where fortune tends to have more influence than bravery' ~ *aut inopia aut superventibus aut terrore melius est hostem domare quam proelio, in quo amplius solet fortuna potestatis habere quam virtus*. See Chapter 11 in this volume.

58 Agathias 5.25.6 Menander fr. 5. For the context, see Koehn (2018) 260–265.

59 Polyaenus I preface 3: ἀνδρεία μὲν γάρ, ὅστις ἀλκῇ χρησάμενος πολεμίων μαχομένων ἐκτήσατεν, εὐβουλία δέ, ἀμαχεὶ τέχνῃ καὶ δόλῳ περιγίγνεσθαι· ὥς ἐστι πρώτη δεινῶν στρατηγῶν σοφία κτᾶσθαι τὴν νικὴν ἀκίνδυνον. For Polyaenus' work on stratagems, see the detailed study of Wheeler (2010), esp. 17f. For a list of stratagems in Leo's *Tactics* which are based on Polyaenus, see Wheeler (2012) 162. The claim of Wheeler (2010) 54 that Polyaenus was becoming 'a prominent figure in Byzantine military thought' might rather be exaggerated (see Rance (2011)); yet he was among the authors recommended by Constantinus Porphyrogennetus to be taken on campaign for consultation.

60 Kaegi (1983) 11–14. In regard to the transmission of certain tactics and stratagems, one has to be cautious to make such a clear distinction as Chapter 2 does, which, in discussing advices for mock battles continuously transmitted from Xenophon and Onasander up to Leo VI and Nikephorus Ouranus, argues that this advice is 'neither evidence for common Greek and Roman nor Byzantine military doctrine. It is only evidence for three authors (Onasander, Leo VI, and Nikephorus Ouranous) relying on the works of their predecessors.' Interdependency and variances of texts and its content and continuity/discontinuity in military doctrine are hard to separate in this way. For a well-balanced recent discussion of tradition vs. innovation and antiquarianism vs. practicability in Byzantine manuals, see Chatzelis (2019).

61 Wheeler (2010) 48–53 has intriguingly argued that Polyaenus' presentation of various Barbarian stratagems in book seven set a precedent for Maurice's and Leo's ethnographical treatment of enemy fighting tactics in their manuals, and that this precisely was 'Polyaenus' original contribution to Western military thought'. What regards the military literature *stricto sensu*, it is certainly true. However, in a practical sense, awareness of military asymmetries and need of military acculturation were always part of Greco-Roman military thought. For example, while the Romans of the first century CE tried to overcome the *bellum silvestre* (Lucretius 5.1245) of their Germanic enemies fighting as light-armed ambushers (cf. Dio Cassius 56.21.4) by converting the

loca iniqua (forests and marshes) into *loca aequa* (cf. Frontinus 1.3.10 with Tacitus, *Annals* 1.68) in order to employ proper close-combat tactics or by adjusting the tactics of their troops (cf. Frontinus 2.3.23: Roman cavalry fighting on foot, with Vegetius 3.6.21). Their descendants of the third century CE were well aware that the Germans had enhanced their skills in confronting them directly; therefore, they took care to have enough light-armed infantry for long-range hit-and-run tactics such as Parthian archers and Maurusian javelin-men: Herodotus 6.7.8–10 and 7.2.1–3 (at the campaign which now is connected with the Harzhorn battle, 236/37 CE). In the case of the latter (auxiliary *numeri Maurorum*), it is interesting to see how knowledge about fighting tactics of other people went into oblivion: when Justinian's troops had to deal with the Maurusian uprising after the conquest of the Vandal kingdom, they were consternated by the effectivity of their javelins, but quickly adopted this weapon (see Koehn (2018) 134–137); it might be that this triggered the transmission of armament and tactics of various foreign potential enemies first in Maurice and then Leo. Yet despite the innovative approach of Byzantine manuals in discerning the peculiarities of their enemies, they still recommend certain tactics as universally applicable for fighting against any people: Maurice 6.4 and Leo 18.10.

62 Cf. Stouratis (2009), who shows that the so-called Themata system designed primarily for a defensive military policy was just an intermediate solution due to the loss of the Eastern territories and was not meant to be a completely new reconfiguration of Byzantium's military strategy. In this context, it is important to note that the introductory remarks in Nicephorus' preface point to the fact that this treatise at the moment it was composed has no relevance anymore for the warfare in the East; Byzantines at that time were busy with conquering enemy territories, not ambushing enemy raids.

Bibliography

Alofs, E. 2014. 'Studies on Mounted Warfare in Asia I: Continuity and Change in Middle Eastern Warfare, c. CE 550–1350: What Happened to the Horse Archer?'. *War in History* 21. 423–444.

Barwick, K. 1948. 'Zu den Schriften des Cornelius Celsus und des alten Cato'. *WJA* 3. 118–132.

Basso, G. and Greatrex, G. 2018. 'How to Interpret Procopius' Preface to the Wars', in C. Lillington-Martin and E. Turquois, eds., *Procopius of Caesarea: Literary and Historical Interpretations*. Abingdon: Routledge. 59–72.

Birkenmeier, J. W. 2002. *The Development of the Komnenian Army 1081–1180*. Leiden: Brill.

Breccia, G. 2001. '"Con assennato coraggio . . .". L'arte della guerra a Bisanzio tra Oriente e Occidente'. *Medioevo Greco* 1. 53–78.

——— 2007–2008. 'Grandi imperi e piccole guerre: Roma, Bisanzio e la guerriglia'. *Medioevo Greco* 7 & 8. 13–68 and 49–132.

Chatzelis, G. 2019. *Byzantine Military Manuals as Literary Works and Practical Handbooks: The Case of the Tenth-Century Sylloge Tacticorum*. London: Routledge.

Chlup, J. 2014. 'Just War in Onasander's ΣΤΡΑΤΗΓΙΚΟΣ'. *Journal of Ancient History* 2. 37–63.

Dagron, G. 1987. 'Ceux d'en face': Les peuples étrangers dans les traités militaires byzantins'. *TM* 10. 207–232.

Dawson, T. 2002. 'Syntagma Hoplon: The Equipment of Regular Byzantine Troops, c. 950 to c. 1204', in D. Nicolle, ed., *A Companion to Medieval Arms and Armour*. Woodbridge: Boydell Press. 81–90.

Graff, D. A. 2016. *The Eurasian Way of War: Military Praxis in Seventh-Century China and Byzantium*. London: Routledge.

Grotowski, P. L. 2010. *Arms and Armour of the Warrior Saints: Tradition and Innovation in Byzantine Iconography (843–1261)*. Leiden: Brill.

Haldon, J. 1999. *Warfare, State and Society in the Byzantine World, 565–1204*. London: Routledge.

———, ed. 2007. *Byzantine Warfare*. Aldershot: Ashgate.

——— 2014. *A Critical Commentary on the Taktika of Leo VI*. Washington: Dumbarton Oaks.

Hanson, V. D. 1989. *The Western Way of War: Infantry Battle in Classical Greece*. Berkeley: University of California Press.

Heather, P. 2018. *Rome Resurgent: War and Empire in the Age of Justinian*. Oxford: Oxford University Press.

Humm, M. 2007. 'Des fragments d'historiens grecs dans l'Ineditum Vaticanum?', in M.-L. Freyburger and D. Meyer, eds., *Visions grecques de Rome – Griechische Blicke auf Rom*. Paris: de Boccard. 277–318.

Janniard, S. 2015. 'Les adaptations de l'armée romaine aux modes de combat des peuples des steppes (fin IVe-début Vie siècle apr. J.-C.)', in U. Roberto and L. Mecella, eds., *Governare e riformare l'impero al momento della sua divisione: Oriente, Occidente*. Rome: Collection de l'École française de Rome. 247–288.

——— 2018. 'Procope, les Huns et les transformations tactiques de la cavalerie romaine', in G. Greatrex and *idem*, eds., *Le monde de Procope: The World of Procopius*. Paris: de Boccard. 205–214.

Kaegi, W. E. 1964. 'The Contribution of Archery to the Turkish Conquest of Anatolia'. *Speculum* 39. 96–108.

——— 1983 [2007]. 'Some Thoughts on Byzantine Military Strategy', in J. Haldon, ed., *Byzantine Warfare*. Aldershot: Ashgate. 251–268.

Kaldellis, A. 2004–2005. 'Classicism, Barbarism, and Warfare: Procopius and the Conservative Reaction to the Later Roman Military Policy'. *AJAH* 3–4. 189–218.

Keegan, J. 1976. *The Face of Battle*. London: J. Cape.

Koehn, C. C. 2016. 'Justinian's Wonder Warriors: Some Observations on the Development of Multi-Role Cavalry in Early Byzantium'. *Mnemon* 16. 155–180.

——— 2018. *Justinian und die Armee des frühen Byzanz*. Berlin: De Gruyter.

Kruse, M. 2017. 'Archery in the Preface to Procopius' Wars: A Figured Image of Agonistic Authorship'. *Studies in Late Antiquity* 1. 381–406.

Kühn, H.-J. 1991. *Die byzantinische Armee im 10. und 11. Jahrhundert. Studien zur Organisation der Tagmata*. Vienna: Verlag Fassbaender.

Lenoir, M. 1996. 'La littérature De re militari', in C. Nicolet, ed., *Les littératures techniques dans l'antiquité romaine*. Geneva: Entretiens Hardt. 77–115.

Luttwak, E. N. 2009. *The Grand Strategy of the Byzantine Empire*. Cambridge, MA: Harvard University Press.

Lynn, J. A. 2003. *Battle: A History of Combat and Culture*. New York: Westview Press. 12–25.

Mauch, O. 1941. *Der lateinische Begriff 'disciplina': Eine Wortuntersuchung*. Diss. Basel. Paulusdruckerei.

McGeer, E. 2008. *Sowing the Dragon's Teeth: Byzantine Warfare in the Tenth Century*. Revised edition. Washington: Dumbarton Oaks.

Montesquieu, Ch. de. 1748 [1890]. *Considérations sur les causes de la grandeur de Romains et leur décadence*. Bielefeld and Leipzig: Jacques Desbordes.

Petitjean, M. 2014. 'Classicisme, barbarie et guerre romaine: l'image du cavalier dans le monde romain tardif'. *AnTard* 22. 255–262.

Phang, S. E. 2008. *Roman Military Service: Ideologies of Discipline in the Late Republic and Early Principate*. Cambridge: Cambridge University Press.

Rance, P. 2007a. 'The Date of the Military Compendium of Syrianus Magister (Formerly the Sixth-Century Anonymus Byzantinus)'. *ByZ* 100. 701–737.

———— 2007b. 'Battle', in P. Sabin, H. van Wees, and M. Whitby, eds., *The Cambridge History of Greek and Roman Warfare Volume II: Rome from the Late Republic to the Late Empire*. Cambridge: Cambridge University Press. 342–378.

———— 2011. Review of Brodersen, K., ed. *Polyainos. Neue Studien-Polyaenus*. New Studies. Berlin. BMCR 2011.06.07.

———— 2017a. 'Maurice's Strategicon and the "Ancients": The Late Antique Reception of Aelian and Arrian', in *idem* and N. Sekunda, eds., *Greek Taktika: Ancient Military Writing and Its Heritage*. Gdansk: Foundation for the Development of Gdansk University for the Department of Mediterranean Archeology. 217–255.

———— 2017b. 'The Reception of Aineias' Poliorketika in Byzantine Military Literature', in M. Pretzler and N. Barley, eds., *Brill's Companion to Aineias Tacticus*. Leiden: Brill. 290–374.

Sheldon, R. M. 2012. *Ambush: Surprise Attack in Ancient Greece*. London: Frontline Books.

Shuvalov, P. V. 2006. *Sekret armii Justiniana. Vostochnorimskaja armija v 491–641 gg.* St. Petersburg: Peterburgskoe Voskokovedenie.

Stouratis, I. 2009. *Krieg und Frieden in der politischen und ideologischen Wahrnehmung in Byzanz (7.-11. Jh.)*. Vienna: Fassbaender Verlag.

Strässle, P. M. 2006. *Krieg und Kriegführung in Byzanz. Die Kriege Kaiser Basileios' II. gegen die Bulgaren (976–1019)*. Cologne: Böhlau.

Traina, G. 1986–1987. 'Aspettando i barbari. Le origine tardoantiche della guerriglia di frontera'. *Romanobarbarica* 9. 247–279.

Whately, C. 2016. *Battles and Generals: Combat, Culture, and Didacticism in Procopius' Wars*. Leiden: Brill.

Wheeler, E. L. 1988. *Stratagem and the Vocabulary of Military Trickery*. Leiden: Brill.

———— 2007. 'Introduction', in *idem*, ed., *The Armies of Classical Greece*. Aldershot: Ashgate. xi–lxiv.

———— 2010. 'Polyaenus: Scriptor Militaris', in K. Brodersen, ed., *Polyainos. Neue Studien-Polyaenus: New Studies*. Berlin: Verlag Antike. 7–54.

———— 2012. 'Notes on a Stratagem of Iphicrates in Polyaenus and Leo Tactica'. *Electrum* 19. 157–163.

Wojnowski, M. 2012. 'Periodic Revival or Continuation of the Ancient Military Tradition? Another Look at the Question of the KATAPHRAKTOI in the Byzantine Army'. *Studia Ceranea* 2. 195–220.

14 'God has sent the thunder'

Ideological distinctives of middle Byzantine military manuals

Meredith L. D. Riedel

In his military manual, the Byzantine emperor Leo VI the Wise (r. 886–912) tells a story about a wise general. The story goes like this: just before battle, as the army was drawing up in formation, there was a sudden loud clap of thunder! Everyone probably jumped. In the face of impending battle, this sort of shock to the system triggers a cascade of reactions both physical and emotional. The soldiers naturally became fearful and wondered if the thunder was an evil omen of imminent death. The wise general, however, sent word to his soldiers that the thunder was not against *his* army or *his* soldiers, but was a warning shot against their enemies.[1] Leo declared: God had sent the thunder. It was God's order to march out against the doomed opposing army.

Of course, this story did not originate with Leo VI. He copied it from Polyaenus, who had written a book on strategy eight centuries earlier, dedicated to Marcus Aurelius (d. 180) and Lucius Verus (d. 169). The much-praised *Stratagems* of Polyaenus is purportedly a book of history. Its author was a rhetorician with a didactic purpose, recounting past victories by Greek and Roman military heroes. Not all of his stories are true; some are fictitious. However, the goal of the work was to help second-century Roman emperors defeat the Parthians by presenting a record of *strategemata*, a word that denotes successful military acts. Written in Greek, it appears also to have been beloved by the Romans of the East, anachronistically called Byzantines, who kept and copied several summary versions of this work. It is not difficult to see the allure of these manuals to a culture that viewed itself as the inheritors of Greek learning and Roman science.

The story that Leo borrows highlights the strategic wisdom of the Theban general Epaminondas (d. 362 BC).[2] However, in Leo's hands, this particular stratagem as written by Polyaenus is edited to reflect a distinctively religious outlook. In Polyaenus' story, there is no mention of 'God' whatsoever, whilst in Leo's rendition, divine activity is explicitly credited: 'God has sent the thunder.' Although Plutarch does refer to τὸν θεόν, the version of Polyaenus used by Leo does not. There is more going on here than mere artistic license.

What was Leo doing and why? The answers to these questions matter because this one emperor singlehandedly oversaw a renaissance of military writing in Byzantium in the tenth century.[3] His influence, in terms of religious language and ideas, shaped the composition of tenth-century manuals on warfighting in Byzantium and arguably contributed to the military successes of the 'Byzantine reconquest', the only period between the rise of Islam and the fall of Constantinople – a period of about eight centuries – during which the eastern empire was able to enlarge its territory. Given that these two rival powers embraced similar monotheistic religions, this paper contends that Leo's articulation of the distinctiveness of Christianity, particularly in the arena of warfare, written to renew military motivation and morale, was visibly demonstrated by the tenth-century successes.

I. Who was Leo VI?

Leo VI is best known among Byzantinists for the scandal of his tetragamy; that is, he married four different women successively (not simultaneously!). Although a widower before each of his subsequent weddings, the scandal arose because Orthodox canon law permits no more than two marriages; Leo's own civil legislation declared third marriages 'contemptible' (καταφρονήσασα).[4] Scholars have written and investigated far more than is necessary about Leo's love life, so more elaboration here would be unnecessarily repetitive. The most interesting thing about the tetragamy scandal is that although Leo was excommunicated, he was also later restored to good standing in the Church and obtained everything he wanted in terms of legitimacy for his son, as well as keeping his fourth wife (whom he had wed against the explicit wishes of the patriarch). He was a man who fought the church and won.

Two further things make Leo a fascinating emperor: first, he was unexpected, and second, he was a prolific writer. As a second son, he had not been destined to rule and therefore was not trained in military arts, but rather received a religious education. In fact, in the mirror of princes addressed to Leo from his father, the founder of the Macedonian dynasty, faith is underlined as of primary importance.[5] Leo's tutor was probably Photios, the patriarch of Constantinople, a brilliant scholar and humanist best known for instigating the schism between the Greek and Latin churches.[6] In other words, Leo was educated (at least in part) by a man of formidable intellect with an equally formidable will plus a lack of concern for the opinions of those who disagreed with him. Lest one be tempted to view Leo as a mild-mannered, slightly bookish, somewhat weak emperor, it is worth remembering that he was unusual in many ways. He reigned for 26 years over an empire where fewer than half of its rulers died peacefully. Leo himself survived three assassination attempts, two revolts and the defection of one of his highest-ranking generals to the Caliphate. Leo died of ill health in 912, still on the throne. I would therefore suggest that his long reign and peaceful death were the result of more than dumb luck. He was clever, he was underestimated both by his contemporaries and by modern scholarship, and he was serious about the role of religion.

Second, he was a prolific writer. A survey of Leo's literary output demonstrates that he took the importance of Orthodox convictions to heart.[7] He wrote things no other emperor had, like homilies and advice for monks. He wrote more legislation than any emperor save Justinian. And despite never serving in the army, he wrote a tactical manual that was not only the first to address the military threat of the

Islamic caliphates, but was read, copied, translated even into Arabic, followed, and later given credit for inspiring the most successful campaigns of the Byzantines in the Islamic era.

It is unlikely that Leo meets the criteria for who is properly considered a 'military author', because he never accrued military experience, on the battlefield or otherwise, and is not overly concerned with technical accuracy.[8] As a young man, he produced the *Problemata*, a text of questions and answers arranged nearly verbatim from the sixth-century *Strategikon* of Maurice.[9] He also compiled extracts on warfare and commissioned a work on imperial military expeditions, but there is no evidence that Leo had any personal experience on a battlefield.[10] His martial expertise, such as it was, remained second-hand; to write the *Tactics* (*Taktika*), he consulted past military dispatches, the records left by his father, Basil I, and reports from his own field generals, stitching the material together in a way that served his own imperial purposes.[11] However, his work did have an official status; he did not write his military manual as a private individual, but as the vicegerent of Christ instructing the generals of a consciously Christian polity on the renewal of military strategy. Leo was also specific that his manual should be accorded the weight of imperial law, calling it another *procheiros nomos*.[12] The precepts of his military writing are not mere guidelines, but obligatory principles.

Leo's *Tactics* is ground-breaking in several ways: it mentions the Muslims for the first time in the context of Byzantine military strategy, carries the force of imperial law, and uses consciously ideological language. Between Justinian I (r. 527–65) and Leo VI (r. 886–912), Byzantine armies regularly faced Byzantium's enemies on the field of battle, and since the defeat of Heraclius (r. 610–641) by Islamic forces at the Yarmuk River in 636, every Byzantine emperor had been forced to reckon most often with the formidable threat of Muslim power. Until Leo, none of them had ever thoughtfully considered in any extant writing how to counter that threat.[13] His riposte, a military manual entitled τῶν ἐν πολέμοις τακτικῶν σύντομος παράδοσις, or more commonly, *Tactical Constitutions*, focuses the attitudes of Byzantine Christians vis-à-vis Islam and, for the first time in Byzantine history, attempts to present a solution for the military threat posed by the Caliphate.[14] Although his handbook, unlike the *Stratagems* of Polyaenus, was not composed for a specific military campaign, it was written with a definite enemy in mind. In the Prologue, Leo bemoans the 'collapse of strategic knowledge' that has brought the condition of the Roman (read: Byzantine) state to the necessity of 'tak[ing] their stand against those nations that want war'.[15] In the Epilogue, Leo clearly references the need to prepare for war 'especially against the Saracen nation now causing us trouble'.[16] In between these, in Constitution 18, Leo introduces a lengthy discussion on this enemy in particular, 'the nation of the Saracens that is presently troubling our Roman commonwealth'.[17] Although widely accepted, this view of the motive for composing the *Tactics* is not universally held by scholars; John Haldon has argued that a particular enemy was not in view when the manual was first compiled, but does agree that 'there was undoubtedly a preoccupation in imperial and military circles during Leo's reign with the threat from its Islamic foes in the east and the duties of a Roman empire in the face of this threat'.[18] He goes on to say that 'it is not always clear that there was a desire to do anything more than defend and maintain the status quo', but I argue that this is precisely the goal of Leo's *Tactics*: defence of Roman territory and its status as the Byzantine Orthodox Christian *oikoumene*.

Leo combines the exigencies of battle with the doctrines of Christianity and casts the conflict in broader terms of religious distinction over against mere political or ethnic difference to a markedly higher degree than did his main source, the emperor-general Maurice (r. 582–602), who wrote a military manual in the late sixth century during the war with the Persians.[19] Although the Byzantine army suffered some serious reverses under Leo's reign, his concern for the renewal of military strategy proved highly effective: the tenth-century reconquest returned lost territory to the empire and expanded the borders to an extent not seen since the initial defeats of the seventh century with the rise of Islam.

This, then, is the seminal author who inspired the writers of the tenth-century military works. He was a man of power, of privilege, of deep learning, and of unusual faith who nevertheless refused to subordinate his perspective to the ecclesiastical authorities of his day.

II. Distinctive characteristics of the *Tactics*

Although his military manual discusses strategy, tactics, received wisdom from earlier eras, and the problem of Saracen military success, Leo's primary interest and expertise lie elsewhere. For him, Byzantine military success finds its true foundation in Christian theology and the advantages that accrue to those who believe in the Orthodox Christian God. Although Greek and Roman military manuals prior to Leo's often included a few exhortations about divine approval for brave warriors, no manual before Leo's *Tactics* took such pains to clarify religious distinctions or exhort soldiers and generals to a pious Christian life. His theological approach and consciously religious language reverberates throughout the manual in an unprecedented fashion. Leo's work expanded the ideological emphasis of armies accustomed to superficial religious battle cries, and extended his military theory to include civilian financial support of a Christian society for Christian soldiers fighting on behalf of Christians.[20] Compiling and annotating the strategies and tactics of previous military manuals provided the vehicle for him to set out a distinctly Byzantine and Christian philosophy of warfare.

First, given the purpose of the book, Leo drew a distinction between Christianity and Islam, characterising the Muslim faith as blasphemy, essentially because it does not agree with orthodox Christian doctrine.[21] Leo notes that Muslim orthodoxy forbids adherents to refer to Jesus Christ as a Saviour, and it does not accord Jesus Christ the status of a deity.[22] Worse, 'they say that God delights in wars – God, who scatters abroad the nations who desire war'.[23] That is, they make God the author of evil.[24] Further, 'they say that everything comes from God, even if it be bad; so if they happen to suffer some reverse, they do not resist it, as being something determined by God; and the strength of their attack fails'.[25] Because the goal of the book is to increase the effectiveness of the Byzantine army, partly by raising its morale, he gives a truncated description of Islamic belief, intended to demolish any purported religious advantage.

Leo's comments reflect his deep and abiding passion for theological solutions. The Byzantines knew that their enemies expected to achieve martyr status if they were killed in battle, but no such honour awaited Christian fighters.[26] The martyrdom of Christian prisoners may have contributed to the problem of battlefield morale, since capture and eventual exchange or martyrdom was preferable to death in battle. Rather than dwell on this inequity, Leo describes the Muslim faith in a way that draws attention to its deficits, perceived from his Christian point of view.[27] Leo is focused on finding a practical solution for a battlefield disadvantage, and he finds it in arguing for the superiority of Christianity over Islam.

More importantly, Leo christianised the ideal general of the ancient military treatises.[28] The attentive reader will no doubt recognise that he was imitating Onasander's emphasis on the character of the general, but with extra sauce: that is to say, in his hands the general's integrity arises from his Christian character. For Leo, the moral rectitude of the empire that is revealed in the writings of Cicero and Livy has in the medieval Byzantine context the Christian faith as its wellspring.

For example, the ancient military treatises often state that luck or fate is one of the factors a general cannot control.[29] Leo, however, never speaks of luck at all; the word τύχη (fate) does not appear in the *Tactics*, except for once in 20.51, where he directly quotes Maurice's *Strategikon*. This preference for biblical vocabulary is particularly evident in Constitution 2, which lists the characteristics of the ideal general. In the section that Leo copies from Onasander, he scrupulously replaces Onasander's references to fate (τύχη) with references to God, thus explicitly christianising a text already deeply concerned with moral behaviour.[30] Leo's use of Onasander and his adaptations to its text reveal Leo's focus on the role of the general, but with Leo's particular twist. He not only changes the first-century pagan references to fate, but he broadens the ethical and moral component, attaching it to Christian piety as a prerequisite for military success.[31] Leo references or quotes biblical material directly, and he also uses biblical language in such a way that, to an attentive reader, the fragrance of the Christian scriptures waft unmistakably from his prose.[32] For example, scripture is central to the counsel in Constitution 2.10:

> We know that a general who is loved by his subjects will be more highly regarded and be very helpful to the men under his command. When men love someone, they are quick to obey his commands, they do not distrust his words and promises and, when he is in danger, they will fight along with him. For love is like this: to lay down one's life on behalf of the person one loves.
>
> (cf. John 15:13)[33]

In Leo's hands, this biblical idea – dying for one's friends – becomes a tool to engender the obedience of soldiers. His interpretation misses the near universal concept in military science that soldiers obey because they have taken an oath of obedience. They are bound by obligation on account of their oath, not by love.

Love does not usually come into it. Most military manuals before Leo's focus on loyalty owed a general because of a soldier's fidelity to his vow, or on account of a general's military skill, personal integrity, or political power. By bringing in the idea of love, Leo is tinkering with the motivational mechanisms of military command theory and practice. This might indicate a problem with army discipline, but it certainly reveals the emperor's view of the use of Christian scripture in the service of what can only be viewed as political goals.

This sustained focus on the general is also a key element of the Byzantine military treatises that followed Leo's *Tactics*. Indeed, one could argue that Leo's emphasis was so thorough and convincing that later writers of military manuals – who were army generals themselves – reprised or expanded his material, implicitly assuming familiarity on the part of their intended readers. The writer of the *De velitatione* makes specific reference to the *Tactics* of Leo VI, and the *Taktika of Ouranos* reprises Leo's material in the first 55 chapters.[34]

Conor Whately has recently examined the implications of classifying military manuals as theoretical or descriptive, with the latter offering the more useful material.[35] Curiously, for many years Leo's *Tactics* was relegated to the ranks of the theoretical and thought to offer little new material useful for his contemporaries. That has recently begun to change with Fr. Dennis' (2010) critical edition and John Haldon's accompanying 2014 commentary. The argument presented here takes the view that the *Tactics* is descriptive, particularly with regard to ideological motivations, and therefore more useful in terms of army morale.

Perhaps it might be time to expand the definition of what makes a military manual useful. It is worth noting that Leo's manual 'was copied more than any other Byzantine military work',[36] despite the fact that Leo did not write original material on the profession of generalship. He took what he found mainly in two earlier Greek manuals and carefully edited, paraphrased and assembled the principles and aphorisms into an ideal with consciously christianised packaging, tailored for the general of the late ninth-century Byzantine army who is addressed throughout.

Leo enjoins his general to offer genuine worship to God (20.47), to revere the churches and those in them (20.70), and to purify the soldiers from sin before they join battle by having them sanctified by the blessing of the priests (20.172). At the same time, he is practical, declaring that

> in time of war it is necessary to offer prayers to God and to invoke him as an ally . . . but at the same time, hold on to your weapons and, *while you fight*, invoke the Divinity as an ally.
>
> (20.77)

At a deeper level, Leo is concerned to contrast Byzantine Christian practice with that of the Muslim Arabs, primarily for the purpose of bolstering morale so that the soldiers would go forward more confidently into battle. The spiritual principles of battle held by Muslims were not unknown to the Byzantines, who had been their perpetual adversaries since the early seventh century. Nor was Leo the first to think about the Christian piety of soldiers. He may have been exploiting a deep cultural trauma, annually commemorated on March 6 in Constantinople, of the sufferings of 42 prisoners of the siege of Amorion; they had been defeated in battle, held prisoner for seven years, and finally on March 6, 845, were publicly beheaded on the banks of the river Tigris.[37] The most famous version of this ecclesial commemoration declared, 'Their profession of soldier, which is for many people the way of damnation, became for them the road which led to salvation. They fought the good fight and received the crown of righteousness'.[38] The implicit message is that all Christians ought to aspire to holy willingness to die for Christ, or at the very least, be willing to sacrifice for the faith. After all, as Evodios and others point out, it is a higher thing to die for God motivated for love of holiness, than to die for a deity who promises mere carnal pleasures (the Byzantine trope of Muslim paradise).[39]

As an emperor thus interested in reinforcing Byzantine Christian conviction, especially among soldiers, Leo undertook to write his manual with the belief that he uniquely was able to combine the requirements of faith, thorough research into military strategy and tactics, and pious leadership in order to produce a tome that would serve a most practical purpose: the reanimation of Byzantine military know-how in the face of Saracen attacks from land and sea. In this, he was undoubtedly successful, if one credits the later Byzantine reconquest victories to his counsel. His theological approach is not one of mere lip service; he claims that all military endeavour should begin and end with prayer, just as his book does. He demands that the general set an example of piety and every Christian virtue, and that the soldiers are also held to a Christian standard of holiness, because they hold the high office of 'defenders of the faith'.

III. Tenth-century Byzantine military manuals

Four military manuals produced in the tenth century bear the marks of Leo's influence: *De velitatione bellica* (περὶ παραδρομῆς τοῦ κυροῦ Νικηφόρου τοῦ βασιλέως),[40] describing eastern border warfare before 960; *Praecepta militaria* (Στρατηγικὴ Ἔκθεσις καὶ Σύνταξις Νικηφόρου Δεσπότου),[41] dealing with eastern border offensives subsequent to 961; *De re militari* (περὶ καταστάσεως ἀπλήκτου),[42] the complementary treatise to the *De velitatione* that focuses on guerrilla warfare in the Balkans in the 990s; and the *Tactics* of Nikephoros Ouranos (Ἐκ τῶν τακτικῶν Νικηφόρου τοῦ Οὐρανοῦ),[43] covering the experience of a Byzantine commander in northern Syria at the beginning of the 11th century. To varying degrees, each of these manuals bears the ideological distinctives of Leo's *Tactics*: invocation of God, mention of Islam, use of scripture or scriptural language, focus on the general, reward for Christian soldiers, and prayers or other religious rituals as part of daily life in the army.

The first two of these manuals carry the name of Nikephoros II Phokas (r. 963–69), an emperor-general who fulfilled Leo's description of the ideal Christian leader in every respect. Known as a devout and ascetic Christian, Nikephoros II had few interests beyond military excellence and spiritual devotion, as his life demonstrates. He was raised to the purple by acclamation in 963, built the first monastery on Mount Athos under the guidance of his spiritual father, and was discovered, on the night he was assassinated, sleeping not in the soft imperial bed but on the floor, faithful to his ascetic practice, wrapped in a bearskin.

In life, Nikephoros epitomised the Byzantine ideal of the warrior-emperor. He was 'a paragon of the personal and imperial virtues'.[44] Gifted in military arts, he proved himself worthy of his name ('Bringer of Victory') in battle – from the recovery of Crete in 961 to the conquest of Antioch in 969 – and therefore commanded the respect of the soldiers who fought under his leadership. Moreover, he was famed for his asceticism and Christian devotion, and sought to portray himself as the Byzantine ideal of the emperor-priest. He has been called 'the epitome of the pious warrior fighting for the Christian people'[45] and 'the φιλομόναχος emperor, the commander who went into battle with the prayers and the presence of monks'.[46] Thus, he embodied the distinctly Byzantine fusion of war and religion, a monkish ascetic with a flair for fighting. Not only does Nikephoros cite the *Tactics* directly, he also appears to have taken Leo's injunctions about the ideal general to heart.[47] After his death, he was celebrated in monastic circles as a martyr and memorialised as a model of pious chastity for future emperors to emulate.[48]

The two manuals that bear Nikephoros' name both concern warfare on the eastern border, that is, the theatre of operations against the Muslim armies of Cilicia and northern Syria. These two treatises 'form a sort of diptych',[49] the first outlining the guerrilla warfare during the annual raids of the Islamic forces, and the second describing the offensive strategies of the later decades when the Byzantine armies reconquered those lands. This makes sense, since Nikephoros and his brother Leo – scions of the Phokades clan based in that region and both formidable warriors – were the architects of the Byzantine reconquest.[50]

These two treatises – *De velitatione bellica* and the *Praecepta militaria* – reflect the determination to preserve valuable and hard-won wisdom for the benefit of posterity. In writing or sponsoring these books, the emperor was fulfilling his role as the commander-in-chief of the Byzantine military, giving direction to those generals under his authority, for the greater glory of Byzantium. As Gilbert Dagron expresses it, 'On attend d'un empereur qu'il dise le droit et qu'il pense la guerre; ses juristes et ses généraux feront le reste, dans un anonymat qui n'est que l'ombre portée de sa légitimité.'[51] Whether it was Nikephoros himself who put pen to paper does not materially change the importance or cultural weight of the text in the eyes of its Byzantine readers. His name on the book gave it credibility in a culture that admired authority and tradition. His record as a highly successful general added plausibility and cachet to the military exhortations of the texts.

Praecepta militaria

The *Military Precepts* (*Praecepta militaria*) was also written ostensibly at the command of Nikephoros II Phokas sometime after the 950s but first completed in 969 after his death. The Greek title is quite long: 'a presentation and composition on warfare of the emperor Nikephoros' (Στρατηγικὴ Ἔκθεσις καὶ Σύνταξις

Νικηφόρου Δεσπότου), undoubtedly referring to the great soldier-emperor of the Phokades family. Brevity prevents it from being a comprehensive composition on warfare. Lacking an introduction or any evident organising principle, the extant text remains fragmentary. The purpose of the *Military Precepts* is to lay out tactics and strategy for offensive warfare in the east against a weakened Arab enemy, evincing the era of the so-called 'Byzantine reconquest' that began in the 960s. The text itself does not categorically articulate its purpose, as does the *De velitatione*, but its emphasis is evidently on more set-piece tactics. The instructions include infantry and cavalry formations, with advice to leave the baggage train 'in our own territory', indicating that assault, not defence, is envisaged. Unlike the *De velitatione*, there is little mention of guerrilla tactics, except for ambushes by mounted scouts (*prokoursatores*). The *Military Precepts* opens with what sort of men should be chosen as foot soldiers and how they should be organised in battle formation. It is a manual of conquest; the second chapter lists the requirements for provisioning the infantry during invasions into enemy territory, and it continues in the same vein. It offers instructions on heavy infantry formations, the heavy cavalry, deployment of the cavalry generally, how to organise a camp, and how to flush out spies. The fourth and sixth chapters also contain significant sections on prayers for the troops and pre-battle and daily religious ritual. The flow of the text is rough, compared with the relatively more polished *De velitatione*, possibly indicating that a general's detailed field notes may have been the original source.

Of the four manuals considered here, the *Military Precepts*, tellingly the most closely associated to the time and authorship of Nikephoros II, contains the most illuminating comments with regard to the Byzantines' animating faith. The dominant themes resound with confidence in the absolute sovereignty of God to determine battles, shot through with an equally strong conviction of the importance of pious observances that have bearing on a soldier's mettle. There is an equal emphasis on the activity of God, of Christ and of Mary.

In the *Military Precepts*, one finds several passages dealing with prayers, rituals, penalties and war cries. Soldiers are instructed to 'say the invincible prayer proper to Christians, 'Lord Jesus Christ, our God, have mercy on us, Amen,' and in this way let them begin their advance against the enemy' (IV.109).[52] The prayer echoes the admonishment of Maurice's sixth-century *Strategikon*,[53] although the author of the *Military Precepts* adds another, longer prayer to be said during the advance:

> Come to the aid of us Christians, making us worthy to rise up and fight to the death for our faith and our brethren by fortifying and strengthening our souls, our hearts, and our whole body, the mighty Lord of battles, incomparable in power, through the intercession of the Mother of God Who bore Thee, and of all the saints, Amen.

(IV.115–120)[54]

Chapter 6 details similar prayers for Vespers and Matins in the camp but adds regulations concerning the penalties for failing to perform these prayers: beating, tonsure, demotion and general public humiliation (VI.28–30). It also establishes the ritual three-day fast concluded by communion on the night before battle as a rite of purification. One wonders whether this fast was consistently observed, because all generals know that soldiers do not fight well on empty bellies. In the words of the late fourth-century Roman tactician Vegetius, 'armies are more often destroyed by starvation than battle, and hunger is more savage than the sword'.[55]

De velitatione

The *De velitatione* is about twice as long as the *Military Precepts*. It opens with a formal prologue followed by a discussion of the frontier, its surveillance and how far apart the watch posts ought to be from one another. The opening chapters discuss the possible points of attack, the nature and direction of Arab incursions and how to prepare for them. Chapter 3 discusses blocking the passes that give access to Byzantine lands with infantry formations. Chapter 4 then states it is sometimes better to attack on the return because of the difficulty in quickly mustering defences in anticipation of a raid. Chapter 5 discusses controlling water supplies, and so it goes. Clearly, this book is focused on the exigencies of defensive warfare in a particular location, and faced with a specific threat. The anomalous Chapter 20 refers specifically to Byzantine victories in Arab lands that occurred during the reconquest era. It advises an invasion coupled with a scorched-earth policy as a means of forcing the Arab army to break off their own invasion in order to return to defend their own country.[56] At the same time, however, the writer advises caution in the case of inferior forces. While siege warfare is mentioned, the instructions are brief in comparison to the fuller tactics enumerated in the *Military Precepts*.

The *De velitatione* seeks to show how a Byzantine commander can come to terms with an opposing army of superior number while his own forces are reduced, weak or otherwise inferior to the enemy's. It was composed in the late 960s at a time when the Byzantine forces were in the ascendant and the Arab threat no longer posed a danger to the Byzantine state, but it refers to the state of affairs that held during the 950s when Muslim forces were regularly ravaging Asia Minor. Although it explicitly acknowledges the changed situation, the material in the handbook assumes the need for aggressive tactics to wage defensive battles. Dennis calls it a treatise on 'hit-and-run warfare'.[57]

In the *De velitatione*, pious observances are less obtrusive, perhaps because it was redacted somewhat later by an editor who wished to dilute the emphasis on religion, possibly because the religious underpinning was already clearly understood – not an unlikely scenario, given that this manual was promulgated decades after Leo's *Tactics* – thus a repetition of the requirements of piety may have seemed unnecessarily redundant. Rather, references to the cooperation or aid or favour of the deity (usually Θεοῦ συνεργίᾳ or βοηθείᾳ Θεοῦ) are sprinkled throughout the text, imitating common Byzantine numismatic slogans, and giving it a generally Christian aroma. The religious outlook of the culture is present in the text in the way that salt is present in seawater, infused throughout. If one causes the non-religious language to 'evaporate', so to speak, one is left with the crystals of Byzantine military Christianity. For example, advice offered to the officer addressed in the text states the common belief relatively clearly: 'The outcome of war is not brought about according to the will of men but, just as the affairs of each one are weighted, by the providence of God on high' (ἀλλ' ὡς ἡ ἄνωθεν τοῦ Θεοῦ πρόνοια).[58] This assertion verges on fatalism; taken out of context, one might be tempted to think that the 'will of men' counts for nothing, while the 'providence of God on high' inexorably determines all. Yet it appears in a manual dedicated to honing the effect of 'the will of men' in battle. Additionally, it comes at the end of chapter 19 and reads like a general conclusion for the entire work, but bizarrely placed nearer the middle than the end of the book. This emphasis on God's provision (πρόνοια) echoes Leo's *Tactics*, who previously made a case against Islamic fatalism whilst distinguishing it from Christian providence.

Further on, one finds the appropriate motivation of Christian soldiers mentioned repeatedly: The soldiers fight for 'all the Christian people' (καὶ παντὸς Χριστιανικοῦ πληρώματος), for the 'freedom and vindication of Christianity' (τῆς τῶν Χριστιανῶν ἐλευθερίας καὶ ἐκδικήσεως) because they are the 'defenders and, after God, the saviors of Christians who, so to speak, die each day on behalf of the holy emperors' (τοὺς ἐκδίκους καὶ, μετὰ Θεόν, σωτῆρας Χριστιανῶν, καὶ ὑπὲρ τῶν ἁγίων βασιλέων καθ' ἑκάστην, ὡς εἰπεῖν, ἀποθνήσκοντας).[59] These repeated statements emphasise the spiritual importance of the army's activity and, therefore, the actions of each individual soldier. God is the defender of Christians, but so is the army. Soldiers fight for Christianity, for Christian people, for freedom – imitating in their deaths the sacrifice of Christ as they 'die on behalf of' others. The writer of the manual is at pains both to acknowledge the providence of God and to praise the indispensability of military action. Here also one finds substantial overlap with the exhortations of Leo's *Tactics*.

Eric McGeer finds four parallels between the *De velitatione* and the *Military Precepts*: context, appeals for divine aid, technical terminology, and similar wording and instructions.[60] The context for both is the eastern frontier conflict against mounted Arab horsemen seeking to regularly raid Roman lands. The need for divine aid 'runs as a leitmotiv through both treatises', reflecting the customary Christian worldview of Byzantine piety.[61]

De re militari

The companion volume to the *De velitatione*, usually known as the *On Military Matters (De re militari)*, was an anonymous text on guerrilla warfare in the mountains to the north of Constantinople; it was not written until later but presents the same approach to the principles of skirmishing. Scholars have disagreed whether it was composed in the reign of Nikephoros II Phokas, or of Basil II (r. 958–1025), because the tactics fit with the campaigning patterns of both emperors.[62] The parallels between the *On Military Matters* and the *De velitatione* are obvious, as are some significant differences: they share a concern for discipline and for discerning the will of God. Both were written by practical, experienced military leaders concerned to preserve military know-how for guerrilla warfare on rugged terrain. Both authors draw from personal experience, not just theoretical knowledge. Both are concerned with fighting at night, siege warfare, scouts, marching through hostile territory, and ambushes. Their subject matter, while similar, is intended to describe warfare in different theatres: the mountains on the eastern border of Asia Minor on the one hand (*De vel.*), and the mountains of Bulgaria (*De re mil.*) on the other. The former focuses on raiding, while the latter has more universal application and includes extensive information on things like how to set up camps, for which it also includes diagrams. The *De velitatione* is written by a general to a general; the anonymous *On Military Matters* is written by a general to an emperor-general, a scenario which seems to fit the state of Byzantine-Bulgarian campaigns with a younger Basil II under the tutelage of an older military colleague in the early 990s. Furthermore, the two texts are found in the same three extant manuscripts of the early 11th century, a likely indicator of their complementary relationship.[63] Finally, both use language that assumes the sovereignty of the Orthodox Christian God over all human endeavour, including warfare, and hint that spiritual rewards belong to those who fight faithfully.

Although the *On Military Matters* does not indicate its author, nor is there a title in the manuscripts, it resonates with the *Tactics* of Leo VI in significant ways. Although this manual is primarily concerned with northern foes like the Pechenegs, Bulgarians and Rus, it also references Arabs and Turks[64] as enemies, and mentions fighting 'in the land of the Agarenes'.[65] However, the ending chapter declines to elaborate further because it is 'superfluous to write about something which everyone already knows'.[66] At the beginning of the tenth century, Leo VI wrote that military science was in disarray and that the 'Saracens' were devastating Byzantine lands; by the 990s, his advice had clearly been found successful.[67]

The religious language of the *On Military Matters* provides further similarities. The general instructed in this manual is exhorted to honour his soldiers 'as defenders of Christians' (ὡς τῶν Χριστιανῶν προμάχους) because they fight 'on behalf of Christians' (ὑπὲρ τῶν Χριστιανῶν).[68] Both of these comments are first made in Leo's *Tactics*; indeed, they are key to building morale and fortifying the courage of those who fight. Meanwhile, he who leads is referred to throughout as the 'holy emperor', emphasising his divinely approved leadership. Beyond this, the help of God (Θεοῦ συνεργίου) is invoked repeatedly, mostly in the chapters that describe dangerous situations like moving through a mountain pass, repelling an ambush or marching at night.[69] Other locutions include prevailing 'with God's favour' (τοῦ Θεοῦ διδόντος), avoiding harm 'by the grace of Christ' (τοῦ Χριστοῦ ἀντιλήψει), and defeating enemies 'by the grace of God' (τοῦ Θεοῦ χάριτι).[70]

Taktika of Nikephoros Ouranos

The *Tactics* (*Taktika*) of Nikephoros Ouranos (d. ca. 1011), written ca. 1000, is tremendously long, and remains unedited; it incorporates the *Tactics* of Leo VI in chapters 1–55, and the works of ancient strategists in chapters 66–178, whilst the ten chapters between them (56–65) contain an expanded version of the *Precepts*, a section from the *On Military Matters*, and material from the first-hand experience of the author.[71] Known for his victories over the Bulgarians, Ouranos here reveals strategic priorities a full century after the promulgation of Leo's *Tactics*, primarily reflecting the shift from conquest to retention of recovered lands. His paraphrase of Leo's work eliminates the *prooimion* and the first constitution, as rightly out of date. Where Leo was writing to address ongoing defeats at the hands of the Muslims, a century later this Byzantine general could not share the same concern. Rather, his work assumes ascendant Christian military power; thus Ouranos determined to 'set down only the methods that our generation currently employs'.[72] This comment refers specifically to principles of siege warfare, not military science generally, but one might extrapolate by Ouranos' inclusion of most of the *Tactics* that it was worthy of repetition. He refers to tactics succeeding 'with the help of God', echoing the Byzantine belief in the providence of God as well as the stance of previous tenth-century military writings. With this manual, Herbert Hunger notes, Ouranos 'schliesst die Reihe der byzantinischen taktischen Sammlungen ab'.[73]

IV. Byzantine views of the enemy

Andrew Cappel has claimed that the tenth century witnessed an abrupt reversal of a ninth-century trend toward sedentarisation among nomadic tribes in the al-Jazira region, resulting in greater instability along Byzantium's eastern border.[74] He argues that the presence of imperial concerns about the Bedouin horsemen in later tenth-century military manuals provides evidence of the problem. The *De velitatione* mentions 'Arabs' (Ἄραβας) as additions to the large invading armies, composed of jihadis from 'Egypt, Palestine, Phoenicia, and southern Syria to Cilicia' that seasonally gathered along the frontier to raid Roman territory.[75] The effect of this nomadic migration, Cappel argues, was mostly negative: 'The *'arab* cut off roads, plundered commercial caravans, attacked Christian and Muslim pilgrims and devastated agricultural districts, some – like al-Ghuta around Damascus – repeatedly.'[76] However, it is doubtful that such strict distinctions can be drawn, given that these texts also use the more general terms 'children of Ishmael' (τῶν τοῦ Ἰσμαὴλ), 'Cilicians' (τῶν Κιλίκων), 'sons of Hagar' (τῆς Ἄγαρ υἱῶν), 'men of Tarsus' (τῶν Ταρσιτῶν), but preponderantly, simply 'enemies' (οἱ πολέμιοι).[77] The writers of these texts do not distinguish between nomads and pastoralists, and the terminology is sufficiently vague that it cannot bear the weight of Cappel's thesis. The most that can be asserted, then, is the existence of fast-moving horsemen who posed a significant concern to tenth-century Byzantine commanders. Whether those mounted raiders were nomads does not matter. In fact, the *De velitatione* indicates that the majority were settled, because the writer advises invading their country to make them break off a siege on the Roman side of the border.[78]

The tone of the *De velitatione* is more urgent, the *Military Precepts* more confident. The *Military Precepts* calmly recommends that

> the best time to seek general engagements with the enemy is when, with the help of God, the enemy has fled once, twice, or three times and are crippled

and fearful, while on the other hand our host is obviously confident and their thoughts of valor have been awakened.[79]

This statement is retained, virtually unchanged, in the *Tactics* of Ouranos. At the same time, the writer of the *Military Precepts* acknowledges the advantages of guerrilla warfare under otherwise inauspicious circumstances: 'If the enemy force far outnumbers our own both in cavalry and infantry, avoid a general engagement or close combat and strive to injure the enemy with stratagems and ambushes.'[80] The difference seems to be that the *Military Precepts* views general engagement as a possibility the general ought to be prepared to initiate, while the *De velitatione* appears primarily though not exclusively concerned with reacting appropriately to Arab incursions.

It is worth noting that the Muslim and Christian populations of the frontier regions were not limited to contact by armed conflict. Throughout the tenth century, commercial contacts were maintained, diplomatic ties waxed and waned but never completely disappeared, and prisoner exchanges were not unusual. In fact, both the *De velitatione* and the *Military Precepts* discuss the use of spies. In the former, they are *trapezitai*, fast-riding warriors used as scouts and raiders, and merchants who are instructed to 'pretend to make friends with the emirs who control the castles in the border regions.'[81] In the latter, the writer reveals an awareness of the danger of Arab spies, giving common-sense instructions for flushing out infiltrators inside the army encampments.[82]

In the opinion of Alphonse Dain, the brevity of the contents of the *Military Precepts* lead the reader to believe it must have continued beyond the six published chapters, apparently with material that is now lost.[83] The *Tactics* of Nikephoros Ouranos may provide some of the missing material; it follows the text of the *Precepts* rather closely but adds more information on setting ambushes, performing counterraids and designing sieges.[84] Both compositions describe warfare on the eastern front, against Arab raiders and Bedouin horsemen. The material in the *Military Precepts*, supplemented by the *Tactics* of Ouranos, thus reveals a fuller picture of the methods and exigencies of the Byzantine reconquest of the 960s, characterised by the use of heavy cavalry charges and effective siege warfare.

The author of the *De velitatione* also makes frequent reference to the Byzantine canon of military literature, such as the *Tactics* of Leo VI and other works.[85] The writer of the *Military Precepts*, by contrast, makes no such explicit references. It is quite possible that the fault here lies with the 14th-century scribe who gave up the work of transcribing a tachygraphic text, especially if the general's shorthand was illegible or sparsely detailed. The presence of older works incorporated into the *Tactics* of Ouranos testifies to the obvious familiarity of Byzantine military writers with older compositions.

V. Conclusion

The first two treatises are written in two registers of literary Koine, the better of which Haldon identifies as 'what might loosely be termed a formalised and practical Constantinopolitan technical and administrative register.'[86] The *Precepts* because of its very roughness is therefore widely considered an original text written by Nikephoros II Phokas, similar to the precursor of the *De velitatione* that was later reworked and polished by a skilled editor. The length of the *Military Precepts*, however, casts doubt on its characterisation as a field manual. Although fragmentary and obviously lacking its full text, the sheer amount of detailed instruction seems too cumbersome for quick reference on campaign. That generals were required to be literate and that they often brought such manuals on campaign with them is evidenced in the tenth-century writings of Leo's son, Constantine VII Porphyrogennetos, which lists the books that ought to belong in a field commander's mobile library, including 'military manuals (βιβλία στρατηγικὰ) . . . [and] historical books (βιβλία ἱστορικὰ), especially Polyaenus and Syrianus'.[87]

McGeer concludes, 'Nikephoros Phokas composed the *Praecepta militaria*, his notes formed the basis for the *De velitatione*, and his call for a treatise on campaigning in the west inspired the *De re militari*.'[88] This summary seems plausible, if a bit too tidy. On the same grounds, one could also argue that Nikephoros composed the *Military Precepts*, a more literate relative with similar combat experience actually wrote the *De velitatione*, and while the emperor's wish may have inspired the *On Military Matters*, it was assuredly written a generation or more after his death.

The literary importance of these military handbooks, according to Dagron, rests in their approach to war as an art, a regular and not occasional part of social life.[89] He views these texts as providing a sort of moral, aesthetic and even technical background, rather than attempting to regulate the ongoing activity of waging a war. His analysis is persuasive, since warfare in the east had been a perennial fact of life for centuries. In effect, these texts are philosophical as well as practical, offering more than advice and revealing more than mere information. However, their primary function was that they 'sauvent de l'oubli' Byzantium's past heritage, while also presenting models for posterity.[90] They succeed in preserving not only knowledge, but a Byzantine faith-infused way of expressing it.

However, the concept of a military manual as a moral background for Byzantine military culture has not been fully exploited. Of course, Dagron was principally referring in his comment to the sections that list qualities desirable in a general, but the frequent references to prayer, ritual, religious disposition and even penalties for failing to fulfil them – these also qualify as rich material for exploring the moral stance of the Byzantines. Morality is inextricably bound up with religion, with what is viewed as right and wrong, and Byzantine culture was inescapably influenced by its Eastern Orthodox religion.

If, as some have defined it, ideology is 'the intellectual legitimation of social domination,'[91] then the dominant social force of Byzantium was surely Orthodox Christianity. This is abundantly evident even in its military tactical manuals. The declarations in the military manuals of the mid-tenth century follow the received wisdom in the *Tactics* of Leo VI, who says that Byzantine soldiers are not only defenders of the empire, but indeed defenders of the faith: 'we fight against this sort of impiety [= Islam] by means of our piety and orthodox faith' (XVIII.111).

Indeed, they are exhorted to consider themselves brothers of all those who fight against the enemies of the true God.[92] It may have ultimately been this conviction, inspired and modelled by a powerful and devout emperor-general, that enabled the Byzantine soldiers to push through the last defences of the frontier emirates and reclaim the offensive in the never-ending battle against the Muslims. To give credit where it is due, spiritual fervour alone of course would not suffice – the genius of centuries of Greek and Roman military strategy fuelled the shift, while ideology and the rhetoric of Christian Orthodoxy gave it the weight of legitimacy.

Notes

All dates are CE.

1 In both Polyaenus' account (2.3.3) as well as that of Plutarch in his *Sayings of the Kings and Commanders*, Epaminondas announces to his frightened soldiers that the thunder was an ill omen for the opposition, whom he described as 'thunderstruck' (ἐμβεβροντῆσθαι): πάλιν δὲ προσάγων τοῖς πολεμίοις, βροντῆς γενομένης καὶ τῶν περὶ αὐτὸν πυνθανομένων τί σημαίνειν οἴεται τὸν θεόν, ἐμβεβροντῆσθαι τοὺς πολεμίους εἶπεν ὅτι τοιούτων χωρίων ἐγγὺς ὄντων, ἐν τοιούτοις στρατοπεδεύουσιν. ~ 'On another occasion, when he was leading his troops against the enemy, there came a thunder-stroke, and when those about him inquired what he thought the god meant to signify by this, he replied, "That the enemy have been thunder-struck out of all sense because, when such places as those are near at hand, they pitch their camp in places such as these"' (Loeb Classical Library translation).
2 Polyaenus 2.3.4: 'When Epaminondas invaded the Peloponnese, he found the enemy encamped at Mount Oneium. A violent storm of thunder happened at the same time, which greatly intimidated his army, and the soothsayer declared against fighting. But Epaminondas said that it was the right time for fighting; because the thunder was clearly directed against the enemy in their camp. His interpretation of the occurrence brought fresh courage to his soldiers; and they advanced eagerly to the attack' ~ Ἐπαμινώνδας ὑπερέβαλλεν ἐς Πελοπόννησον. οἱ πολέμιοι κατὰ τὸ Ὄνειον (ἐμβαλόντες) ἐστρατοπεδεύοντο. βροντὴ γίγνεται, καὶ φόβος αἱρεῖ τοὺς στρατιώτας. ὁ μὲν μάντις ἐπισχεῖν ἐκέλευεν. Ἐπαμινώνδας δὲ 'μηδαμῶς', ἔφη· τοὺς πολεμίους γὰρ ἐν τοιούτῳ χωρίῳ στρατοπεδεύσαντας ἐμβεβροντῆσθαι. ὁ λόγος τοῦ στρατηγοῦ θάρσος ἐνέδωκε τοῖς στρατιώταις, καὶ προθύμως ἠκολούθησαν.
3 Dain and Foucault (1967) 354.
4 French edition: Dain and Noailles (1944). Greek edition: Troianos (2007).
5 Haldon (2014) 33.
6 Dvornik (1948, repr. 1970). Turner (2016) 475–489. For a different view of the schism, see Zymaris (2001) 345–362.
7 A good and comprehensive summary of Leo's literary output can be found in Antonopoulou (1997) 16–23.
8 See Chapter 2.
9 Dain (1935). See also Dain and Foucault (1967) 354.
10 Tougher (1997) 167–168.
11 Leo 18.117. Dennis (2014) 481.
12 Dennis (2010) 6–7. Ὥσπερ οὖν ἄλλον τινὰ πρόχειρον νόμον ὑμῖν, ὡς εἴρηται, στρατηγικὸν τὴν παροῦσαν πραγματείαν ὑπαγορεύοντες προσεχῶς τε καὶ ἐπιπόνως ἀκούειν ὑμῶν παρακελευόμεθα. (Prologue 6)
13 Certainly previous emperors had written military manuals, but Leo VI is the first who wrote a treatise on tactics that specifically considered the military challenge of the Caliphate.
14 Riedel (2018a).

15 Dennis (2010) 3–5.
16 Dennis (2010) 643.
17 Dennis (2010) 475.
18 Haldon (2014) 24.
19 Leo's book appears to use primarily three identifiable sources: *The General* of Onasander; the *Tactica theoria* of Aelian for definitions; and the *Strategikon* of Maurice for the arrangement of the material. These three indicate the essential library of the early tenth-century strategist. Haldon has identified several further possible sources, including a derivative compilation of Polyaenus (for Constitution 20 primarily) and possibly three texts of Syrianus *magistros* in his *Commentary* (2014) 39–55.
20 Riedel (2018a) chapters 2–4. Dennis (2010) 561.
21 18.111. δοκοῦσιν εὐσεβεῖν, βλασφημίαν δὲ τὴν αὐτῶν δοκοῦσαν εὐσέβειαν.
22 18.111. Χριστὸν μὲν τὸν ἀληθινὸν Θεὸν καὶ τοῦ κόσμου σωτῆρα καλεῖν Θεὸν οὐκ ἀνέχονται.
23 18.111. καὶ πολέμοις χαίρειν λέγουσι τὸν Θεὸν τὸν διασκορπίζοντα ἔθνη τὰ τοὺς πολέμους θέλοντα.
24 18.111. παντὸς δὲ καὶ κακοῦ ἔργου τὸν Θεὸν εἶναι αἴτιον ὑποτίθενται.
25 18.117. ὡς ἀπὸ Θεοῦ γὰρ τὸ πᾶν, εἰ καὶ κακὸν εἴη, λέγοντες εἶναι, εἰ συμβῇ αὐτοὺς ἐναντίον τι παθεῖν, ὡς ἀπὸ Θεοῦ ὁριζομένου οὐκ ἀντιπίπτουσιν, ἀλλὰ τῇ προσβολῇ σφαλέντες χαλῶσι τὸν τόνον.
26 For more on the church's refusal to grant martyrdom to soldiers, see Riedel (2015) 121–147.
27 Nicephorus II Phokas proposes an even more radical solution: automatic granting of military martyrdom.
28 The *Advice and Anecdotes* of Kekaumenos in the late eleventh century is a parainetic text that in parts most closely resembles Leo's perorations on the ideal general; it is also addressed to a son and to an emperor. See Roueché (2003) and (2013).
29 For more on this idea, see Chapter 5.
30 Leo identifies Onasander by name at 14.112. Constitution 2.1–38 follows very closely the material in Onasander's *Strategikos*, chapter 1, albeit in paraphrase. See also Chlup (2014).
31 These themes are repeated briefly in the Epilogue and more fully in Constitution 20.
32 Riedel (2018b).
33 Καὶ φιλούμενον δὲ παρὰ τῶν ὑπηκόων τὸν στρατηγὸν εὐδοκιμώτερον ἴσμεν γίνεσθαι καὶ γὰρ μεγάλα τοὺς ἀρχομένους ὠφελήσειεν. ὅντινα γὰρ ἄνθρωποι φιλοῦσι, τούτῳ ἐπιτάττοντι μὲν ταχὺ πείθονται, λέγοντι δὲ καὶ συντιθεμένῳ οὐκ ἀπιστοῦσι, κινδυνεύοντι δὲ συναγωνίζονται τοιοῦτον γὰρ ἡ ἀγάπη, τὸ τιθέναι τὴν ψυχὴν ὑπὲρ τοῦ φιλουμένου.
34 *De vel.*, 20:11–12. See also Darkó (1916–17) 129–146.
35 Whately (2016), citing Krenz and Wheeler, McGeer, and Rance in scholarly literature.
36 Dennis (1985a) xix.
37 Kolia-Dermitzaki (2002) 141–162.
38 This is a reference to 2 Tim 4:7–8, 'I have fought the good fight, I have finished the race, I have kept the faith. Henceforth there is laid up for me the crown of righteousness, which the Lord, the righteous judge, will award to me on that Day, and not only to me but also to all who have loved his appearing.' The text that refers to this verse is from the prologue to Evodios, 'Life and struggle of the 42 holy martyrs' (βίος καὶ ἄθλησις τῶν ἁγίων τεσσαράκοντα δύο μαρτύρων) [*AASS mart.6* (Venice 1668), vol. I (Greek: cols. 887–893; Latin: 460–466]. Evodios' version is the one that appears in the *Synaxarion*. The *Synaxarion* is a 13th-century list of the feasts and stational churches of the city of Constantinople, based on a number of manuscripts. Delehaye (1902).
39 Evodios repeatedly mentions the inducements of physical pleasures in sections 17, 19, 20, 21, 25, 27 and 40. Of course the martyrs resisted them all.
40 English edition: Dennis (1985b) 144–239. French edition: Dagron and Mihăescu (1986) 28–135.
41 McGeer (1995) 12–59.

42 Dennis (1985b) 246–335.
43 McGeer (1995) 88–164.
44 Morris (1988) 84.
45 Laiou (2006) 35.
46 Laiou (1998) 399.
47 *De vel.*, XX.8–10, XXI.8–13. See also Riedel (2015) 121–147.
48 *Theognosti Thesaurus*, Munitiz (1979) 196–203.
49 McGeer (1995) 228.
50 Cheynet (1986), 289–315.
51 Dagron and Mihǎescu (1986) 162.
52 Κύριε Ἰησοῦ Χριστέ, ὁ Θεὸς ἡμῶν, ἐλέησον ἡμᾶς, ἀμήν. McGeer (1995) 44. This same prayer is copied verbatim in the *Tactics* of Ouranos.
53 Maurice counsels prayers before battle, rather than a noisy approach to battle. His recommended prayers are the *Kyrie eleison* and just before joining with the enemy, an officer shouting 'Deus nobiscum' (II.18). Dennis (1985a) 33–34. This last is also exhorted by Vegetius 3.5.
54 χριστιανοὺς ἡμᾶς παράλαβε, ἀξίους ποιῶν ὑπὲρ τῆς πίστεως καὶ τῶν ἀδελφῶν ἡμῶν ἀναστῆναι καὶ ἀγωνισθῆναι μέχρι θανάτου, ῥωννύων καὶ ἐνισχύων τὰς ψυχὰς καὶ τὰς καρδίας καὶ τὸ ὅλον ἡμῶν σῶμα, ὁ κραταιὸς ἐν πολέμοις Θεὸς καὶ ἐν ἰσχύϊ ἀνείκαστος, πρεσβείαις τῆς τεκούσης σε Θεοτόκου καὶ πάντων τῶν ἁγίων, ἀμήν. McGeer (1995) 44. Also copied verbatim in the *Tactics* of Ouranos.
55 *Qua ratione famem collecti patiuntur hostes, dispersi uero crebris superuentibus facile uincuntur.* Vegetius, *Epitome*, tr. N.P. Milner, III.3, 67. Although written in Latin, the *pi Epitome of Military Science* was probably translated into Greek at the same time as his volume on veterinary medicine, that is, in the fifth century.
56 *De vel.*, XX.45–72. Cf. Dennis (1985b) 220–221.
57 Dennis (1985b) 139.
58 *De velitatione*, 17.135. Cf. Dennis (1985b) 210–211.
59 *De vel.*, 19.34, 19.46, 19.52–54.
60 McGeer (1995) 174–175. Cf. Dagron and Mihǎescu (1986) 161–165.
61 McGeer (1995) 174.
62 Kulakovskij and Dennis plump for Basil II, while Vári thought it was composed during the lifetime of Nikephoros II. Cf. Dennis (1985b) 242–243.
63 The manuscripts are *codex Vaticanus graecus 1164* (V), *codex Scorialensis graecus 281 (Y-III-11)* (S), and *codex Barberinanus graecus II 97 (276)* (B). According to Dennis, S is a copy of V, and B comes from a lost manuscript deriving from the same exemplar as V. V contains only chapters 1–14, while S and B contain chapters 1–32.
64 Likely meaning Hungarians or Magyars of the Balkans. Cf. Moravcsik (1958) 131–145.
65 Dennis (1985b) 302–303 and 326–327 (chs. 21 and 32).
66 Dennis (1985b) 327 (ch. 32).
67 Prologue 3: 'Everyone can clearly see, with his own eyes, how the collapse of strategic knowledge has cast all the affairs of the Romans down to such a degree as we experience at this very moment'.
68 Dennis (1985b) 318–321 (ch. 28).
69 Variations on this same vocabulary translated by Dennis as 'God's assistance' (300, ch. 20), 'with God working on our side' (300, ch. 20), 'with God aiding us' (302, ch. 20), 'with the cooperation of God' (304, ch. 21 and 312, ch. 25).
70 Chapters 12, 23, 24, and 26, respectively.
71 McGeer (1995) 80. Critical edition of chapters 56–65: McGeer (1995) 88–163. Critical edition of chapters 63–74: Foucault (1973) 281–312; critical edition of chapters 54 and 119–123 in Dain (1943) 69–104; the classic study of the sources and manuscripts: Dain (1937).
72 McGeer (1995) 163.
73 Hunger (1978) 2.337.

74 Cappel (1994) 114.
75 *De vel.*, chapter 7. Dennis, 162–163.
76 Cappel (1994) 116.
77 *De vel.*, prooimion and chapters 7, 15, 20. The manual attributed to Syrianos magistros, formerly thought to be sixth-century but recently dated to perhaps the later ninth century, sadly does not help in this regard, as it contains no mentions of these people groups. There is one sole mention of Arabs, distinguishing them from Romans, but not in the context of describing an enemy. See Rance (2007) 711–714.
78 *De vel.*, chapter 20.
79 McGeer (1995) 51 (IV.19).
80 McGeer (1995) 51 (IV.19).
81 *De vel.*, II and VII. Cf., Dennis, 163.
82 Ch VI. Cf. McGeer (1995) 57.
83 Dain and Foucault (1967) 370.
84 Cf. chapters 63–65.
85 Primarily in chapters 20 (direct reference to Leo) and 21 (generally to books on strategy and tactics).
86 Haldon (1990) 70.
87 Haldon (1990), Text C.196–202.
88 McGeer (1995) 178.
89 Dagron and Mihăescu (1986) 139.
90 Dagron and Mihăescu (1986) 140.
91 Larraín (1979) 17.
92 Vieillefond (1935) 323.

Bibliography

Antonopoulou, T. 1997. *The Homilies of Leo VI*. Leiden: Brill.
Cappel, A. J. 1994. 'The Byzantine Response to the 'Arab (10th–11th Centuries)'. *Byzantinische Forschungen* 20. 114.
Cheynet, J. C. 1986. 'Les Phocas', in G. Dagron and H. Mihăescu, eds., *Le traité sur la guérilla de l'empereur Nicéphore Phocas (963–969)*. Paris: Centre National de la Recherche Scientifique. 289–315.
Chlup, J. T. 2014. 'Just War in Onasander's ΣΤΡΑΤΗΓΙΚΟΣ'. *Journal of Ancient History* 2. 37–63.
Dagron, G. and Mihăescu, H. 1986. *Le traité sur la guérilla de l'empereur Nicéphore Phocas (963–969)*. Paris: Centre National de la Recherche Scientifique.
Dain, A. 1935. *Leonis VI sapientis problemata*. Paris: Les Belles Lettres.
——— 1937. *La 'Tactique' de Nicéphore Ouranos*. Paris: Les Belles Lettres.
——— 1943. *Naumachica*. Paris: Les Belles Lettres.
——— and Foucault, J.-A. 1967. 'Les stratégistes byzantins'. *Travaux et Mémoires* 2. 317–392.
——— and Noailles, P., eds. 1944. *Les novelles de Léon VI le Sage*. Paris: Les Belles Lettres.
Darkó, E. 1916–17. 'Die Glaubwürdigkeit der Taktik des Leo Philosophus'. *Ungarische Rundschau für historische und soziale Wissenschaften* 5. 129–146.
Delehaye, H., ed. 1902. *Synaxarium Ecclesiae Constantinopolitanae*. Brussels: Socios Bollandianos.
Dennis, G. T. 1985a. *Maurice's Strategikon*. Philadelphia: Penn Press.
——— 1985b. *Three Byzantine Military Treatises*. Washington: Dumbarton Oaks.
——— 2010, repr. 2014. *The Taktika of Leo VI*. Washington: Dumbarton Oaks.

Dvornik, F. 1948, repr. 1970. *The Photian Schism: History and Legend*. Cambridge: Cambridge University Press.

Foucault, J.-A. 1973. 'Douze chapitres inédits de la *Tactique* de Nicéphore Ouranos'. *TM* 5. 281–312.

Haldon, J., ed. 1990. 'Constantine Porphyrogenitus, Three Treatises on Imperial Military Expeditions'. *CFHB* 28. 196–202.

———— 2014. *A Critical Commentary on the Taktika of Leo VI*. Washington: Dumbarton Oaks.

Hunger, H. 1978. *Die hochsprachliche Literatur der Byzantiner*. Munich: Beck.

Kolia-Dermitzaki, A. 2002. 'The Execution of the Forty-Two Martyrs at Amorion: Proposing an Interpretation'. *Al-Masāq* 14. 141–162.

Laiou, A. 1998. 'The General the Saint: Michael Maleinos and Nikephoros Phokas'. *Byzantina Sorbonensia* 16. 399–412.

———— 2006. 'The Just War of Eastern Christians and the Holy War of the Crusaders', in R. Sorabji and D. Rodin, eds., *The Ethics of War: Shared Problems in Different Traditions*. London. 30–43.

Larraín, J. 1979. *The Concept of Ideology*. London: Hutchinson.

McGeer, E. 1995. *Sowing the Dragon's Teeth: Byzantine Warfare in the Tenth Century*. Washington: Dumbarton Oaks.

Moravcsik, G. 1958. *Byzantinoturcica*. Vol. 1. Berlin: Akademie-Verlag. 131–145.

Morris, R. 1988. 'The Two Faces of Nikephoros Phokas'. *Byzantine and Modern Greek Studies* 12. 83–116.

Munitiz, J. A., ed. 1979. 'Theognosti Thesaurus'. *Corpus Christianorum, Series Graeca, No. 5*. 196–203.

Rance, P. 2007. 'The Date of the Military Compendium of Syrianus Magister (Formerly the Sixth-Century *Anonymus Byzantinus*)'. *Byzantinische Zeitschrift* 100. 702–737.

Riedel, M. L. D. 2015. 'Nikephoros II Phokas and Orthodox Military Martyrdom'. *Journal of Medieval Religious Cultures* 41.2. 121–147.

———— 2018a. *Leo VI and the Transformation of Byzantine Christian Identity: Writings of an Unexpected Emperor*. Cambridge: Cambridge University Press.

———— 2018b. 'Biblical Echoes in the *Taktika* of Leo VI', in C. Rapp and A. Külzer, eds., *The Bible in Byzantium: Appropriation, Adaptation, Interpretation*. Göttingen. 23–38.

Roueché, C. 2003. 'The Rhetoric of Kekaumenos', in E. Jeffreys, ed., *Rhetoric in Byzantium*. Aldershot. 23–38.

———— 2013. *Kekaumenos, Consilia et Narrationes*. London. Online publication (www.ancientwisdoms.ac.uk/library/kekaumenos-consilia-et-narrationes/).

Tougher, S. 1997. *The Reign of Leo VI (886–912) Politics and People*. Leiden: Brill.

Troianos, S. N. 2007. *Οι Νεαρές Λέοντος Ϛ΄ του Σοφού. Προλεγόμενα, κείμενο, απόδοση στη νεοελληνική, ευρετήρια και επίμετρο*. Athens: εκδόσεις Ηρόδοτος.

Turner, J. 2016. 'Was Photios an Anti-Latin? Heresy and Liturgical Variation in the *Encyclical to the Eastern Patriarchs*'. *Journal of Religious History* 40.4. 475–489.

Vieillefond, J.-R. 1935. 'Les pratiques religieuses dans l'armée byzantine d'après les traités militaires'. *Revue des études anciennes* 36. 322–330.

Whately, C. 2016. 'The Genre and Purpose of Military Manuals in Late Antiquity', in G. Greatrex and H. Elton, eds., *Shifting Genres in Late Antiquity*. Abingdon. 249–261.

Zymaris, P. 2001. 'Neoplatonism, the Filioque and Photios' Mystagogy'. *Greek Orthodox Theological Review* 46. 345–362.

Epilogue

Is war an art? The past, present, and future of Greek, Roman, and Byzantine military literature

Immacolata Eramo

I. In the beginning, there was Homer

'Homer was the first to write on the tactical theory in war' ~ Ὅτι Ὅμηρος πρῶτος περὶ τῆς ἐν τοῖς πολέμοις τακτικῆς θεωρίας ἔγραψεν. This is the fourth time in this volume that this passage of Aelian appears; Homer is therefore a thread that binds how to think about military texts.[1] Presenting his *Tactics*, Aelian briefly mentions his predecessors, with whom he compares himself favourably, underlining proudly his own experience in the field of *mathemata*. Just as Aelian's predecessors were his own point of reference, Aelian himself wishes to be the same for his successors, who will prefer his writing to those who preceded him.[2] Among these predecessors, Aelian puts Homer in first place: Homer discovered tactical theory and had great appreciation for experts in tactics, such as Menestheus.[3] Furthermore, tactical theory was born in Homer's time. Finally, the authors of tactics 'wrote about tactics following from Homer'.[4] Moreover, it is not surprising that Aelian began the list of his predecessors with Homer. Aelian was perfectly integrated in a cultural tradition that identified Homer as the 'first inventor' (πρῶτος εὑρετής) of each literary framework.[5] Just as Strabo considered Homer to be the first geographer, according to Aelian (but also Arrian and before them their common source), Homer was the first military author in western literature.[6]

As Conor Whately observes in his chapter, Homer's authority in the literature and history of military thought lasted centuries; and as noted more than once in this volume, the examples of Alexander and Philopoemen described by Plutarch in his *Parallel Lives* clearly show this. Alexander kept a copy of the *Iliad*, 'a guidebook of the virtue in war', under his pillow. The great Macedonian general 'once said that the *Iliad* and the *Odyssey* accompanied him on his campaigns and acted as a relief from strain or as a companion during moments of pleasant idleness'.[7] Philopoemen, who was Achaean *strategos* on eight occasions, often resorted to the reading of passages taken from Homeric poems in order to excite and stimulate his soldiers' bravery and imagination.[8] Even though Homer's primacy for the *paideia* and behaviour of the Greeks had been questioned by the teaching of the Sophists and finally desacralised by Plato, Homeric poems were also considered an essential point of reference for subsequent ages.[9] Supplying a general although synthetised picture of Homer's *Fortleben* in western military literature and thought is a challenging task and is not our aim here. It will be enough to cite a few examples. Homer is the only author explicitly mentioned by Onasander in his *The General*, which is even more noteworthy if we consider that Onasander never cites his sources.[10] The same thing occurs in manuals on tactics, generally without literary references. The only exception is Homer himself. Indeed, both Aelian and Arrian explicitly cite Homeric passages in underling the need for order and silence in the army, so that soldiers can pay attention to their general's orders.[11] Two verses taken from the *Iliad* are reported in Maurice's *Strategicon* among the maxims, as if they were merely a decorative tribute, not useful to explain the precept that these verses accompany.[12]

Nicholas Sekunda demonstrates how authors after the Homer engaged with his poems – possibly even after the first actual military manuals entered circulation. The influence of Homer is also evident – thanks to Aelian's mention – among the pages of modern military essays, where his absence is at times as telling as his presence. This can be seen in the 'Companion of military knowledge' (*Syntagma de studio militari*) by Gabriel Naudé (1600–1653) and in the edition of all Greek and Latin military authors that Friedrich Haase (1808–1867) intended to publish. Gabriele Naudé (1600–1653), Cardinal Mazarin's librarian, who was a great admirer of Machiavelli and the author of the first treaty on the coup d'état, was also the first writer to collect a well-reasoned repertoire of military writings, manuscripts, and printed books, and to include it in his *Syntagma de studio militari*, dedicated to Count Ludovico Guidi di Bagno and published in Rome in 1637.[13]

According to Naudé, Homer, like Vegetius, is a supreme example of a military author without experience of war.[14] In the introduction, the author claims he should not be accused of incompetence, openly admitting that he does not have experience in the military field.[15] Naudé is well aware that he is taking on a work of an unfair and unusual weight, so much so that he might even appear ridiculous, like Phormio who dared to talk of military matters in the presence of Hannibal.[16] Therefore, even though his Muses are lovers of peace, and his habits are far from the turmoil of war, the author states that he is able to carry out his work and deal with military matters using texts by ancient authors. These include firstly works of history, from which he can draw the hidden thoughts and secrets of command, closed within as if they were in a secret hiding place; secondly, the biographies of important generals; and finally, the authors of military texts (*auctores de re militari*): Greeks, Romans, and 271 'more recent authors' (*recentiores*), classified into seven categories.[17] In short, many and almost countless writers have tried to illustrate military matters, even though in a crude and disordered way. Homer is one of these. It is a fact that Homer's mention indicates a precise choice, directly connected, as Naudé himself admits, to the list of Aelian's military predecessors.[18]

In the editorial project devised but never carried out by Haase, Homer's fate was different. In an 1835 article published in the *Neue Jahrbücher für Philologie und Pädagogik*, Haase complained that there are not enough Greek sources writing before the Peloponnesian war, and above all before the numerous wars of Alexander's successors, in order to reconstruct a theory of the *Kriegsthaten*.[19] Regarding times before these events, Haase admitted that it is necessary to resort to other sources, to Homer for the older ages, and to the historians for the more recent ones. Therefore, Haase too believed that the history of Greek military thought cannot be separated from either Homer or from the historians; even though these authors are not 'military' in the broadest sense, they filled a gap in a specialised literary form. However, just a few years later, he changed his mind. Indeed, when Haase came up with the ambitious project of publishing all the Greek and Latin military authors, the author he omitted was Homer himself. We do not know the reasons for this choice; it was not to concentrate exclusively on military treaties, since he kept the military sections from historical works in the project. He probably preferred to focus on unknown or lesser-known authors rather than select – as he did with the historians – the Homeric sections dealing with war. In this case, his choice would have been anything but selective and would have involved considerable effort, making the project even more ambitious than it already was.

II. Military history and military literature

The story of Haase's project, or rather of his failed editorial enterprise, deserves to be briefly told, since it offers food for thought regarding military history and its relationship with ancient military ideas. In 1833, Haase published his edition of Xenophon's *On the Constitution of the Spartans*.[20] According to Carl Rudolf Fickert, Haase's biographer as well as his disciple and friend, this edition was a noteworthy work, not only from a philological point of view – Haase was of course a student of the well-known philologist Karl Lachmann – but also because of the great attention paid to military questions. The more difficult they were to understand, the more application they required. However, Haase was not a man to leave his work incomplete. Therefore, driven by the desire to focus more deeply upon these subjects, Haase began to study ancient military authors, to draw tactical manoeuvres, and to consult the experts in this field.[21]

This research was the beginning of the thirty years of work which Haase devoted to military authors, resulting in his extensive library and his idea to publish a companion. The scholar persevered with this project until his death but was not able to carry it out, since he never found an editor willing to publish the collection. Indeed, he succeeded only in publishing – more than ten years later and at his own expense – a brochure, the 'On the edition of all Greek and Roman military writers in progress' (*De militarium scriptorum Graecorum et Latinorum omnium editione instituenda*), where he exposed the structure, contents, and characteristics of his work.[22] Firmly convinced that philology was a single discipline, and that no text was of greater worth than another, Haase believed he was carrying out a precious and useful work by publishing an edition of military writers, for which he requested also the collaboration of other scholars in searching for manuscripts or published books and in editing the texts, clearly hoping above all for some financial help.[23]

As we have already seen, this ambitious work never saw the light of day. Haase's book illustrates the different phases of his work, some of which had already been carried out. These phases included the collection of all Greek and Latin military writings up to the fifteenth century, the search for manuscripts and emendation of texts, the drafting of the edition with critical apparatus, commentary, introduction, indices, military technical lexicon, and Latin translation of Greek writings.

The collection should have comprised nine volumes, organised in a logical and chronological structure: 1. ancient authors and those not included in other volumes (among these: Xenophon, *Cavalry Commander* and *On Horsemanship*, Aineias the Tactician, Onasander), but also texts from Thucydides and Xenophon's historical works, 'and others on Greek military art' and from Polybius, both his work on the Roman army and other military fragments; 2. Writers of tactics; 3–4. Authors of mechanics; 5. Collection of stratagems; 6–7. Byzantine military treaties; 8. Undefined fragments, Latin authors (Frontinus, Vegetius, the so-called Modestus, the *De rebus bellicis*); and 9. Various writings (book 10 of Vitruvius' *De architectura*, Ps.-Hyginus, Medieval authors).

III. Military thought and literature

A quick look at the content of Haase's plan suggests that he used an extensive criterion for collection, including not only military treaties *stricto sensu*, but also passages taken from different sources, *in primis* the histories, providing they dealt with 'the military art of the Greeks' (*de Graecorum arte militari*).[24] Ultimately, Haase made a choice regarding which texts to include. He could do little else, if we consider that war is directly or indirectly present in most Greek, Roman, and Byzantine literature, being an essential part of life for the ancients. Two passages from ancient Greek authors – Heraclitus and Plato – that the introduction cites are important here: according to Heraclitus of Hephesus, 'war is father of all (beings) and king of all, and so he renders some gods, others men, he makes some slaves, others free'.[25] Plato believed that the word 'peace' was actually an empty name, because a state of undeclared and nonstop war existed between all the *poleis*.[26] The situation is not so different if we shift focus to Rome, which was able to build an empire thanks to the force of arms, and to New Rome, which defended its borders with arms for more than a thousand years.[27]

These circumstances were crucial conditions for the birth and evolution of a military praxis and theory, which developed above all thanks to the comparison with 'the other': hoplite tactics evolved into the Macedonian phalanx, which in turn had an influence on the Roman legion; Pyrrhus taught the Romans how to set up a camp; Scipio Aemilianus conquered Carthage with the art of stratagems learnt from Hannibal (and the skills he acquired from reading – and re-reading – Xenophon's *Education of Cyrus*). James T. Chlup's chapter serves as a reminder that Hannibal never lurks far from discussion of military strategy and history.[28] Therefore, the history of the ancient world is also the history of war; and the best writers of military history are military experts:

> a military historian cannot be a sort of divinely inspired master of the art of war. Comments, interpretations, and conclusions worth advancing can only be offered when they are based on the writer's own personal military

experience . . . *or* on his reading of good military reports from the past, *or* on his contact and discussion with expert military men.[29]

Hans Michael Schellenberg would appear to agree based upon his *caveat* in his chapter.

Thucydides was a *strategos* in Thrace in 424–423 BC and therefore not only described the Peloponnesian war, 'great and worthier of recording than any previous conflict (μέγαν τε ἔσεσθαι καὶ ἀξιολογώτατον τῶν προγεγενημένων)', but also *how* Athenians and Spartans fought (ὡς ἐπολέμησαν πρὸς ἀλλήλους).[30] Xenophon was a mercenary soldier led by Cyrus the Younger but was above all 'an able tactician, as is clear from his writings' (τακτικός, ὡς ἐκ τῶν συγγραμμάτων δῆλον).[31] It is most probable that Diogenes Laërtius here refers not only to *Cavalry Commander* and *On Horsemanship*, but also to the *Education of Cyrus*, which contains a long digression on military precepts, considered also in the early modern period as a point of reference for anyone wishing to learn the art of command, as Conor Whately introduces and Jeffrey Rop explores in this volume; the shadow of this text, almost as large as Homer's, unequivocally confirms its importance in the study of the military text. Indeed, in Spring 1444, Alfonso V of Aragon asked Giovanni Aurispa to translate the collection of military writers sold to him by Aurispa himself. In the summer of the same year, Aurispa replied that it would be more useful to translate the *Education of Cyrus*, of which he was already in possession.[32] Among the military historians who were expert in war, a pivotal place is occupied by Polybius, firstly a *hipparchus* of the Achaean League, before going on to follow Scipio Aemilianus in the campaign in Numantia in 134 BCE.[33]

The problem of the relationship between military history and war experience became crucial in historiographic thought and its methods. In the aftermath of Lucius Verus' victory against the Parthians (166 CE), the increasingly great number of 'amateur' historians led Lucian of Samosata to write a polemical pamphlet, *How to Write History*. Lucian states that most people believe it is perfectly simple to write about history and that anyone can do it if only he is able to express what he thinks. On the contrary, it is not so easy, because the best writer on history should have political understanding and power of expression, but also military ability. He therefore should have a knowledge of generalship, he should be one who has at some time been in a camp and seen soldiers training or marching and knows arms and machines, what 'in column' and 'in line' mean, how the infantry and cavalry units are manoeuvred, what 'lead-out' means, 'in short, not a stay-at-home [the historian] who only believes what people tell him'. According to Lucian, direct experience of war allows the writer of history to tell the truth, which is the main aim of a historical work: 'it was Thucydides himself to lay down this law' (ὁ δ' οὖν Θουκυδίδης εὖ μάλα τοῦτ' ἐνομοθέτησεν).[34]

The extremely close relationship between wars and the histories that describe them ensures that histories themselves are the best source not only to reconstruct events, but also to try to outline the history of ancient military thought. This is something very different to the aims of military manuals. As we will see, these manuals are subject to different rules to histories, having their own rules entirely. Considering military treaties to be military histories has caused confusion and misunderstanding, which is also responsible for confining these works to the rank of 'secondary' or 'minor' literature, being accused of 'antiquarianism', 'pedantry', or 'abstractness'.[35] Aelian's list allows us to clarify this misconception. Along with Homer, Aelian cites authors who he considers to be his predecessors for the tactical genre. Indeed, Homer himself is not considered to be the first military author, but the first author writing on tactics. Unfortunately, the state of the textual tradition of these works provides us with only few clues. Among the cited authors, some are little more than names (Stratokles, Hermias, Cineas the Thessalian, Alexander Pyrrhus' son, Clearchus, Evangelus, Eupolemus, Bryon).[36] Others are known but not because they are writers of military works; these include Pyrrhus, Pausanias, Iphicrates, Poseidonius.[37] And still others wrote *de re militari* works, but not on tactics or generalship, or better, their writings on these subjects have not been handed down: Frontinus is the author of the *Strategemata*, along with a treaty on aqueducts, but probably he wrote a larger military work, as we will see. Aineias the Tactician wrote not only the *Poliorketika*, but also a military compendium, divided into several sections, to whom the author himself refers.[38]

Polybius seems to be an exception to Aelian's criterion, since he was a military historian and not a writer of tactics.[39] In actual fact, Polybius himself cites his own treaty of tactics, the περὶ τὰς τάξεις ὑπομνήματα (*Tactics*), which according to some scholars could be the source of the treaties by Asclepiodotus, Aelian, and Arrian, through Posidonius.[40]

Following this line of reasoning, Polybius takes on a pivotal role. We can infer from his *Histories* that Polybius was an expert both in tactics and generalship.[41] We do not know if he collected the military thoughts, which he conceived during his account of the Roman wars, later in the ὑπομνήματα or, more probably, if he inserted the thoughts already expressed in the ὑπομνήματα in the key points of his *Histories*, as could be inferred from what 'I have already explained in more elaborate detail' (ἀκριβέστερον δεδήλωται). Although the loss of this work and the absence of every other clue mean that we cannot make a precise judgement, we might believe, however, that the ὑπομνήματα belonged to a different literary genre, being not a military history but a military manual.[42]

The birth of the technical manual, which regarded several fields of knowledge, from agriculture to medicine and mathematics, was a key event in Greek culture at the end of the fifth century BCE, when the increasing diffusion of more specialised crafts created the problem of how to pass on technical knowledge. It was above all Plato who understood and expressed this need. He believed that the acquisition of knowledge does not require notions of a certain skill to be passed on from an expert to one who has no experience. Instead, knowledge requires a more complex process, able to unite theory, codified in written form, with supervised training and then experience.[43]

This type of literature did not in any way stand as a substitute for the traditional channels of knowledge, the word of the teacher and practice, but it was a precious help. On one side, it provided in advance everything that could be learnt during personal experience through precepts, rules, and general information; on the other, it addressed the past, because it collected the results of others' experience. Vitruvius highlights this important function of technical literature in the Augustan age while dealing with the subject of architecture: a different field, but one which has many things in common with the military ambit. According to Vitruvius, the ancestors wisely and usefully handed down their knowledge and scientific theories (*cogitata*) for posterity, through written evidence, so that they would not be lost. Since these theories became richer and richer with the passing of time, they were published in volumes in order to achieve the highest degree of knowledge. Vitruvius concludes that we should therefore be grateful to the ancestors, as they did not neglect to spread their results because they did not pass on in envious silence, but were instead committed to handing down in written form the result of each scientific acquisition.[44]

The first surviving examples of this type of literature were Xenophon's *Cavalry Commander* and *On Horsemanship*, and, as Conor Whately notes in his chapter, Aeneas the Tactician's *How to Survive under Siege*.[45] The fact that they are the first explains some of their stylistic features, like a lack of homogeneity, repetition, the treatment of subjects one after the other without a precise general structure, and the influence of spoken language.[46] On the other hand, as Felmingham-Cockburn's chapter suggests, with respect to language there may, in fact, be more at play. Indeed, regarding Xenophon's writings, the didascalic tone is associated with philosophical reflections (the relationship between the thing to mould and the artist's will in *Cavalry Commander* 6.1; the power given by gods to men of instructing one another by word in *On Horsemanship* 8.13), in the same way that technical descriptions are mixed with images from daily life (the blade of a knife in *Cavalry Commander* 2.3; the rows of seats in a theatre in 2.7; hawks and wolves in 4.18–20; a child's game in 5.10; the foundations of a house in *On Horsemanship* 1.2; the dancer in 11.6.).

IV. The work of a military writer

When technical knowledge was written down, it was able to leave the restricted field of practice and receive a wider and more general audience. Crucially, it could travel far in space and time. In this way, because this technical knowledge was written down in literary form, it was not limited to a mere reproduction and transmission among a small circle of professionals, but instead became 'another' type of knowledge, susceptible to analysis and reflection and therefore inserted into wider technical and cultural areas, giving rise to repetition, comparison, distinction, and integration.[47]

Nadya Williams in her chapter raises important questions about authors and readers of military texts. This choice to aim at a wider readership of professionals in each field is also evident in the military ambit.[48] Aelian concludes the list of his predecessors in tactics stating that their writings were addressed to experts. This inconvenience created some difficulty, because Aelian himself could not find anyone able to explain the most difficult topics to him. For this reason, he chose to illustrate clearly the principles of tactics, resorting, when necessary, to diagrams. He also decided to use the same technical terms employed by the ancients, in order to familiarise his readers with their use and to make it easier for them to understand the works of his predecessors, thanks to his simple treatment of the topics.[49]

Aelian was not the first to do this. A little less than a century before, Hero of Alexandria stated that his *Engines of War* (*Belopoeica*) had been written with a technical, but not a specialised, language. He too chose a simple and essential language, which was detailed and accessible to any reader. In contrast to his predecessors who wrote for an expert readership, Hero considers that taking information from these writers and dealing with the same subjects, even though they are sometimes old-fashioned, and presenting them in such a way as to be understood by anyone, make a work worthy of praise. Therefore, Hero states that in the manual, he is dealing with the construction of machines and their parts, dwelling on names, composition, construction, dimensions, and uses, after having described the differences between the machines and their evolution.[50] The intended readership of the *Belopoeica* were not artillery experts but were instead officials and generals wishing to complete their military training with precepts on this subject. For this reason, Hero, even more than Aelian, because of the specialised nature of his subject, considers it appropriate to insert some drawings in his text, illustrating various mechanisms as a didactic aid, since this was of great help to the reader understanding the text. These illustrations are present in the miniatures of the handed-down codices.[51]

The need for clarity is of great importance when a military author wants to address a large and unidentified readership, helping them to understand some key concepts that ensure the transmission of knowledge. Vitruvius had been a military engineer in Caesar's army dealing with the supply and repair of ballistae, scorpions, and other artillery.[52] For this reason, he decided to include in his *On Architecture* a section on military mechanics. Justifying his choice, Vitruvius provides his reader with an interesting reflection on the characteristics and aims of military manuals. Indeed, he explains that the machines which he describes cannot be used in the same places and in the same way, since fortification systems and adversaries are always different: each reader can choose the most suitable information for his needs. In the same way, it is pointless to describe defence systems, 'since the enemy does not construct their defences on the basis on our writings'.[53]

The aim of addressing a large and heterogeneous readership in some way leads the military author to not dwell upon in-depth details, which are linked to each specific situation, but to provide the reader with generic teachings, drawing from tradition or his own experience only what he believes to be useful for this type of readership and for both present and future conduct. Once again, one might reflect that the caution articulated by Schellenberg in his chapter may be warranted.

The question of readership and characters of the military manual genre can be well understood through an examination of the preface to Onasander's *The General* (*Strategikos*).[54] This work is addressed to Quintus Veranius, *legatus Augusti* between 42 and 43 CE in the province of Lycia and later (57–58) governor of

Britain. Along with Veranius, Onasander also identifies magistrates *cum imperio*, senators, and, generally, all Romans as the ideal users of a manual on generalship, a readership as numerous as it is undefined, as is also evident from a comparison of the potential readership of manuals on other disciplines: in contrast to military writings, treaties on riding, hunting, fishing, and agriculture should be addressed to those who want to engage in these activities.[55]

In the preface, Onasander also explains the criterion he uses in writing the manual and his aims. He then outlines his own role as a collector, composer, and writer of strategic precepts. Firstly, Onasander wished to 'deploy' a 'deployment' – a play on words by the author – of precepts so that it would be training for good generals and a delight for old commanders.[56] Secondly, he wanted to explain why some commanders have made mistakes and failed, whereas others have achieved victory and glory. Ultimately, he aimed to highlight Roman values, since the Romans did not extend their empire to the boundaries of the world by chance, but thanks to war. Illustrating the ambitious aim of his collection, Onasander defines his role as a writer: not someone who is too young and inexperienced in war, which are indispensable conditions to being considered credible, since people put their trust in one who shows himself to be an expert, even though he might not write elegantly from a literary point of view. On the other hand, people do not lend credence to non-experts, even if they write about precepts which can be carried out. Onasander states that all his subjects come from the great feats of the Romans, not his own; nevertheless, he justifies his approach saying that if a general composes a manual on generalship telling of the great enterprises of others along with reflections based on his own experience, he cannot be considered as an unreliable witness. In the same way, not everything included in his manual is the result of his own mind.

Actually, no reflection and precept from *(The General) Strategicus* can be directly linked to detailed and well-known episodes, despite the best efforts of modern scholarship to find elements of inspiration in them.[57]

The potential but also the limits of this type of research are clearly demonstrated by the fact that we can encounter many equally valid examples for each precept. Manuals on strategy and tactics differ from works on stratagems because they are theoretical in nature and do not necessarily require examples. Indeed, while Frontinus' *Stratagems* (*Strategemata*) and Polyaenus' *Stratagems* (*Strategica*) are military writings, they are not strictly speaking real manuals, since they do not feature the characteristics typical of this genre: didactic style, precepts, admonitions, and descriptions.[58] They are instead collections of anecdotes. In practice, they cannot substitute the military precepts of military manuals, but are an addition to these precepts.[59] The chapters of Aaron L. Beek and James T. Chlup, however, combine to challenge this assumption by suggesting that there may be broader, overarching objectives in these texts: they may be more than their individual pieces suggest. Indeed, Frontinus, the first representative figure of this genre (or subgenre), had already engaged in the writing of a military manual before collecting the *Stratagems*. In the preface to his work, Frontinus proudly claims to have been the only author to have organised the rules of military knowledge.[60] As he believes to have achieved this aim, he consequently summarises in convenient sketches the skilled enterprises of the generals (*sollertia ducum facta*) that Greeks call 'stratagems'. In this way, generals – who evidently constitute the readership of Frontinus' collection – will have access to models of wisdom and experience, useful to give them a greater ability to design and carry out similar enterprises and to not worry about the results of their stratagems, since they can compare these with the successful experiences of the past.[61] This underlines what Wrightson explores in his chapter, though with a twist of reading about stratagems to understand how to counteract them. Along with the problem of direct references to identified episodes, the research and identification of sources for this type of literature are no less problematic and always provide open and often contradictory results.

We have seen that, except for Homer, Onasander never cites his sources, but instead puts together a work that does not include concrete examples. The result is a manual which is generic bordering on banality. This characteristic determined his fortune in Byzantine military literature but also condemned him to the criticism of modern scholarship. Alphonse Dain defined Onasander as 'une aimable graeculus, nullement versé dans l'art militaire, prodigue de flatteries'.[62] Collecting and writing useful precepts for future generals is Onasander's declared aim. In other words, knowledge of war is essential, as nobody is afraid of putting into practice what he has well learnt (*nemo facere metuit quod se bene didicisse confidit*).[63]

This principle is the basis of Vegetius' *Epitome of Military Science* (*Epitoma rei militaris*), including *Regulae bellorum generales* at the end of the book (3.26.1– 33) – certainly written by Vegetius and central to the conceptualising of the text, as Jonathan Warner explains in his chapter, which, importantly, had great fortune in the subsequent military tradition (for example, Maurice and Machiavelli's *Art of War* (*L'arte della guerra*).[64] Here, the historical context is completely different from the years of the empire when Onasander lived. Vegetius wrote in the aftermath of the defeat at Adrianople (378 CE), a moment of great disruption for the collective mindset of the Romans.[65] Vegetius identifies military structures as being one of the weaknesses of the empire, leading to defeat. According to the writer, Roman military structures were weak due to a long period of inactivity and were completely unsuitable to defeating the barbarians who attacked the borders of the empire. In order to remedy this problem, Vegetius proposes his own solution to the emperor – to whom his work is addressed[66] – consisting in the *Epitome*[67] of a compendium of military theories from the past. He was convinced that he was undertaking a difficult and tiring operation which would be of great use to everyone, with the aim of reforming the military structures of the time.[68] Therefore, Vegetius' plan includes a return to the glorious ancient legion, or rather, to the values which it represents, through the application of its pivotal principles: the recruitment of Roman citizens and careful selection of soldiers, hard training, strict discipline, and knowledge of tactics and strategy for the generals. At first glance, Vegetius' proposal has a mere antiquarian value (see also earlier, 270).

Indeed, this was the case. However, we should not forget that Vegetius' solution had its origin in literature and was intended to remain there. This concept is the key characteristic and at the same time the most evident limit of this type of literature. Onasander and Vegetius, like Vitruvius before them, highlight the limitations of precepts in a military manual, which cannot include the whole repertory of possible events in war, but only offer guidelines, suggestions from one's own or another's experience, and also – in the case of stratagematic literature – examples and anecdotes as models to follow or to avoid.

V. The role of Byzantium in the tradition of military texts

Clear ideas, which can be easily understood by non-experts, general precepts, which can be used in various circumstances, elementary notions: this dimension without time and space determined the fortune of military manuals in Byzantine culture and, indeed, their survival – their 'future', or rather, the survival of some manuals to the detriment of others. For example, the presence of an articulated and in-depth compendium like Vegetius' *Epitome* was probably decisive for the loss of the previous writings, which, in part, Vegetius himself cites and uses (Cato the Elder, Cornelius Celsus, Frontinus, Tarruntenius Paternus: 1.8.10–11; 15.4; 2.3.6–7). We have only clues about these, and, as Murray Dahm's chapter suggests, speculating about a 'lost' text is useful in filling in the history of genre and demonstrating its appeal to ancient readers and generals. Indeed, if we exclude the accidents of the manuscript tradition – for example, regarding a corpus of writings on seamanship – the transmission of Greek military literature was the result of a process of selection which occurred in Byzantium.[69]

There was also a widespread activity in elaborating *sylloges*, epitomes, and excerpts.[70] To comment on each of these writings and on the complex relationship they have with each other and with ancient military writings would be long and complex. It is enough to take a brief look at the titles of Alphonse Dain's studies over more than twenty years, which give us a clear idea of how these anonymous companions made use of the military literature of the previous centuries.[71]

This was not only a literary operation. Besides the fact that 'the compulsiveness of the Byzantines' archaistic bent makes it frequently difficult for students of the works of Byzantine Greek literature to diagnose whether a writer is recording contemporary facts or is retailing some conventional tradition',[72] the Byzantine approach towards war was a pivotal element in order to know of the military writings of the past and their reuse: Clemens Koehn provides a useful frame of reference here. Referring to an ancient *topos*,[73] the Byzantines considered war to be the worst and absolute evil, a barbarous act which the civil man should avoid:[74] conflicts are created by the devil, who uses sin in order to arm one man against another.[75] Although obliged to defend its own borders with arms, the Eastern Roman Empire never managed to conceive a positive theory on war, which could enhance or at least justify the use of arms. Therefore, Byzantium developed a war *techne* which was strongly based on the precepts of the ancients, but richer, more complex and codified, giving priority to strategic aspects, even at the expense of avoiding, whenever possible, the use of arms. They therefore valued reason over blind courage, organisation and discipline over number and strength.[76]

The retrieval of past military tradition is clearly shown in a compendium of an *armchair general* such as Syrianus Magister, particularly in the section devoted to tactics.[77] There Syrianus uses Aelian's *Tactics*, from which he draws the description of various manoeuvres, without providing changes or updates and actually proposing abstract models, which certainly were no longer used or usable in either war or training.[78] In his 1980 article, Vladimir Kučma underlined that Syrianus applied to his compendium a method of 'minimum transformativity', characteristic of a 'transformative-reductive' work.[79] In practice, Syrianus collects the precepts of the ancients modifying them just enough to make them relevant, through a process of simplification. The author's aim is evidently not to give practical advice to generals or officials who are actively involved in the field of battle, but instead to carry out a cultural operation, engaging both the highest-ranking officers and officials or men close to the emperor. In short, his aim is to provide a *Kriegsbildung*.

Past scholarship highlighted the 'antiquarian' value of Syrianus' compendium, particularly when compared with Maurice's *Strategicon*.[80] As a few contributors in this volume have suggested (Whately, Chlup, and Caldwell), antiquarianism is a recurring perceived feature that recurs throughout the genre; in fact, perhaps it is what 'tethers' the genre. Maurice's text, however, is the most original and creative product of Byzantine military literature. Maurice was personally engaged in wars against the Persians and intended to write (or have someone write) a handbook which would be, above all, useful and therefore easily and clearly written – 'we have paid no attention to the correctness of the writing or the sound of the words' – a modest manual of elementary notions or an introduction for those devoting themselves to the art of generalship (μετρίαν τινὰ στοιχείωσιν ἤτοι εἰσαγωγὴν τοῖς εἰς τὸ στρατηγεῖν ἐπιβαλλομένοις).[81] Nevertheless, in the same text where Maurice illustrates the aim and characteristics of his work, he admits his debt to ancient military authors: 'we have resolved to write on this subject, as best as we can, briefly and simply, drawing in part on ancient authors and in part on our limited experience'.[82] In doing this, Maurice declares that he does not expect to introduce new elements or try to improve what the ancients have already done. Nevertheless, from his point of view, the ancients addressed themselves to experts and for this reason dealt with subjects which were difficult for the common reader, omitting necessary and fundamental elements, those which were, in turn, useful for his times.[83]

It is undeniable that Maurice reworked materials taken from ancient authors in order to create an original compendium, where he gives great attention to reflections based on his own and others' recent experience. In the same way, his treatment of tactics and military organisation is contemporary. Furthermore, in book 11 of his *Strategicon*, Maurice introduces a novelty regarding both content and method. Indeed, for the first time, a military manual devotes particular attention to the enemy, described as an abstract entity in the manuals of the past,[84] considering the characteristics of each (the Persians, Scythians, Franks and Longobards, and Slaves) both military and ethnographic, on the basis of ἁρμόζεσθαι, the need to adapt to the enemy in order to defeat him.[85]

In Byzantium, the *Strategicon* was the most important and pivotal manual, so much so that when, in the tenth century, Leo VI the Wise wished to collect a compendium *de re militari*, he simply referred to Maurice's work (*Strategicon*), reproducing the majority of its content, often *verbatim*, reworking and enriching it, in order to achieve his own aims.[86] Unlike Maurice, Leo was not an emperor-general and probably never took part in a military campaign or saw an army in battle in his life.[87] Like some of his predecessors (Syrianus, for example) or successors (as is the case of Naudé seen previously), Leo knew his Arab enemy from a distance, through his readings, his father's memories, and reports of his generals.[88] Besides the need to fit available military knowledge to practical necessities, and, as Meredith L. D. Riedel discusses in her chapter, insert a Christian frame, Leo wished to provide his readership with a consolidated repertoire of precepts. For this reason, he not only explained, paraphrased, and updated subjects from the *Strategicon* through linguistic revision, distributing them in a larger and more organised companion than the model adding to each subject his own personal reflections, but also enriched his work with sections taken from ancient authors, both on tactics (Aelian and Arrian) and strategy (Onasander) and on stratagems (Polyaenus).

The result is not a handbook that is ready to be applied in war, but a collection of precepts in order to educate the ruling class of a very militarised society. With Leo's *Tactica*, the tradition of ancient military manuals withstands and overcomes innovation, and theory prevails over practice. Presenting his work, Leo immediately states his starting point, the characteristics, the method, and aims. He believes that the fatal mistakes made in war are not so much due to a lack of discipline and courage on the part of the soldiers, or to the inexperience and cowardice of the generals, as to the lack of a comparison with military writers of the past, due to their obscure characteristics. Therefore, Leo decided to undertake the enterprise of collecting and organising the 'ancient and recent methods' of generalship and tactics, but also stories and descriptions of other types, in order to give his subjects a great benefit, offering them, as synthetically and clearly as possible, a 'regulatory manual', a πρόχειρος νόμος, a kind of rulebook for officers and men with military roles.[89]

VI. Is war an art?

The organisation of military knowledge in a compendium by Leo VI perfectly reflects the literary and cultural canons of the age of the Macedonian dynasty, which are expressed with the word 'encyclopaedism'.[90] Based on his experience in the study of Byzantine military literature, with this definition Alphonse Dain identified two different tendencies, the creation and compilation of extracts.[91]

Both these forms found their best expression thanks to Leo's son, Constantine VII Porphyrogennetos. During his reign, he patronised a historical collection (*Scriptores post Theophanem*) and above all *excerpta* of a moral-historical (*Excerpta historica*) and political nature (*De Cerimoniis*, *De Thematibus*, *De Administrando imperio*), and technical compendia, such as *Geoponica* (agriculture), *Iatrika* (medicine), and *Basilika* (law).[92] In our field, Constantine VII was responsible for the recovery of the Hellenistic and Roman military tradition, in order to collect and transmit ancient military knowledge and make this available in the imperial library of Constantinople.

The most important result of this activity is the *Laurentianus* LV.4. This codex is the official copy of writings on tactics and strategy circulating in the *scriptorium* of Constantinople in the age of Constantine VII and the most important witness to the direct tradition of Greek military manuals. In Alphonse Dain's words, this manuscript is 'a military encyclopaedia, where one finds a complete volume of different kinds of writing related to the art of war'.[93] The structure of the *Laurentianus* itself is noteworthy, since it is a corpus obtained from the composition of three different corpora. The first and the last are composed of Byzantine treaties. The first corpus presents the *Strategicon* at the beginning, in second place.

The core consists of the ancient military writings (Asclepiodotus, Aelian, Aeneas, Arrian's *Tactics* and *Expedition against the Alans*, Onasander). With a perfect *Ringkomposition*, the corpus is bookended by two small handbooks attributed to Constantine VII: at the beginning the so-called *Praecepta imperatori observanda*, which is really a collection of notes taken from Leo Katakylas' work.[94] At the end, we find the *De moribus diversarum gentium*, mutilous without an end due to the loss of the final folio of the manuscript, which is in fact an excerpt taken from Maurice's *Strategicon*, regarding the military uses of the Persians and Scythians. Likewise, the *How the Saracens Fight* (*Quomodo Saracenis debelletur*), which is positioned immediately before this, is an *excerptum* from Leo's *Tactics*.[95] This collection reveals an important and decisive date for the transmission of ancient military manuals. Constantine not only organised military knowledge, but, following in Leo's footsteps, began to collect notes on military subjects, probably wishing to give them a literary form later. In this way, the transmission of knowledge went hand in hand with the production of the same knowledge in a field that we moderns would expect to be closer to contemporary times. Byzantine military writers referred to the authority and precepts of the ancients, first as followers of a tradition of military thought and then as makers of a process of transmission of this knowledge, establishing a relationship of continuity which was valuable despite the existing military structures, social conditions, and historical situations.

So, is war an art or a science?[96] The Byzantines would have had no doubt when answering this question. Syrianus considers tactics as an ἐπιστήμη which allows the general to array and manoeuvre a body of armed men in an orderly way.[97] Defining the different classes in a state and why they have been established, Syrianus considers ἐπιστήμη and τέχνη (or rather τεχνικόν) as 'a way of carrying out something with minimum effort in a proper and long-lasting manner', so that an activity done 'τέχνη' is an activity which will be more easily completed and long-lasting.[98] If ἐπιστήμη identifies itself only in part with τέχνη, it cannot however be reduced to a simple 'knowledge' (γνῶσις), as Syrianus states defining the ἱερατικὴ τέχνη.[99] Considering war as a science, then, and as such, something to be known, communicated, preserved, codified, and transmitted, determined the survival of manuals by ancient polemographers, who were men of literature and science. Or rather, they were *also* men of literature and science.

Notes

1 See the Introduction (1–16) and Chapters 1 (17–38) and 4 (78–98).
2 Aelian, Preface 1: ἐμαυτὸν δὲ πείθων ἠβουλήθην ταύτην συντάξαι τὴν θεωρίαν, ὅτι τοῖς ἡμετέροις οἱ μεθ᾽ ἡμᾶς πρὸ τῶν ἀρχαιοτέρων προσέξουσι συγγράμμασιν.
3 *Iliad* 2.551–554. Still of great use is Cantarelli (1974), esp. 460–471.
4 Aelian, Preface 1: 'Tactical theory among the Greeks goes back to the age of Homer' ~ Τὴν παρὰ τοῖς Ἕλλησι τακτικὴν θεωρίαν ἀπὸ τῶν Ὁμήρου χρόνων τὴν ἀρχὴν λαβοῦσαν; 1.1–2: 'those who write on tactics following Homer' ~ καὶ περὶ τῆς καθ᾽ Ὅμηρον τακτικῆς συγγραφεῦσι.
5 Vela Tejada (2004) 130–131. In general, regarding so-called Greek 'heuremathography', see Zhmud (2006) 23–34; however, he does not deal with military matters.
6 Strabo 1.1.2; 1.11. On Homer's influence on the collections of *exempla*, see Wheeler (2010) 24–27.
7 Plutarch, *Alexander* 8.2; *On the Fortune or Virtue of Alexander* 327F–328A.
8 Plutarch, *Philopoemen* 4.6–7. See also the Introduction (2) and Chapter 9 (177). On Plutarch's attention to military matters in his *Parallel Lives*, see Gazzano and Traina (2014) with further bibliography; Jacobs (2018).
9 Plato, *Ion*, esp. 536d–539d (see earlier Chapter 4, 79); *Republic* 606e; see also Kahn (1993); Ford (2002) 201–208; and Hunter (2004) 246–249.
10 *Odyssey* 7.36 at 1.7 and *Iliad* 13.122 at 23.1. See Petrocelli (2008) 142–143 n.32 and Dueck (2011) 379–380.
11 Aelian 41, who cites (in order): *Iliad* 4.428–431; 3.8–9; 2.459–463; 4.436–437; 3.1–2; 3.8–9. Arrian at *Tactics* 31.5–6 cites *Iliad* 4.428–431; 2.459–460; 4.436–437; 3.8–9.
12 Maurice, *Strategikon* 8.2.82. It is *Iliad* 11.802–803 = 16.44–45.
13 This list of writers *de re militari* was then published posthumously, in 1683 in Jena, by Georg Schubart, with the title *Bibliographia militaris*. See Cochetti (1989) 91–93 and Bianchi (1996) 203–207.
14 'It is one thing to talk about war, it is another thing entirely to defeat the enemy and scare him off. Even though Flavius Vegetius wrote most elegantly about the Roman army, he cannot be compared to generals such as Sertorius or Marius or to soldiers like Lucius Dentatus or Marcus Sergius. Likewise, neither can Homer, *vates ille pauper, ac caecus, . . . et suis versibus pangendis idoneus*'.
15 Naudé (1637) 3–4.
16 Cicero, *On Public Speaking* 2.75–76.
17 Naudé (1637) 2 and 510; 514–550 for the *auctores de re militari*, on which see Ilari (2012) 142–144.
18 Naudé (1637) 514.
19 Haase (1835) 88–91.
20 *Xenophon. De Republica Lacedaemoniorum*, emendavit et illustravit Fr. Haase Magdeburgensis. Accedunt verborum index locupletissimus et rerum tacticarum figurae, Berolini: Dümmler 1833.
21 Fickert (1868) 18.
22 Haase (1847).
23 Haase (1847) 1–6 *passim*.
24 Haase (1847) 7.
25 Heraclit. fr. 53 Diels-Kranz (= 29 Marcovich 2001²), transl. M. Marcovich. See the Introduction (1).
26 Plato, *Laws* 626a. See Vela Tejada (2004) 129–130 and the Introduction (1).
27 The essays by Hölkeskamp (1997), esp. 481–489; Hornblower (2007), esp. 22–39; and Koder and Stouraitis (2012) 10–14 highlight the critical points of this matter.
28 On Pyrrhus, see Livy 35.14.8–9. On Hannibal, see Eramo (2013–14) 87–91, with additional bibliography.

29 Marsden (1974) 274 dealing with Polybius (see further later).
30 Thucydides 1.1.1. See Canfora (2006) 11–13. On ὡς ἐπολέμησαν πρὸς ἀλλήλους, see Petrocelli (2012) 16–19.
31 Diogenes 2.56. See Lee (2017) 26–29 with additional bibliography.
32 *Iniunxisti mihi ut opus, in codice quodam, graecum, De re militari transferre in latinum . . . neque hoc dico quod laborem fugere velim, sed menti habeo, si iusseris, Xenophontem De institutione regis Cyri et de omni eius vita scribentem in linguam nostram vertere; in quo opera magnam, ut spero, voluptatem legentibus feram et maiestati regiae, si quid illi gloriae addi potest, gloriam faciam. Habeo iam opus in manu et id pertracto*: Ferrara, July-August 1444, in Sabbadini (1931) 109–110. On this episode, see Eramo (2006) 169–171. In general, on the fortune of the *Education of Cyrus*, see also Humble (2017) 416–419.
33 Polybius 28.6.9. See Meißner (2013) with additional bibliography. In this volume, Whately fills a gap in critical literature, highlighting Polybius' specific tactical competences.
34 *How to Write History* 5; 34; 37; 41–42, with a commentary by Porod (2013) 285–287; 472–476; 490–493; 516–523.
35 See also observations by Rance (2017a) 10–13. On antiquarianism, see also later, 276, Chapter 10 and Chapter 12.
36 On Stratokles, see Chapter 4 (86–87).
37 On Pyrrhus' *Tactics*, see Wheeler (2010) 21. Some reports on these cited authors can be found in Rance (2017a) 15–16.
38 Works that Aineias mentions include: *On preparations* (*Paraskeuastike biblos*): 7.4, 8.5, 21.1, 40.8, *On Procurement* (*Poristike biblos*): 14.2, and *On Encampment* (*Stratopedeutike biblos*): 21.2. He identified the last of these as a work still to be written. Indeed, Polybius defines Aineias as an author of 'treatises on tactics' (τὰ περὶ τῶν Στρατηγικῶν ὑπομνήματα, 9.44.1), and Aelian states that Aineias 'composed a considerable number of books on strategy' (στρατηγικὰ βιβλία ἱκανὰ συνταξάμενος, 1.2). See Vela Tejada (1993) 83–85. Regarding Aineias' reference to ἀκούσματα that were written before, see Eramo (2008) 129.
39 An analysis of Polybius' military lexicon is in Meißner (2013) with additional bibliography.
40 Polybius 9. 20.4: 'On this subject, I have already written in more detail in my "Notes of Tactics"' ~ ὑπὲρ ὧν ἡμῖν ἐν τοῖς περὶ τὰς τάξεις ὑπομνήμασιν ἀκριβέστερον δεδήλωται. Devine (1995) exposes the hypothesis of Polybius as the common source of Asclepiodotus, Aelian, and Arrian through Posidonius, but see previously Müller (1896) col. 1640.
41 See, for example, the digressions on generalship at 9.12–20 or on fire-signalling at 10.41–47 and above all on the Roman military system in 6.19–42. On Polybius as a military historian, see Marsden (1974); Poznanski (1978), (1980a), (1980b), (1993), and (1994); and Devine (1995).
42 On the use of ὑπομνήματα in the military field, see Eramo (2018a) 13–14 with additional bibliography. More generally, see Potter (1999) 30–33.
43 Plato, *Symposium* 175d: 'How fine it would be, Agathon . . . if wisdom were a sort of thing that could flow out of the one of us who is fuller into him who is emptier, by our mere contact with each other, as water will flow through wool from the fuller cup into the emptier' ~ εὖ ἂν ἔχοι, φάναι, ὦ Ἀγάθων, εἰ τοιοῦτον εἴη ἡ σοφία ὥστ' ἐκ τοῦ πληρεστέρου εἰς τὸ κενώτερον ῥεῖν ἡμῶν, ἐὰν ἁπτώμεθα ἀλλήλων, ὥσπερ τὸ ἐν ταῖς κύλιξιν ὕδωρ τὸ διὰ τοῦ ἐρίου ῥέον ἐκ τῆς πληρεστέρας εἰς τὴν κενωτέραν.
44 Vitruvius 7 Preface 1. *Maiores cum sapienter tum etiam utiliter instituerunt, per commentariorum relationes cogitata tradere posteris, ut ea non interirent, sed singulis aetatibus crescentia voluminibus edita gradatim pervenirent vetustatibus ad summam doctrinarum subtilitatem. itaque non mediocres sed infinitae sunt his agendae gratiae, quod non invidiose silentes praetermiserunt, sed omnium generum sensus conscriptionibus memoriae tradendos curaverunt.* See Romano (2017) 53–56.

45 See Vela Tejada (1993) 89–92 and (2004) 140–142; Formisano (2009) 358–359; and Rance (2017a) 14–15 with additional bibliography; Shipley (2018) 57–64 and Dillery (2017) 196–199, 209–213; see here Chapters 1, 5, and 7. Regarding the chronological relationship between *How to Survive under Siege* and the *Cavalry Commander*, see also Bettalli (1990) 264. Besides the authors on Aelian's list, Diogenes Laërtius (9.48) cites two writings by Democritus, *On fighting in armour* (Τακτικόν) and *On tactics* (Ὁπλομαχικόν). We do not know if they are spurious (Leszl (2007) 62) or original (Jouanna (2015) 35–36).

46 Whitehead (1990) 17–21 and Vela Tejada (2018) 115–119.

47 Vela Tejada (1993) 87–89; Cambiano (1992) 525–528; in general, Zhmud (2006) 45–54, who nevertheless does not say anything regarding the military field.

48 Rance (2017a) 13–15.

49 Aelian 1.3–6.

50 *Engines of war* 73–74. Vd. Meißner (1999) 69–71; 250–251; Cuomo (2002); Tybjerg (2005) 212–213 and Cortney (2016) 86–87, 222–223.

51 The drawings are present in the miniatures of the Vat. gr. 1605 (eleventh century), the first exemplar of the tradition of Hero's text: Dain (1933) 25–33. It is evident that the reliability and adherence to the original illustrations is an open and difficult problem for this writing as it is for all illustrated technical manuals. See Weitzmann (1959) 7–10; Meißner (1999) 69–71 and 250–251 and Whitehead (2008) 149–150.

52 Vitruvius I Preface 2.

53 *Non enim ad nostra scripta hostes comparant res oppugnatorias*: *Vitruvius* 10.16.1–2.

54 Regarding the *Strategicus*, see Schellenberg (2007); the introduction of Petrocelli (2008) and Formisano (2011) and (2017) 21–24. On the readership of military manuals, see some general considerations by Whately (2015) and Chapter 3.

55 Onasander prae 1. Besides Chlup (2014) 39–43, see also Chapter 3. On the addressees of Onasander's preface, see Eramo forthcoming.

56 Differently to the rest of the work, the preface to *The General* is stylistically elaborated and interwoven with military metaphors referring to the activity of composition itself: τὰ παρ' ἐμοῦ συντεταγμένα . . . στρατηγικὰς συνεταξάμην ὑφηγήσεις . . . τὸ δὲ σύνταγμα θαρροῦντί μοι: prae. 3–4. See Petrocelli (2008) 132–133 n.11.

57 See, for instance, 10.22–24, 'On the secret plans of the general', which is linked to a long military tradition with many examples: Petrocelli (2008) 204–207.

58 According to *Suda* π 1956, Polyaenus was the author of *Tactica* in three books, but this information is anything but reliable. See also o 386 for Onasander, on which Wheeler (2010) 33 n.103 and Rance (2017a) 20.

59 See here Chapter 6. On the characteristics of the writings on stratagematics, see Wheeler (2010) 19–36.

60 The existence of a military manual by Frontinus in addition to the *Stratagems* is corroborated by Aelian's list (1.1, on which see earlier) with the citation by Vegetius (1.8.11; 2.3.7, see later).

61 Frontinus, Preface to book 1.

62 Dain and de Foucault (1967) 328–329; but see also Ambaglio (1981) 354 and Le Bohec (1998) 169. The judgement of Charles Guischardt, one of his translators in a 1758 edition, is merciless: 'pédant comme il l'étoit, et sans lumiéres, il fit son choix sans jugement; il confondit les époques, et la milice des Romains avec celle des Grecs, et il n'entra dans aucun des details, qui auroient rendu son Livre digne de son titre' (50–51).

63 Vegetius 1.1.7.

64 Vegetius 1.1.7, who adds 'scientific knowledge in warfare nurtures courage in battle' ~ *scientia enim rei bellicae dimicandi nutrit audaciam*. See Formisano (2002) 113–114.

65 Lenski (1997) and (2002) 320–367.

66 Various proposals have been put forward regarding the identification of the emperor as the addressee of the *Epitome*: see Reeve (2004) viii–x; Charles (2007) and Zecchini

(2008) 196–197 and n. 8 with further bibliography. It is evident that the emperor was not the only real reader of Vegetius' work, assuming that he was really able to read it. See here Chapters 3 and 11 for further discussion here.

67 *Epitoma rei militaris* or *Epitoma institutorum rei militaris* is the *intitulatio* present in the authoritative manuscripts. On this question, see Reeve (2004) v–vi and 1.

68 Vegetius, Preface to book 4 (§8).

69 Asclepiodotus, Aelian, and Arrian refer to a nautical section, although it is unclear if it is a new work or part of the same manual. Aelian announces a subsequent and separate work (2.1), Asclepiodotus and Arrian merely point out the two different fields in which war occurs. On the other hand, Aineias the Tactician's *How to Survive under Siege*, as it is handed down, is interrupted *ex abrupto* exactly in the passage of a writing regarding a nautical subject (40.8): see Eramo (2017) 140–141. On the other hand, we have the nautical appendix of Vegetius' *Epitome* (4.31–46): see here Chapter 12.

70 See Rance (2018) on the reception of Aeneas. For a general idea, see also Kaldellis (2012) on the role of Byzantium in the survival of the corpus of ancient Greek historiography.

71 See above all Dain (1930), (1931), (1933), (1934a), (1934b), (1935), (1937), (1939), (1940), (1942), (1946), (1946–47), (1953), and (1967). Regarding the long-lasting research by Dain on Greek military writings, see Grabar (1964) 237–238 and Rance (2017a) 31–32.

72 Definition of Toynbee (1973) 297.

73 See, for instance, Polybius 4.31.3; cf. Xenophon, *Hellenica* 6.3.6.

74 Syrianus, *On Strategy* 4: μέγα κακὸν εὖ οἶδ' ὅτι ὁ πόλεμος καὶ πέρα κακῶν.

75 Leo, *Tactics* Preface 4: 'But since the devil, the killer of men from the beginning, the enemy of our race, has made use of sin to bring men to the point of waging war against their own kind, it becomes entirely necessary for men to wage war making use of contrivances of the devil, developed through men and, without flinching, to take their stand against those nations that want war' ~ ἐπειδὴ δὲ ὁ ἀπ' ἀρχῆς ἀνθρωποκτόνος διάβολος καὶ τοῦ γένους ἡμῶν ἐχθρὸς διὰ τῆς ἁμαρτίας ἰσχύσας κατὰ τῆς ἰδίας φύσεως ἀντιστρατεύεσθαι τοὺς ἀνθρώπους παρεσκεύασεν, πᾶσα ἀνάγκη ταῖς αὐτοῦ γινομέναις διὰ τῶν ἀνθρώπων μηχαναῖς ἀνθρώπους ἀντιστρατεύεσθαι καὶ τοῖς ἐθέλουσι πολέμους ἔθνεσι μὴ εὐχειρώτους καθίστασθαι. See Laiou (1993) 166–168; Oikonomides (1995); and Stouraitis (2012).

76 Syrianus, *On Strategy* 6: τὸ εἰρηνεύειν αἱρούμεθα, κἂν τύχῃ ζημίαν τινὰ ἡμῖν ἐντεῦθεν ἐπάγεσθαι. The approach of Byzantium towards war sometimes assumed ambiguous characteristics: see Kaegi (1983); Breccia (2001) 53–72 and (2009) lxxv–lxxx; Stouraitis (2009); and Haldon (2014) 5–6.

77 See Eramo (2009), (2011), and (2012) with additional bibliography.

78 See Eramo (2018b).

79 Kučma (1980) 68–69.

80 See, for example, Dain and de Foucault (1967) 343.

81 Maurice, Preface 21–31. On the readership of Maurice, see Whately (2015) 253–254.

82 Maurice, Preface 14–17: 'We have resolved, therefore, to do some writing on this subject, as best we can, succinctly and simply, drawing in part on ancient authors and in part on our limited experience of active duty, with an eye more to practical utility than to fine words' ~ ἔκ τε τῶν ἀρχαίων λαβόντες καὶ μετρίαν πεῖραν ἐπὶ τῶν ἔργων εὑρόντες ταύτην συγγραφῇ παραδοῦναι, κατὰ τὸ ἡμῖν δυνατόν, συντόμῳ τε καὶ ἁπλῇ, τὴν ὠφέλειαν ἐπὶ τῶν πραγμάτων ἐχούσῃ μᾶλλον ἤπερ ἐν λέξεσι. For an in-depth commentary on this section, see Rance (2017b), esp. 222–226. In general, regarding Byzantine military literature, see Hunger (1978) 321–340; Cosentino (2009); and Sullivan (2010).

83 It is evident that here Maurice refers to Aelian's observation, which probably had become a type of *topos* in military literature (1.3: 'we know that these writers address to those whom are already acquainted with this subject' ~ ἐπέγνων δέ, ὡς ἔπος εἰπεῖν,

τοὺς συγγραφεῖς ὡς εἰδόσι τὰ πράγματα τοῖς ἀνθρώποις συντεταχότας). See also Vegetius 1.8.9 and observations by Rance (2017b) 225–226.

84 A 'military ethnography' is already present in Polyaenus, Preface 7: see Wheeler (2010) 48–54. However, we have seen that writings on stratagems cannot be considered as either manuals in a strict sense or compendia, but simply stratagem collections: Wheeler (2010) 19–34.

85 On this term, see also Chapter 13 (228–237).

86 Breccia (2011) 113–114.

87 Tougher (1997) 166–168 and Eramo (2018c) 159–162 with additional bibliography.

88 Leo, *Tactics* 18.103–125, on which see Dagron (1983) 219–224 and Haldon (2014) 352–374 with additional bibliography. A slightly different version to this section is appended as a separate text at the end of the *Laurentianus* LV.4.

89 Leo, *Tactics* Preface 5–6 and 9. See Haldon (2014) 125–127. Regarding the aims of Leo's *Tactics* and its relationship with military writings of the past, above all Maurice's treatise, see Tougher (1997) 168–172 and Haldon (2014) 39–55 with additional bibliography.

90 See Dain (1953) and Lemerle (1971) 148–154. The limits and the weak points of this definition are perfectly highlighted by Odorico (1990) 1–8 and (2011) 89–92; see also Piccione (2003) 54–56 and Magdalino (2011) 143–151 and (2013).

91 Dain (1953) 80: 'A dire vrai, cet encyclopédisme a revêtu deux formes. La première est spécifiquement constantinienne et consiste en fabrication d'extraits: on réduisit la production antérieure, et notamment la tradition héritée des anciens, à des formules plus brèves, formules qui, par conséquence, entraînaient l'oubli et la perte des ouvrages que l'on avait pu jusque là conserver *in extenso*. – L'autre forme este celle de la compilation, ordinairement paraphrasée, qui s'applique plus volontiers aux travaux modernes qu'à la littérature ancienne: de sources antérieures diverses et diversement combinées, on fait un ouvrage nouveau, passible à son tour de rajeunissement et de nouvelles compilations'. See also Trombley (1997) 261–269.

92 Regarding the cultural activities of Constantine VII, see Toynbee (1973) 575–605 and Breccia (2011) 111–114.

93 Dain (1954) 144; see Rance (2017a) 32–33 and (2018) 302–311 with additional bibliography. On the content and structure, see Eramo (2018a) 43–44.

94 Dain and de Foucault (1967) 361 and Haldon (1990) 45–53.

95 Maurice 11.2 and 3: 'On the costumes of different peoples' ~ Στρατηγικὸν περὶ ἐθῶν διαφόρον ἐθνῶν; Leo, 18.103–125: 'How to defeat the Saracens' ~ Πῶς δεῖ Σαρακηνοῖς μάχεσθαι (on which see earlier); see Dain and de Foucault (1967) 362, 384–385.

96 See also the Introduction of this volume (4–7).

97 Syrianus, *On Strategy* 14. Cf. Polyaenus, Preface 2.

98 Syrianus, *On Strategy* 14 2: 'Technicians make sure that projects will be carried out with a minimum of effort, in the proper manner, and with due regard for durability. Work done in a professional manner will be more easily completed and prove more solid' ~ τὸ δὲ τεχνικὸν διὰ τὸ ῥᾷον καὶ ὡς ἂν δέοι γίγνεσθαι τὰ γινόμενα καὶ διαρκεῖν τῷ χρόνῳ. τὸ γὰρ τέχνῃ τελούμενον οὐ μόνον ῥαδίως, ἀλλὰ καὶ ἀσφαλῶς γίνεται. See Meißner (1999) 337.

99 'Those who devote themselves to the sciences and the arts should possess the natural qualifications for their specialties' ~ τοὺς δὲ περὶ τὰς ἐπιστήμας καὶ τέχνας ἀπησχοληνένους φύσεως εὖ ἔχειν πρὸς τὸ σπουδαζόμενον: Syrianus, *On Strategy* 3; 'priestly service might also be classed as a profession, but I do not think it should be, nor should it be listed as a science' ~ τὴν δέ γε ἱερατικὴν δόξειε μὲν ἄν τις ἴσως τέχνην εἶναι, ἡμῖν δὲ οὐ τοῦτο δοκεῖ, ἀλλ' οὐδὲ ἐπιστήμην: *On Strategy* 1. Regarding these reflections by Syrianus, which is probably a valuable clue in order to date the compendium, see Eramo (2011) 219–222. On the relationship between ἐπιστήμη and τέχνη in the Aristotelian theory of science, see Zhmud (2006) 122–133.

Bibliography

Ambaglio, D. 1981. 'Il trattato "Sul comandante" di Onasandro'. *Athenaeum* 59. 353–377.

Bettalli, M. 1990. *Enea Tattico. La difesa di una città assediata* (Poliorketika). Pisa: ETS.

Bianchi, L. 1996. *Rinascimento e libertinismo. Studi su Gabriel Naudé*. Naples: Bibliopolis.

Breccia, G. 2001. '"Con assennato coraggio . . .". L'arte della guerra a Bisanzio tra Oriente e Occidente'. *MEG* 1. 53–78.

———— 2009. *L'arte della guerra. Da Sun Tzu a Clausewitz*. Turin: Einaudi.

———— 2011. 'I trattati tecnici e l'enciclopedia di Costantino VII Porfirogenito: arte militare e agronomia, in *Voci dell'Oriente. Miniature e testi classici da Bisanzio alla Biblioteca Medicea Laurenziana*, in M. Bernabò, ed., *Catalogo della mostra* (*Firenze, 4 marzo-30 giugno 2011*). Florence: Biblioteca Medicea Laurenziana. 133–142.

Cambiano, G. 1992. 'La nascita dei trattati e dei manuali', in G. Cambiano, L. Canfora, and D. Lanza, eds., *Lo spazio letterario della Grecia antica*. I, *La produzione e la circolazione del testo*. 1, *La polis*, Roma: Salerno Editrice. 525–553.

Canfora, L. 2006. 'Biographical Obscurities and Problems of Composition', in A. Rengakos and A. Tsakmakis, eds., *Brill's Companion to Thucydides*. Leiden: Brill. 3–31.

Cantarelli, F. 1974. 'Il personaggio di Menesteo nel mito e nelle ideologie politiche greche'. *RIL* 108. 459–505.

Charles, M. B. 2007. *Vegetius in Context: Establishing the Date of the* Epitoma Rei Militaris. Stuttgart: Franz Steiner.

Chlup, J. T. 2014. 'Just War in Onasander's Στρατηγικός'. *JAH* 2. 37–63.

Cochetti, M. 1989. 'Gabriel Naudé, *Mercurius philosophorum*'. *Il bibliotecario* 22. 61–104.

Cortney, R. 2016. *Technical Ekphrasis in Greek and Roman Science and Literature: The Written Machine between Alexandria and Rome*. Cambridge: Cambridge University Press.

Cosentino, S. 2009. 'Writing about War in Byzantium'. *Revista de História das Ideias* 30. 83–99.

Cuomo, S. 2002. 'The Machine and the City: Hero of Alexandria's *Belopoeica*', in C. J. Tuplin and T. E. Rihill, eds., *Science and Mathematics in Ancient Greek Culture*. Oxford: Oxford University Press. 165–177.

Dagron, G. 1983. *Byzance et le modèle islamique au Xe siècle. À propos des* Constitutions Tactiques *de l'empereur Léon VI*, CRAI 127.2. 219–243 [= Id., *Idées byzantines*. Paris, 2012. 329–352].

Dain, A. 1930. *Les manuscrits d'Onésandros*. Paris: Les Belles Lettres.

———— 1931. 'Les cinq adaptations byzantines des Stratagèmes de Polyen'. *REA* 33. 321–345.

———— 1933. *La tradition du texte d'Héron de Byzance*. Paris: Les Belles Lettres.

———— 1934a. 'Les manuscrits d'Asclépiodote le Philosophe'. *RPh* 60. 341–360.

———— 1934b. 'Les Manuscrits des Traités tactiques d'Arrien'. *Annuaire de l'Institut de philologie et d'histoire orientales* 2 (Mélanges Bidez). 157–184.

———— 1935. 'Les manuscrits d'Énée le Tacticien'. *REG* 48. 1–32.

———— 1937. *La Tactique de Nicéphore Ouranos*. Paris: Les Belles Lettres.

———— 1939. *Le Corpus perditum*. Paris: Les Belles Lettres.

———— 1940. *La collection florentine des tacticiens grecs: essai sur une entreprise philologique de la renaissance*. Paris: Les Belles Lettres.

———— 1942. *L'extrait tactique tiré de Léon le Sage*. Paris: Les Belles Lettres.

———— 1946. *Histoire du texte d'Élien le Tacticien. Des origines à la fin du moyen âge*. Paris: Les Belles Lettres.

———— 1946–47. 'Inventaire raisonné des cent manuscrits des *Constitutions tactiques* de Léon VI le Sage'. *Scriptorium* 1. 33–49.

———— 1953. 'L'encyclopédisme de Constantin Porphyrogénète'. *Bulletin de l'Association Guillaume Budé. Lettres d'humanité* 12. 64–81.

———— 1954. 'La transmission des textes littéraires classiques de Photius à Constantin Porphyrogénète'. *DOP* 8. 31–47.

———— 1959. *Titres et travaux de Marie-Alphonse Dain*. Limonges: Bontemps.

———— and de Foucault, J. A. 1967. 'Les stratégistes byzantins'. *T&MByz* 2. 317–392.

Devine, A. M. 1995. 'Polybius' Lost *Tactica*: The Ultimate Source for the Tactical Manuals of Asclepiodotus, Aelian, and Arrian?'. *AHB* 9. 40–45.

Dillery, J. 2017. 'Xenophon: The Small Works', in M. A. Flower, ed., *The Cambridge Companion to Xenophon*. Cambridge: Cambridge University Press. 195–219.

Dueck, D. 2011. 'Poetry and Roman Technical Writing: Agriculture, Architecture, Tactics'. *Klio* 93. 369–384.

Eramo, I. 2006. '"Un certo tractatello de l'officio del buon capitanio". Ludovico Carbone traduttore di "opere pellegrine"'. *Paideia* 51. 153–195.

———— 2008. 'Omero e i Maccabei: nella biblioteca di Siriano Μάγιστρος'. *AFLB* 51. 123–147.

———— 2009. '*Romaioi* e *Arabes* a battaglia? Nota al *De re strategica* di Siriano Magistros'. *InvLuc* 31. 95–104.

———— 2011. 'Sul compendio militare di Siriano Magister'. *RSA* 41. 201–222.

———— 2012. 'Composition and Structure of Syrianus Magister's Military Compendium'. *Classica et Christiana* 7. 97–116.

———— 2013–14. 'I triboli di Annibale. Nota a Servio *ad georg.* 1,153'. *InvLuc* 35–36. 85–91.

———— 2017. 'Syrianus' *Naumachiae*. Tactics, Strategy, and Strategies of Composition'. *HiMA* 5. 139–154.

———— 2018a. *Appunti di tattica* (De militari scientia). Besançon: Presses universitaires de Franche-Comté.

———— 2018b. 'Les écrits tactiques byzantins et leurs sources: l'exemple du *De re strategica* de Syrianos Magister'. *Rivista di diritto romano* n.s. III, 18. 159–174 (www.ledonline.it/rivistadirittoromano/).

———— 2018c. 'Pirati a Bisanzio. Una minaccia alla talassocrazia della Nuova Roma', in I. G. Mastrorosa, ed., *Latrocinium maris. Fenomenologia e repressione della pirateria nell'esperienza romana e oltre*. Rome: Aracne. 143–170.

Fickert, K. R. 1868. *Fickert, Friderici Haasii memoria*. Breslau: Grass, Barth and Company.

Ford, A. 2002. *The Origins of Criticism: Literary Culture and Poetic Theory in Classical Greece*. Princeton: Princeton University Press.

Formisano, M. 2002. 'Strategie da manuale: l'arte della guerra, Vegezio e Machiavelli'. *QS* 55. 99–127.

———— 2009. 'Strategie di autorizzazione. Enea Tattico e la tradizione letteraria dell'arte della guerra'. *Euphrosyne* 37. 349–361.

———— 2011. 'The *Strategikós* of Onasander: Taking Military Texts Seriously'. *Technai* 2. 39–52.

———— 2017. 'Introduction: The Poetics of Knowledge', in *idem* and P. Van der Eijk, eds., *Knowledge, Text and Practice in Ancient Technical Writing*. Cambridge: Cambridge University Press. 12–26.

Gazzano, F. and Traina, G. 2014. 'Plutarque, historien militaire?'. *Ktèma* 39. 347–370.

Grabar, A. 1964. 'Éloge funèbre de M. Marie-Alphonse Dain, membre libre de l'Académie'. *Comptes rendus des séances de l'Académie des Inscriptions et Belles-Lettres* 108. 234–238.

Guischardt, C. 1758. *Mémoires militaires sur les Grecs et les Romains* [. . .]. La Haye: Pierre De Hondt.

Haase, F. 1835. 'Ueber die griechischen und lateinischen Kriegsschriftsteller'. *Neue Jahrbücher für Philologie und Pädagogik* 5.14.1. 88–118.

Haase, F. 1847. *De militarium scriptorum Graecorum et Latinorum omnium editione instituenda narratio*. Berolini: Verlag Trautwein.

Haldon, J. 1990. *Constantine Porphyrogenitus: Three Treatises on Imperial Military Expeditions*. Vienna: Verlag der Österreichischen Akademie der Wissenschaften.

——— 2014. *A Critical Commentary on the Taktika of Leo VI*. Washington: Dumbarton Oaks.

Hölkeskamp, K. J. 1997. 'La guerra e la pace', in S. Settis, ed., *I Greci. Storia Cultura Arte Società*. Turin. 481–539.

Hornblower, S. 2007. 'Warfare in Ancient Literature: The Paradox of War', in P. Sabin, H. van Wees, and M. Whitby, eds., *The Cambridge History of Greek and Roman Warfare. Volume I: Greece, the Hellenistic World and the Rise of Rome*. Cambridge: Cambridge University Press. 22–53.

Humble, N. 2017. 'Xenophon and the Instruction of Princes', in M. A. Flower, ed., *The Cambridge Companion to Xenophon*. Cambridge: Cambridge University Press. 416–434.

Hunger, H. 1978. *Die hochsprachliche profane Literatur der Byzantiner*. II, *Philologie, Profandichtung, Musik, Mathematik und Astronomie, Naturwissenschaften, Medizin, Kriegswissenschaft, Rechtsliteratur*. Munich: C. H. Beck.

Hunter, R. 2004. 'Homer and Greek Literature', in R. Fowler, ed., *The Cambridge Companion to Homer*. Cambridge: Cambridge University Press. 235–253.

Ilari, V. 2012. 'Tra bibliografia ed epistemologia militare. Introduzione allo studio degli scrittori militari italiani dell'età moderna'. *Rivista di studi militari* 1. 141–170.

Jacobs, S. 2018. *Plutarch's Pragmatic Biographies: Lessons for Statesmen and Generals in the 'Parallel Lives'*. Leiden: Brill.

Jouanna, J. 2015. 'Guerre et philosophie en Grèce ancienne: aux origines de l'art de la guerre', in Ph. Contamine, J. Jouanna, and M. Zink, eds., *Colloque La Grèce et la guerre. Actes du XXVe colloque de la Villa Kérylos, 3–4 octobre 2014*. Paris: Inscriptions. 29–44.

Kaegi, W. E. 1983. *Some Thoughts on Byzantine Military Strategy*, Brookline, MA: Hellenic College Press. (reprinted in J. Haldon, ed., *Byzantine Warfare*. Aldershot: Ashgate, 2007. 251–268).

Kahn, C. H. 1993. 'Plato's Ion and the Problem of *Techne*', in R. M. Rosen and J. Farrell, eds., *Nomodeiktes: Greek Studies in Honor of Martin Ostwald*. Ann Arbor: University of Michigan Press. 369–378.

Kaldellis, A. 2012. 'The Byzantine Role in the Making of the Corpus of Classical Greek Historiography: A Preliminary Investigation'. *JHS* 132. 71–85.

Kim, L. 2010. *Homer between History and Fiction in Imperial Greek Literature*. Cambridge: Cambridge University Press.

Koder, I. and Stouraitis, I. 2012. 'Byzantine Approaches to Warfare (6th–12th Centuries): An Introduction', in *idem*, eds., *Byzantine War Ideology Between Roman Imperial Concept and Chtistian Religion: Akten des Internationalen Symposiums (Wien, 19.-21. Mai 2011)*. Vienna: Verlag der Österreichischen Akademie der Wissenschaften. 9–15.

Kučma, V. V. 1980. '"Vizantijskij Anonim VI v.": osnovnye problemy istočnikov i soderzanija'. *VizVrem* 41. 68–91.

Laiou, A. E. 1993. 'On Just War in Byzantium', in J. S. Langson, ed., *Τὸ Ἑλληνικόν: Studies in Honor of Speros Vryonis Jr.* New Rochelle, NY: Caratzas. 153–177.

Le Bohec, Y. 1998: 'Que voulait Onesandros?', in Y. Burnand, *idem*, and J.-P. Martin, eds., *Claude de Lyon empereur romain. Actes du Colloque Paris-Nancy-Lyon, Novembre 1992*. Paris: Presses de L'Université de Paris-Sorbonne. 169–179.

Lee, W. I. 2017. 'Xenophon and His Times', in M. A. Flower, ed., *The Cambridge Companion to Xenophon*. Cambridge: Cambridge University Press. 15–36.

Lemerle, P. 1971. *Le premier humanisme byzantin. Notes et remarques sur enseignement et culture à Byzance des origines au Xe siècle*. Paris: Presses Universitaires de France.

Lenoir, M. 1996. 'La littérature *de re militari*', in C. Nicolet, ed., *Les littératures techniques dans l'antiquité romaine. Statut, public et destination, tradition*. Geneva: Entretiens Hardt. 77–115.

Lenski, N. 1997. '*Initium mali Romano imperio*: Contemporary Reactions to the Battle of Adrianople'. *TAPA* 127. 129–168.

——— 2002. *Failure of Empire: Valens and the Roman State in the Fourth Century A.D.* Berkeley: University of California Press.

Leszl, W. 2007. 'Democritus' Work: From Their Titles to Their Contents', in A. Brancacci and P.-M. Morel, eds., *Democritus: Science, the Arts, and the Care of the Soul: Proceedings of the International Colloquium on Democritus (Paris, 18–20 September 2003)*. Leiden: Brill. 11–76.

Magdalino, P. 2011. 'Orthodoxy and History in Tenth-Century Byzantine "Encyclopedism"', in P. van Deun and C. Mace, eds., *Encyclopedic Trends in Byzantium? Proceedings of the International Conference held in Leuven, 6–8 May 2009*. Leuven: Peeters. 143–159.

——— 2013. 'Byzantine Encyclopaedism of the Ninth and Tenth Centuries', in J. König and G. Woolf, eds., *Encyclopaedism from Antiquity to the Renaissance*. Cambridge: Cambridge University Press. 219–231.

Marsden, E. W. 1974. 'Polybius as a Military Historian', in E. Gabba, ed., *Polybe*. Geneva: Entretiens Hardt. 267–301.

Meißner, B. 1999. *Die technologische Fachliteratur der Antike. Struktur, Überlieferung und Wirkung technischen Wissens in der Antike (ca. 400. v. Chr.-ca. 500 n. Chr.)*. Berlin: De Gruyter.

——— 2013. 'Polybios als Militärhistoriker', in V. Grieb and C. Kohen, eds., *Polybios und seine Historien*. Stuttgart: Franz Steiner. 127–157.

Müller, K. K. 1896. *Asklepiodotos* [10]. *RE* II.2, coll. 1637–1641.

Naudé, G. 1637. *Gabrielis Naudaei Syntagma de studio militari. Ad Illustrissimum Iuvenem Ludovicum ex Comitibus Guidiis a Balneo*. Rome: Facciottus.

Odorico, P. 1990. 'La cultura della Συλλογή. 1) Il cosiddetto enciclopedismo bizantino. 2) Le Tavole del sapere di Giovanni Damasceno'. *ByzZ* 83. 1–21.

——— 2011. 'Cadre d'exposition/cadre de pensée – la culture du recueil', in P. van Deun and C. Mace, eds., *Encyclopedic Trends in Byzantium? Proceedings of the International Conference held in Leuven, 6–8 May 2009*. Leuven: Peeters. 89–107.

Oikonomides, N. 1995. 'The Concept of "Holy War" and Two Tenth-Century Byzantine Ivories', in T. S. Miller and J. Nesbitt, eds., *Peace and War in Byzantium: Essays in Honor of George T. Dennis, S.J.* Washington: Dumbarton Oaks. 62–68.

Petrocelli, C. 2008. *Onasandro. Il generale*. Bari: Edizioni Dedalo.

——— 2012. 'Racconti di guerra. Figure della narrazione nelle Storie di Tucidide', in V. Maraglino, ed., *Scienza antica in età moderna. Teoria e immagini*. Bari: Cacucci Editore. 15–33.

Piccione, R. M. 2003. 'Scegliere, raccogliere e ordinare. Letteratura di raccolta e trasmissione del sapere'. *Humanitas* 58. 44–63.

Porod, R. 2013. *Lukians Schrift "Wie man Geschichte schreiben soll". Kommentar und Interpretation.* Vienna: Phoibos Verlag.

Potter, D. S. 1999. *Literary Texts and the Roman Historian.* London: Routledge.

Poznanski, L. 1978. '"Le traité de Tactique" de Polybe d'après le livre II des Histoires'. *LEC* 46. 205–212.

——— 1980a. 'Essai de reconstitution du Traité de tactique de Polybe d'aprèe le livre III des Histoires'. *AC* 49. 161–172.

——— 1980b. 'A propos du "Traité de tactique" de Polybe'. *Athenaeum* 58. 340–352.

——— 1993. 'Commander, contrôler, communiquer. Polybe, de la tradition à la modernité'. *LEC* 61. 205–220.

——— 1994. 'La polémologie pragmatique de Polybe'. *JS* 1. 19–74.

Rance, P. 2017a. 'Introduction', in P. Rance and N. Sekunda (eds.), *Greek Taktika: Ancient Military Writing and its Heritage.* Gdansk: Akathina. 9–64.

———. 2017b. 'Maurice's *Strategicon* and the "Ancients": The Late Antique Reception of Aelian and Arrian', in P. Rance and N. Sekunda, eds., *Greek Taktika: Ancient Military Writing and Its Heritage.* Gdansk: Foundation for the Development of Gdansk University for the Department of Mediterranean Archeology. 217–255.

———. 2018. 'The Reception of Aineias' *Poliorketika* in Byzantine Military Literature', in M. Pretzler and N. Barley, eds., *Brill's Companion to Aineias Tacticus.* Leiden: Brill. 290–374.

Reeve, M. D. 2004. *Epitoma rei militaris.* Oxford: Oxford University Press.

Romano, E. 2005. 'Il difficile rapporto fra teoria e pratica nella cultura romana', in F. Bessone and E. Malaspina, eds., *Politica e cultura in Roma antica. Atti dell'incontro di studio in ricordo di Italo Lana, Torino, 16–17 ottobre 2003.* Bologna: Pàtron. 81–99.

——— 2017. '*Si qui voluerit.* Vitruvius on Architecture and "the Art of the Possible"', in M. Formisano and P. Van der Eijk, eds., *Knowledge, Text and Practice in Ancient Technical Writing.* Cambridge: Cambridge University Press. 53–67.

Sabbadini, R. 1931. *Carteggio di Giovanni Aurispa.* Rome: Tip. del Senato.

Schellenberg, H. M. 2007. 'Einige Bemerkungen zum Strategikos des Onasandros', in L. de Blois and E. Lo Cascio, eds., *The Impact of the Roman Army (200 BC-AD 476): Economic, Social, Political, Religious and Cultural Aspects: Proceedings of the Sixth Workshop of the International Network Impact of Empire (Roman Empire, 200 B.C.-A.D. 476).* Leiden: Brill. 181–191.

Shipley, D. G. J. 2018. 'Aineias Tacticus in His Intellectual Context', in M. Pretzler and N. Barley, eds., *Brill's Companion to Aineias Tacticus.* Leiden: Brill. 49–67.

Stouraitis, I. 2009. *Krieg und Frieden in der politischen und ideologischen Wahrnehmung in Byzanz (7.-11. Jahrhundert).* Vienna: Fassbaender Verlag.

——— 2012. '"Just War" and "Holy War" in the Middle Ages: Rethinking Theory through the Byzantine Case-Study'. *JÖByz* 62. 227–264.

Sullivan, D. F. 2010. 'Byzantine Military Manuals: Prescriptions, Practice and Pedagogy', in P. Stephanson, ed., *The Byzantine World.* London: Routledge. 149–161.

Tougher, S. 1997. *The Reign of Leo VI (886–912): Politics and People.* Leiden: Brill.

Toynbee, A. 1973. *Constantine Porphyrogenitus and His World.* London: Duckworth.

Trombley, F. 1997. 'The Taktika of Nikephoros Ouranos and Military Encyclopaedism', in P. Binkey, ed., *Pre-Modern Encyclopaedic Texts: Proceedings of the Second COMERS Congress, Groningen.* Leiden: Brill. 261–274.

Tybjerg, K. 2005. 'Hero of Alexandria's Mechanical Treatises: Between Theory and Practice', in A. Schürmann, ed., *Physik/Mechanik*. Stuttgart: Franz Steiner. 204–226.

Vela Tejada, J. 1993. 'Tradición y originalidad en la obra de Eneas el Táctico: la génesis de la historiografía militar'. *Minerva* 7. 79–92.

——— 2004. 'Warfare, History and Literature in the Archaic and Classical Periods: The Development of Greek Military Treatises'. *Historia* 53. 129–146.

——— 2018. 'Creating Koine: Aineias Tacticus in the History of the Greek Language', in M. Pretzler and N. Barley, eds., *Brill's Companion to Aineias Tacticus*. Leiden: Brill. 96–122.

Weitzmann, K. 1959. *Ancient Book Illumination*. Cambridge, MA: Harvard University Press.

Whately, C. 2015. 'The Genre and Purpose of Military Manuals in Late Antiquity', in G. Greatrex, H. Elton, and L. McMahon, eds., *Shifting Genres in Late Antiquity*. Farnham-Burlington: Ashgate. 249–261.

Wheeler, E. L. 2010. 'Polyaenus: *Scriptor Militaris*', in K. Krodersen, ed., *Polyainos. Neue Studien. Polyaenus. New Studies*. Berlin: Verlag Antike. 7–54.

Whitehead, D. 1990. *Aineias the Tactician: How to Survive under Siege*. Oxford.

——— 2008. 'Fact and Fantasy in Greek Military Writers'. *AAHung* 48. 139–155.

Yeide, H. 2011. *Fighting Patton: George S. Patton Jr. through the Eyes of His Enemies*. Minneapolis and Osceola: Zenith Press.

Zaloga, S. J. 2010. *George S. Patton: Leadership, Strategy, Conflict*. Oxford: Osprey Publishing.

Zecchini, G. 2008. 'Utopie militari tardoantiche?', in C. Carsana, ed., *Utopia e utopie nel pensiero storico antico*. Rome: "L'Erma" di Bretschneider. 195–206.

Zhmud, L. 2006. *The Origin of the History of Science in Classical Antiquity*. Translated from the Russian by A. Chernoglazov. Berlin: De Gruyter.

Index